T0309212

FRESH-CUT FRUITS AND VEGETABLES

FRESH-CUT FRUITS AND VEGETABLES

Technologies and Mechanisms for Safety Control

Edited by

MOHAMMED WASIM SIDDIQUI

Academic Press is an imprint of Elsevier
125 London Wall, London EC2Y 5AS, United Kingdom
525 B Street, Suite 1650, San Diego, CA 92101, United States
50 Hampshire Street, 5th Floor, Cambridge, MA 02139, United States
The Boulevard, Langford Lane, Kidlington, Oxford OX5 1GB, United Kingdom

Library of Congress Cataloging-in-Publication Data
A catalog record for this book is available from the Library of Congress

British Library Cataloguing-in-Publication Data
A catalogue record for this book is available from the British Library

ISBN 978-0-12-816184-5

For information on all Academic Press publications visit our website at https://www.elsevier.com/books-and-journals

Publisher: Charlotte Cockle
Acquisition Editor: Patricia Osborn
Editorial Project Manager: Vincent Gabrielle
Production Project Manager: Omer Mukthar
Cover Designer: Miles Hitchen

Typeset by SPi Global, India

Contents

Contributors

A. Aguayo-Acosta Department of Microbiology, Faculty Biological Sciences, Autonomous University of Nuevo Leon, San Nicolas de los Garza, Mexico

Bindvi Arora Division of Food Science and Postharvest Technology, ICAR-Indian Agricultural Research Institute, New Delhi, India

Omar Bashir Division of Food Science and Technology, Sher-e-Kashmir University of Agricultural Sciences and Technology, Kashmir, India

Usawadee Chanasut Department of Biology, Faculty of Science, Chiang Mai University, Chiang Mai, Thailand

Kartikey Chaturvedi National Institute of Food Technology Entrepreneurship and Management, Sonipat, India

Sindhu Chinnaswamy Division of Food Science and Postharvest Technology, ICAR-Indian Agricultural Research Institute, New Delhi, India

Larissa Morais Ribeiro da Silva Department of Food Engineering, Federal University of Ceará, Fortaleza, Brazil

Aamir Hussain Dar Department of Food Technology, Islamic University of Science and Technology, Awantipora, India

J.E. Dávila-Aviña Department of Microbiology, Faculty Biological Sciences, Autonomous University of Nuevo Leon, San Nicolas de los Garza, Mexico

Andréa Cardoso de Aquino Department of Food Engineering, Federal University of Ceará, Fortaleza, Brazil

Luciana de Siqueira Oliveira Department of Food Engineering, Federal University of Ceará, Fortaleza, Brazil

Vinayak Deshi Department of Food Science and Postharvest Technology, Bihar Agricultural University, Bhagalpur, India

Kaliana Sitonio Eça Department of Food Engineering, Federal University of Ceará, Fortaleza, Brazil

Gajanan Gundewadi Division of Food Science and Postharvest Technology, ICAR-Indian Agricultural Research Institute, New Delhi, India

Alka Joshi Division of Food Science and Postharvest Technology, ICAR-Indian Agricultural Research Institute, New Delhi, India

Jumina Chemo and Biosensors Group, Faculty of Pharmacy, University of Jember, Jember; Department of Chemistry, Faculty of Mathematics and Natural Sciences, Gadjah Mada University, Yogyakarta, Indonesia

Shafat Khan Department of Food Technology, Islamic University of Science and Technology, Awantipora, India

Wilawan Kumpoun Science and Technology Research Institute, Chiang Mai University, Chiang Mai, Thailand

Bambang Kuswandi Chemo and Biosensors Group, Faculty of Pharmacy, University of Jember, Jember; Department of Chemistry, Faculty of Mathematics and Natural Sciences, Gadjah Mada University, Yogyakarta, Indonesia

Ishrat Majid Lovely Professional University, Phagwara, India

Hilal A Makroo Department of Food Technology, Islamic University of Science and Technology, Awantipora, India

Swarajya Laxmi Nayak Division of Food Science and Postharvest Technology, ICAR-Indian Agricultural Research Institute, New Delhi, India

Burhan Ozturk Department of Horticulture, Faculty of Agriculture, Ordu University, Ordu, Turkey

Uma Prajapati Division of Food Science and Postharvest Technology, ICAR-Indian Agricultural Research Institute, New Delhi, India

Ovais Shafiq Qadri Department of Bioengineering, Integral University, Lucknow, India

A. Ríos-López Department of Microbiology, Faculty Biological Sciences, Autonomous University of Nuevo Leon, San Nicolas de los Garza, Mexico

Qurat ul Eain Hyder Rizvi Eternal University, Baru Sahib, India

Shalini Gaur Rudra Division of Food Science and Postharvest Technology, ICAR-Indian Agricultural Research Institute, New Delhi, India

Pankaj Preet Sandhu Center of Innovative and Applied Bioprocessing, Mohali, India

Shruti Sethi Division of Food Science and Postharvest Technology, ICAR-Indian Agricultural Research Institute, New Delhi, India

Rafiya Shams Sher-e-Kashmir University of Agricultural Sciences and Technology, Jammu, India

Ram Roshan Sharma Division of Food Science and Postharvest Technology, ICAR-Indian Agricultural Research Institute, New Delhi, India

Mohammed Wasim Siddiqui Department of Food Science and Postharvest Technology, Bihar Agricultural University, Bhagalpur, India

Gisha Singla Center of Innovative and Applied Bioprocessing, Mohali, India

L.Y. Solís-Soto Department of Microbiology, Faculty Biological Sciences, Autonomous University of Nuevo Leon, San Nicolas de los Garza, Mexico

Sarana Rose Sommano Plant Bioactive Compound laboratory (BAC), Department of Plant and Soil Sciences, Faculty of Agriculture, Chiang Mai University, Chiang Mai, Thailand

Sucheta Center of Innovative and Applied Bioprocessing, Mohali, India

Cherakkathodi Sudheesh Department of Food Science and Technology, Pondicherry University, Puducherry, India

Kappat Valiyapeediyekkal Sunooj Department of Food Science and Technology, Pondicherry University, Puducherry, India

Aseeya Wahid Department of Processing and Food Engineering, Punjab Agricultural University, Ludhiana, India

Basharat Yousuf Department of Post Harvest Engineering and Technology, Faculty of Agricultural Sciences, Aligarh Muslim University, Aligarh, India

1

Fresh-cut fruits and vegetables: Quality issues and safety concerns

Basharat Yousuf*, Vinayak Deshi†, Burhan Ozturk‡, Mohammed Wasim Siddiqui†

**Department of Post Harvest Engineering and Technology, Faculty of Agricultural Sciences, Aligarh Muslim University, Aligarh, India. †Department of Food Science and Postharvest Technology, Bihar Agricultural University, Bhagalpur, India. ‡Department of Horticulture, Faculty of Agriculture, Ordu University, Ordu, Turkey*

1 Introduction

Fruit and vegetables are living entities that comprise a vast group of plant foods that differ greatly in content of energy and nutrients. They are a main source of nutrients such as carbohydrates, fibers, protein, vitamins, and minerals. Fruit and vegetables form an important part of the human diet. They are considered good for health and contain important vitamins, minerals, and other phytochemicals that can help in smooth functioning of the human body and can also help in protecting from some diseases (Van Duyn & Pivonka, 2000). Consuming adequate quantities of fruit and vegetables will benefit an individual in maintaining a healthy lifestyle. Many vegetables are a rich source of dietary fiber. Fruit and vegetables are not only an important part of the diet, but also form the basis of a healthy diet. Hence, recommendations for a balanced diet must involve the consumption of fresh fruits and vegetables. Many countries have dietary recommendations that include fruits and vegetables. There are ever increasing demands for fruits and vegetables due to the recommendations by nutritionists and researchers, and the promotion of their consumption through campaigns at a governmental level. Minimally processed or fresh-cut fruit and vegetables help to fulfill such consumer needs.

Minimal or fresh-cut processing of fruits and vegetables has gained importance over recent decades due to the availability of convenience products to consumers. The International

Fresh-cut Fruits and Vegetables. https://doi.org/10.1016/B978-0-12-816184-5.00001-X

Fresh-cut Produce Association (IFPA) defines fresh-cut produce as “any fruit or vegetable or combination thereof that has been physically altered from its original form, but remains in a fresh state” (IFPA & PMA, 1999). Peeled, sliced, shredded, trimmed, and/or washed fruits and vegetables may be regarded as fresh-cut produce (Francis et al., 2012). More specific examples of minimally processed vegetables are sliced potatoes, shredded lettuce and cabbage, mixed salads, washed and trimmed spinach, cauliflower and broccoli florets, cleaned and diced onions. Fig. 1

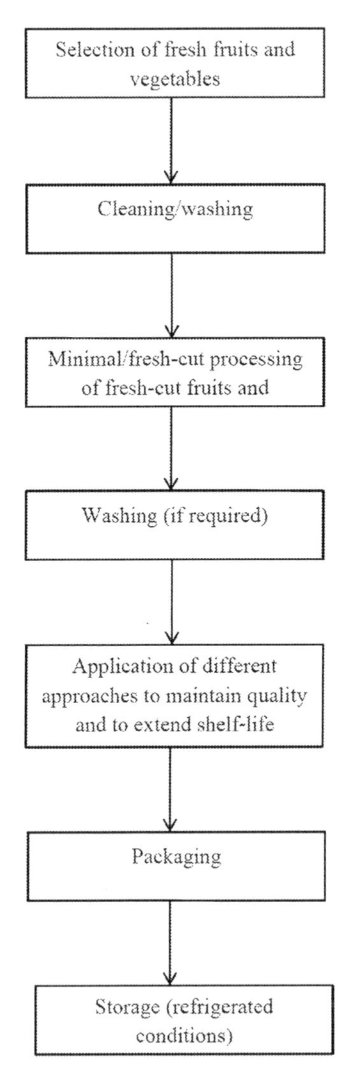

Fig. 1 General flow sheet for production of fresh-cut fruits and vegetables.

represents a general flow sheet for production of fresh-cut fruits and vegetables.

In general, the deterioration in quality and shelf-life of fresh foods is due to water loss, microbial growth, oxidation, texture and flavor deterioration, and increase in respiration rate and ripening process (Martn-Belloso, Soliva-Fortuny, & Oms-Oliu, 2006). Food preservation is essential for keeping the global food supply safe and available for consumers at every time of the year. It is well established that fresh produce may contain a high contamination level after harvest. Further, fresh-cut processing makes it more prone to contamination and subsequently results in rapid deterioration. Therefore, maintaining the quality of such products is a challenging task as minimal processing leads to increased ethylene production and respiration rates, with the consequent loss of quality. Nevertheless, like production and processing, quality and safety of fresh-cut products are critical parameters, which demand considerable attention.

2 Scope and importance

The present global interest in healthier lifestyles has resulted in a rise in the demand for convenient fresh foods that are free from additives and have good nutritional value. The availability of fresh-cut fruits and vegetables in the marketplace throughout the world has also been increasing. The fresh-cut sector is constantly evolving and innovating to enhance product safety and quality attributes that are generally valued by consumers. Fresh-cut or minimally processed products offer a range of advantages, as they reduce the preparation time and provide access to food that retains high nutritional and sensory quality (Graça et al., 2015). Fresh-cut products have almost the same properties as that of the whole, intact produce and have a uniform quality. They are attractive and appealing to the consumers and at the same time, fresh-like characteristics of the raw materials are preserved. While fresh-cut vegetables may require further processing such as cooking before final consumption, fresh-cut fruits are mostly consumed as ready-to-eat products. Fresh-cut fruits and vegetables are very attractive to consumers looking for healthy and convenient meals, as they are neither heat-treated, nor any chemical preservatives added. Fresh-cut produce includes a range of products named specifically after a particular fruit/vegetable or a particular cutting operation. Fresh-cut fruits and vegetables available in the markets of some developed counties include pineapple, papaya, melons, cantaloupe, watermelon, mango, jackfruit, grapefruit, fruit mixes, shredded leafy vegetables and salad mixes, vegetables for cooking like peeled baby carrots, baby corn, broccoli and cauliflower florets, cut celery stalks, shredded cabbage, cut asparagus, cut sweet potatoes, and many more (James & Ngarmsak, 2010).

3 Some basic requirements for preparation of fresh-cut fruits and vegetables

Preparation of fresh-cut fruits and vegetables is of critical importance considering their delicate nature. Therefore, the process of fresh-cut product processing and preparation demands extreme care. Following are some of the essential considerations while fresh-cut fruits and vegetable are being prepared.

1. The raw materials for preparation of fresh-cut products should be of good quality.
2. Strict hygiene and good manufacturing practices should be considered.
3. Careful cleaning and/or washing should be done before and after peeling.
4. Mild processing aids in wash water for disinfection or prevention of browning and texture loss should be used.
5. Sharp knives on cutters should be used to minimize physical/mechanical damage during fresh-cut processing operations such as peeling, cutting, slicing, and/or shredding.
6. Excess moisture should be gently removed/drained.
7. Appropriate packaging material and packaging methods should be applied.
8. Appropriate temperature control should be applied at every stage including processing, storage, distribution, and handling.

4 Quality and safety concerns

Despite a large number of benefits derived from production and consumption of fresh-cut fruits and vegetables, quality and safety of such products is still a major concern. Fresh fruits and vegetables are highly perishable as they contain 80%–90% water by weight. The natural protections of skin and cuticle on the surface of fresh fruits and vegetables are often damaged during harvest, transportation, and mechanical operations. The risk of deterioration further escalates due to fresh-cut processing. Although fresh-cut products are ready to eat, they are not processed by any thermal or chemical preservation method. Thus, they are more susceptible to microbiological contamination during different stages, from food processing to distribution and commercialization (Francis et al., 2012). Quality and safety of fresh-cut produce largely depends on their associated microbiological load.

Quality of fresh-cut produce may sometimes also be influenced by the condition of the raw material (i.e., fresh fruits and vegetables). The raw material should always be of good quality, free from any defects or microbial load. Each step from production and preparation to final consumption will influence the microbiology of fresh-cut produce. Improper handling at any stage will lead to compromise in quality and safety of fresh-cut produce. During the last decades, minimally processed fruits and vegetables have emerged as new vehicles for the transmission of food-borne diseases. There can be different concerns related to quality and safety of fresh-cut fruits and vegetables. Improper sanitization, improper or inadequate fresh-cut processing operations, faulty packaging, and unsuitable storage environment may also result in serious quality and safety concerns. Moreover, using some techniques to extend shelf-life may sometimes increase the risk of safety problems and therefore need to be carefully evaluated. Some of the important factors for safe production of fresh-cut produce are, screening materials entering the processing chain, suppressing microbial growth, reducing the microbial load during processing, and preventing postprocessing contamination (Artés & Allende, 2005). Consumers' perception regarding fresh-cut fruits is of critical importance because:

(1) fresh-cut fruits are consumed raw, without any thermal treatment, as in the case of cooked products;
(2) fresh-cut processing wounds the tissue and the wounded tissue provides more feasible conditions for microorganisms to spoil the product and to pose human health risks;
(3) fresh-cut fruits are not subjected to any lethal or kill (like thermal processing) step during their preparation; and
(4) no preservatives are added to fresh-cut fruits that would ensure their safety to the consumer.

We have tried to explain some of the factors that can critically influence the quality and safety of fresh-cut products. Fig. 2 represents various quality and safety concerns related to fresh-cut fruits and vegetables.

4.1 Contamination from wash water

Among all the factors to be considered while addressing safety issues for fresh-cut produce, the potential for pathogen contamination and food-borne outbreaks is receiving most attention.

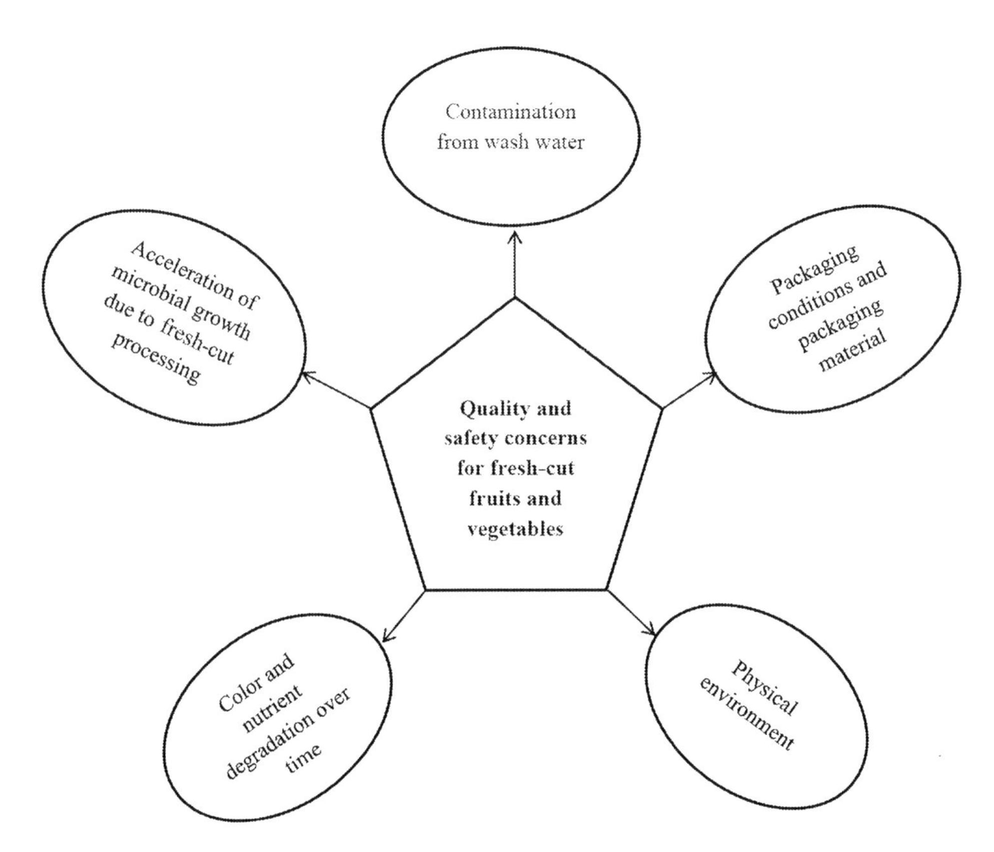

Fig. 2 Different quality and safety concerns of fresh-cut fruits and vegetables.

The best way to eliminate pathogens from fresh produce is to prevent contamination, since washing, even with disinfectants, cannot eliminate pathogens; rather it will only reduce them (Parish et al., 2003). A number of methods have been used to reduce populations of microorganisms on both whole and fresh-cut produce. During fresh-cut processing on a large scale, disinfection is one of the important steps to ensure quality of produce. For instance, chlorine is widely employed as disinfectant. Conventional chlorine-based disinfectants like calcium hypochlorite and sodium hypochlorite are used to wash fruits and vegetables. Sodium hypochlorite is a very potent disinfectant with powerful oxidizing properties, and most commonly used by the food industry for sanitizing both products and equipment of the processing line. However, in the past, disease outbreaks linked with fresh-cut product consumption have raised doubts about the effectiveness of chlorine in ensuring safe and good quality products (Ölmez & Kretzschmar, 2009). In addition, it is established that the reaction of chlorine with other organic constituents results in the development of carcinogenic halogenated disinfection by-products, such as trihalomethanes and haloacetic acids (Hua & Reckhow, 2007). Efficacy of the chlorine sanitizers increases with increase in the concentration of available chlorine, but high levels may cause product tainting and mineral residue on the product and equipment.

Contamination from wash water is of critical importance. The water that is used as a means to reduce the contamination may sometimes contaminate the produce, particularly when the water is reused. It is thus of utmost importance to maintain the quality of wash water to provide a barrier against cross contamination. Due to such problems, the processing sector is seeking new and safe alternatives. The alternatives can be the emerging technologies such as irradiation and ozone for fresh-cut fruits and vegetables. In order to obtain fresh-cut products with fresh-like quality, safety, and high nutritional value, the food processor needs to employ improved approaches by using sustainable techniques, especially standard procedures for sanitation.

4.2 Acceleration of microbial growth due to fresh-cut processing

Attention to microbiology is very important in maintaining shelf stability and safety of fresh-cut fruits and vegetables. Microbiological stability is a critical factor to maintain the commercial marketability of fresh-cut produce. There are growing concerns over microbial safety of fresh-cut fruits and vegetables. Since fresh-cut fruits and vegetables are wounded tissue, there are increased chances of microbial growth, which may cause both decay and safety problems. Further acceleration in microbial contamination and growth is due to the fact that fresh-cut products are neither heat-treated nor any chemical preservatives added (Berger et al., 2010; Olaimat & Holley, 2012). It is extremely difficult to guarantee the microbiological safety of fresh-cut fruits and vegetables. Fresh-cut fruits have nonsterile cut surfaces, which are physiologically active and rich in nutrients and water. Hence, both spoilage and pathogenic microorganisms have a perfect medium to grow and multiply rapidly. Moreover, fresh-cut fruits and vegetables can act as vehicles of the transmission of food-borne pathogens like *Escherichia coli, Listeria monocytogenes,* and *Salmonella* spp. These pathogenic microorganisms can cause severe food-borne disease outbreaks. Spoilage often results in a low organoleptic profile, short shelf-life, and high economic losses, as well as increased public health risks. Therefore, it is imperative to reduce the risk of both types of microorganisms. Thus, microbiological quality of fresh-cut produce is of concern, not only from the food safety point of view, but also due to the spoilage involved, which consequently reduces the shelf-life and causes huge economic losses (Johnston et al., 2006).

In spite of the higher microbiological risks associated with fresh-cut fruits and vegetables, few studies have aimed to evaluate the microbiological quality of fresh-cut products available in global markets. However, in the recent decades a number of techniques have been investigated to reduce microbial growth on fresh-cut products. Fresh-cut fruit and vegetable processors should ensure that good manufacturing practices are followed, which will result in fresh-cut products with lower populations of microorganisms. This will, to some extent, help in increasing the overall quality.

4.3 Rapid nutrient and color degradation due to fresh-cut processing

Processing promotes rapid deterioration because of tissue damage, which leads to an increase in physiological activity and other physico-chemical changes, such as enzymatic browning and softening. Stress caused by operations such as peeling, cutting, shredding, or slicing increase tissue respiration significantly, giving rise to biochemical deterioration like browning, texture breakdown, off-flavors, and risk of microbial development. During fresh-cut processing, contamination with both spoilage and pathogenic microorganisms may occur and then the nutrients inside the fruit contribute to their growth. This causes both deterioration/decay and safety issues. In addition, large numbers of other deteriorative changes may result due to fresh-cut processing.

Nutrients such as ascorbic acid and various phenolic compounds may be lost during the processing. Sensory aspects including appearance, color, flavor, and texture may be altered. Desiccation/shrinkage is one of the prevalent problems in fresh-cut produce, and can result in other quality defects such as change in texture. Texture is a critical quality attribute and loss of firmness can be correlated with tissue degradation. Fresh-cut processing has a remarkable influence on the fruit volatiles profile, as aroma compounds are lost during the processing. Furthermore, browning is one of the main causes of quality deterioration in fresh-cut produce. One of the key causes of quality loss and the major challenge in fresh-cut fruits and vegetables is enzymatic browning (Eissa, Fadel, Ibrahim, Hassan, & Elrashid, 2006), because color is one of the most important noticeable quality attributes of fresh-cut fruits and vegetables. Browning occurs as a result of oxidation of phenolic compounds catalyzed by the polyphenol oxidase enzyme, giving rise to colorless quinones, which later on get polymerized to form melanins. These substances then impart brown, reddish, or black coloration to the product. During fresh-cut processing, there are many cutting operations that expose the surface and increase the risk of browning. Browning is a very common problem in apples, pears, artichokes, and potatoes.

4.4 Packaging conditions and packaging material

Food packaging is an essential step for fresh-cut produce processing that allows the products to be protected and safely distributed to final consumers. When packaging is done appropriately and in an optimized manner, it increases the convenience, safety, and quality of food. Synthetic plastic packaging, which is mostly nonbiodegradable, is an important element in food preservation. The domination and popularity of plastic in the food packaging market is due to low production cost, mechanical resistance, heat sealability, and shape flexibility. However, plastic packaging has the disadvantage of being nonbiodegradable, causing harm to the environment and has adverse impacts on the health of terrestrial and aquatic animals (Thompson, Swan, Moore, & vom Saal, 2009). Fresh-cut fruits and vegetables are often packed in polymeric bags with selective permeability to carbon dioxide and water vapor, where a passive or active modified atmosphere is generated. Inappropriate packaging can sometimes cause development of off-flavors and off-odors, which result in depleted quality in fresh-cut fruits and vegetables (Artés, Gómez, & Artés-Hernández, 2006). Packaging is also considered as one of the important factors influencing the microbiological quality of fresh-cut products.

4.5 Physical environment

Environmental conditions can have great influence on the microorganism populations on fresh-cut fruit surfaces. Contamination in fresh-cut products can occur any time during the entire process including packaging, transportation, storage, and distribution. Fresh fruits and vegetables have natural protective coverings that can protect from the harsh physical conditions surrounding them. However, in fresh-cut processing, this protective covering is often removed. This makes fresh-cut fruits and vegetables more vulnerable to being affected by the surrounding physical conditions and increases the chances of contamination. Microorganisms will grow and multiply rapidly due to the easy availability of nutrients on the injured tissue. Moreover, a harsh physical environment may also adversely affect various quality attributes of fresh-cut fruits and vegetables. The combination of packaging with cold chain management preserves the quality of fresh products, along with extension of shelf-life (Nicola & Fontana, 2014).

5 Different strategies to maintain quality of fresh-cut fruits and vegetables

As stated earlier, fresh-cut products, being wounded tissue, are more vulnerable to deterioration than whole fruits and vegetables. The wounds/injuries incurred during processing operations cause reduction in shelf-life. Hence, various strategies are being investigated and implemented to maintain quality and to extend shelf-life. These strategies mostly include the use of innovative and emerging food processing technologies. Such technologies should also be effective and economically feasible. Modified and controlled atmosphere packaging, chemical treatments such as calcium dip, and use of edible coatings are some of the methods to extend the shelf-life of fresh-cut fruits and vegetables. The effect of modified atmosphere packaging in quality preservation is related mostly to the reduction of respiration rate (Sanchís et al., 2017). It also influences ethylene biosynthesis and action, water loss, phenolic oxidation, and aerobic microbial count. More recent techniques include use of irradiation, ozone, cold plasma, pulsed light, and so on. These and some other techniques have been described in detail in the subsequent chapters in this book. Table 1 shows different strategies to maintain quality and safety of fresh-cut fruits and vegetables with recent examples. Some of the techniques are still at an early stage of their use and exist only at laboratory scale. Therefore, there is a need to evaluate these techniques on pilot-level and larger-scale commercial levels. In addition, technologies ensuring food safety should be preferred over convenience-oriented technologies.

Table 1 Strategies to maintain quality and safety of fresh-cut fruits and vegetables with recent examples

Technique	Fresh-cut fruit/vegetable	Specific effect found	Reference
Modified atmosphere packaging	Fresh-cut lettuce	Strong effect of antibrowning, inhibition of microbial growth	López-Gálvez et al. (2015)
	Fresh-cut cantaloupe and honeydew melons	Preservation of aroma volatiles	Amaro, Beaulieu, Grimm, Stein, and Almeida (2012)
	Fresh-cut pineapple	Strong inhibition effect on the growth of yeasts and their production of volatile organic compounds	Zhang et al. (2013)
Use of edible coatings	Fresh-cut kajari melon	Retention of quality attributes and extension in shelf-life by using soy protein isolate coating	Yousuf and Srivastava (2017)
	Fresh-cut potato	Extension of shelf-life	Wu (2019)
	Fresh-cut strawberries	Reduction in microbiological counts	Tomadoni, Moreira, Pereda, and Ponce (2018)
Ozone	Fresh-cut bell pepper	Reduction in microbial populations	Alwi and Ali (2014)
	Fresh-cut papaya	Significant reduction in total bacteria counts	Yeoh, Ali, and Forney (2014)
	Fresh-cut spinach	Reduced yellowing and maintained compositional characteristics	Papachristodoulou, Koukounaras, Siomos, Liakou, and Gerasopoulos (2018)
Gamma radiation	Fresh-cut lettuce	Polyphenol oxidase activity of fresh-cut lettuce was significantly inhibited	Zhang, Lu, Lu, and Bie (2006)
	Fresh-cut celery	Polyphenol oxidase and respiration rate was greatly inhibited. Moreover, vitamin C, soluble solids, total sugars, and the sensory quality were maintained	Lu, Yu, Gao, Lu, and Zhang (2005)
Electron beam	Fresh-cut cabbage	4.0 log to 7.0 log reduction of *Escherichia coli*	Grasso, Uribe-Rendon, and Lee (2011)
	Fresh-cut cantaloupe	Significant decrease in *Salmonella enterica*	Palekar, Taylor, Maxim, and Castillo (2015)
UV light	Fresh-cut kailan-hybrid broccoli	Significant inhibition of *E. coli*, 495 S. enteritidis, and *L. monocytogenes*	Martínez-Hernández et al. (2015)
	Fresh-cut apples	Significant reduction of browning reactions	Chen, Hu, He, Jiang, and Zhang (2016)

Continued

Table 1 Strategies to maintain quality and safety of fresh-cut fruits and vegetables with recent examples—Cont'd

Technique	Fresh-cut fruit/vegetable	Specific effect found	Reference
	Fresh-cut paprika	Maintenance of antioxidant activity (DPPH activity), total phenolic compound, and vitamin C contents	Choi, Yoo, and Kang (2015)
Pulsed light	Fresh-cut apple	Effective in inhibition of browning and microbial growth	Gómez et al. (2012)
	Fresh-cut mangoes	Maintenance of the firmness, color, and nutritional quality	Charles, Vidal, Olive, Filgueiras, and Sallanon (2013)
	Fresh-cut avocado, watermelon and mushroom	More than 5 log reductions of *L. innocua* and *E. coli* throughout storage	Ramos-Villarroel, Martín-Belloso, and Soliva-Fortuny (2015)
Cold plasma	Fresh-cut apples	65% decrease of browning area	Tappi et al. (2014)
	Fresh-cut kiwifruit	Improved color retention and reduced darkening	Ramazzina et al. (2015)
Ultrasound	Fresh-cut pineapple	Results suggest it can counter oxidative stress	Yeoh and Ali (2017)
	Fresh-cut cucumber	Ultrasound treatment maintained good quality attributes	Fan, Zhang, and Jiang (2019)
	Fresh-sliced button mushrooms	Results demonstrated that US could be effective in maintaining quality and prolonging shelf-life	Wu et al. (2018)

6 Conclusion

The fresh-cut fruit and vegetables industry is expected to expand more in the coming years and, considering the increasing demand of consumers for such products, the use of emerging technologies could be an alternative in their quality maintenance and shelf-life extension. However, fresh-cut preparation is linked with contamination and subsequent acceleration in deterioration. Therefore, with the rise of the fresh-cut produce market, quality and safety concerns demand considerable attention. Lack of standard quality and safety of fresh-cut fruits and vegetables could result in serious disease outbreaks. Therefore, the development of novel and safe strategies that improve fresh-cut produce safety is of the utmost importance. Safety must be of primary concern in any food, including fresh-cut products. Some of the emerging technologies such as ozone, irradiation, pulsed light, and cold plasma have been found to be effective in preserving fresh-cut fruits and vegetables for longer periods of time. It is the responsibility and a challenging task for food scientists and food processors to make the industry able to meet the ever-growing demand of fresh-cut fruits and vegetables while maintaining the highest standards of quality and safety.

References

Alwi, N. A., & Ali, A. (2014). Reduction of *Escherichia coli* O157, *Listeria monocytogenes* and *Salmonella enterica* sv. Typhimurium populations on fresh-cut bell pepper using gaseous ozone. *Food Control, 46*, 304–311.

Amaro, A. L., Beaulieu, J. C., Grimm, C. C., Stein, R. E., & Almeida, D. P. F. (2012). Effect of oxygen on aroma volatiles and quality of fresh-cut cantaloupe and honeydew melons. *Food Chemistry, 130*, 49–57.

Artés, F., & Allende, A. (2005). Minimal fresh processing of vegetables, fruits and juices. In D. W. Sun (Ed.), *Emerging technologies in food processing (pp. 675–715), Vol. 26*. London: Academic Press.

Artés, F., Gómez, P. A., & Artés-Hernández, F. (2006). Modified atmosphere packaging of fruits and vegetables. *Stewart Postharvest Review, 5*(2), 1–13.

Berger, N. C., Sodha, V. S., Shaw, K. R., Griffin, M. P., Pink, D., Hand, P., et al. (2010). Fresh fruit and vegetables as vehicles for the transmission of human pathogens. *Environmental Microbiology, 12*, 2385–2397.

Charles, F., Vidal, V., Olive, F., Filgueiras, H., & Sallanon, H. (2013). Pulsed light treatment as new method to maintain physical and nutritional quality of fresh-cut mangoes. *Innovative Food Science & Emerging Technologies, 18*, 190–195.

Chen, C., Hu, W., He, Y., Jiang, A., & Zhang, R. (2016). Effect of citric acid combined with UV-C on the quality of fresh-cut apples. *Postharvest Biology and Technology, 111*, 126–131.

Choi, I.-L., Yoo, T., & Kang, H.-M. (2015). UV-C treatments enhance antioxidant activity, retain quality and microbial safety of fresh-cut paprika in MA storage. *Horticulture, Environment, and Biotechnology, 56*, 324–329.

Eissa, H., Fadel, H., Ibrahim, G., Hassan, I., & Elrashid, A. (2006). Thiol containing compounds as controlling agents of enzymatic browning in some apple products. *Food Research International, 39*, 855–863.

Fan, K., Zhang, M., & Jiang, F. (2019). Ultrasound treatment to modified atmospheric packaged fresh-cut cucumber: Influence on microbial inhibition and storage quality. *Ultrasonics Sonochemistry, 54*, 162–170.

Francis, G. A., Gallone, A., Nychas, G. J., Sofos, J. N., Colelli, G., Amodio, M. L., et al. (2012). Factors affecting quality and safety of fresh-cut produce. *Critical Reviews in Food Science and Nutrition, 52*, 595–610.

Gómez, P. L., García-Loredo, A., Nieto, A., Salvatori, D. M., Guerrero, S., & Alzamora, S. M. (2012). Effect of pulsed light combined with an antibrowning pretreatment on quality of fresh cut apple. *Innovative Food Science & Emerging Technologies, 16*, 102–112.

Graça, A., Santo, D., Esteves, E., Nunes, C., Abadias, M., & Quintas, C. (2015). Evaluation of microbial quality and yeast diversity in fresh-cut apple. *Food Microbiology, 51,* 179–185.

Grasso, E. M., Uribe-Rendon, R. M., & Lee, K. (2011). Inactivation of *Escherichia coli* inoculated onto fresh-cut chopped cabbage using electron-beam processing. *Journal of Food Protection, 74,* 115–118.

Hua, G., & Reckhow, D. A. (2007). Comparison of disinfection by product formation from chlorine and alternative disinfectants. *Water Research, 41,* 1667–1678.

IFPA (International Fresh-cut Produce Association), & PMA (The Produce Marketing Association). (1999). *Handling guidelines for the fresh-cut produce industry:* (3rd ed., p. 5). Alexandria, VA: IFPA.

James, J. B., & Ngarmsak, T. (2010). *Processing of fresh-cut tropical fruits and vegetables: A technical guide. Available from http://www.fao.org/docrep/014/i1909e/i1909e00.htm.*

Johnston, L. M., Jaykus, L. A., Moll, D., Anciso, J., Mora, B., & Moe, C. L. (2006). A field study of the microbiological quality of fresh produce of domestic and Mexican origin. *International Journal of Food Microbiology, 112,* 83–95.

López-Gálvez, F., Ragaert, P., Haque, M. A., Eriksson, M., van Labeke, M. C., & Devlieghere, F. (2015). High oxygen atmospheres can induce russet spotting development in minimally processed iceberg lettuce. *Postharvest Biology and Technology, 100,* 168–175.

Lu, Z., Yu, Z., Gao, X., Lu, F., & Zhang, L. (2005). Preservation effects of gamma irradiation on fresh-cut celery. *Journal of Food Engineering, 67*(3), 347–351.

Martínez-Hernández, G. B., Huertas, J.-P., Navarro-Rico, J., Gómez, P. A., Artés, F., Palop, A., et al. (2015). Inactivation kinetics of foodborne pathogens by UV-C radiation and its subsequent growth in fresh-cut kailan-hybrid broccoli. *Food Microbiology, 46,* 263–271.

Martn-Belloso, O., Soliva-Fortuny, R., & Oms-Oliu, G. (2006). Fresh-cut fruits. In *Handbook of fruits and fruit processing* (pp. 129–144). Ames, IA: Blackwell Publishing. https://doi.org/10.1002/9780470277737.ch8.

Nicola, S., & Fontana, E. (2014). Fresh cut produce quality: Implications for a systems approach. In W. Florkowski, R. Shewfelt, B. Breuckner, & S. Prussia (Eds.), *Postharvest handling: A systems approach* (pp. 217–273). San Diego, CA: Academic Press.

Olaimat, A. N., & Holley, R. A. (2012). Factors influencing the microbial safety of fresh produce: A review. *Food Microbiology, 32,* 1–19.

Ölmez, H., & Kretzschmar, U. (2009). Potential alternative disinfection methods for organic fresh-cut industry for minimizing water consumption and environmental impact. *LWT-Food Science and Technology, 42*(3), 686–693.

Palekar, M. P., Taylor, T. M., Maxim, J. E., & Castillo, A. (2015). Reduction of *Salmonella enterica* serotype Poona and background microbiota on fresh-cut cantaloupe by electron beam irradiation. *International Journal of Food Microbiology, 202,* 66–72.

Papachristodoulou, M., Koukounaras, A., Siomos, A. S., Liakou, A., & Gerasopoulos, D. (2018). The effects of ozonated water on the microbial counts and the shelf life attributes of fresh-cut spinach. *Journal of Food Processing and Preservation, 42*(1).

Parish, M. E., Beuchat, L. R., Suslow, T. V., Harris, L. J., Garrett, E. H., Farber, J. N., et al. (2003). Methods to reduce/eliminate pathogens from fresh and fresh-cut produce. *Comprehensive Reviews in Food Science and Food Safety, 2,* 161–173.

Ramazzina, I., Berardinelli, A., Rizzi, F., Tappi, S., Ragni, L., Sacchetti, G., et al. (2015). Effect of cold plasma treatment on physico-chemical parameters and antioxidant activity of minimally processed kiwifruit. *Postharvest Biology and Technology, 107,* 55–65.

Ramos-Villarroel, A. Y., Martín-Belloso, O., & Soliva-Fortuny, R. (2015). Combined effects of malic acid dip and pulsed light treatments on the inactivation of *Listeria* innocua and *Escherichia coli* on fresh-cut produce. *Food Control, 52,* 112–118.

Sanchís, E., Ghidelli, C., Sheth, C. C., Mateos, M., Palou, L., & Pérez-Gago, M. B. (2017). Integration of antimicrobial pectin-based edible coating and active modified atmosphere packaging to preserve the quality and microbial safety of fresh-cut persimmon (Diospyros kaki Thunb. cv. Rojo Brillante). *Journal of the Science of Food and Agriculture, 97*(1), 252–260.

Tappi, S., Berardinelli, A., Ragni, L., Dalla Rosa, M., Guarnieri, A., & Rocculi, P. (2014). Atmospheric gas plasma treatment of fresh-cut apples. *Innovative Food Science & Emerging Technologies, 21,* 114–122.

Thompson, R. C., Swan, S. H., Moore, C. J., & vom Saal, F. S. (2009). Our plastic age. *Philosophical Transactions of the Royal Society, B: Biological Sciences, 364*(1526), 1973–1976. https://doi.org/10.1098/rstb.2009.0054.

Tomadoni, B., Moreira, M. R., Pereda, M., & Ponce, A. G. (2018). Gellan-based coatings incorporated with natural antimicrobials in fresh-cut strawberries: Microbiological and sensory evaluation through refrigerated storage. *LWT-Food Science and Technology, 97,* 384–389.

Van Duyn, M. A. S., & Pivonka, E. (2000). Overview of the health benefits of fruit and vegetable consumption for the dietetics professional: Selected literature. *Journal of the American Dietetic Association, 100*(12), 1511–1521.

Wu, S. (2019). Extending shelf-life of fresh-cut potato with cactus Opuntia dillenii polysaccharide-based edible coatings. *International Journal of Biological Macromolecules, 130,* 640–644.

Wu, S., Nie, Y., Zhao, J., Fan, B., Huang, X., Li, X., et al. (2018). The synergistic effects of low-concentration acidic electrolyzed water and ultrasound on the storage quality of fresh-sliced button mushrooms. *Food and Bioprocess Technology, 11*(2), 314–323.

Yeoh, W. K., & Ali, A. (2017). Ultrasound treatment on phenolic metabolism and antioxidant capacity of fresh-cut pineapple during cold storage. *Food Chemistry, 216,* 247–253.

Yeoh, W. K., Ali, A., & Forney, C. F. (2014). Effects of ozone on major antioxidants and microbial populations of fresh-cut papaya. *Postharvest Biology and Technology, 89,* 56–58.

Yousuf, B., & Srivastava, A. K. (2017). A novel approach for quality maintenance and shelf life extension of fresh-cut Kajari melon: Effect of treatments with honey and soy protein isolate. *LWT-Food Science and Technology, 79,* 568–578.

Zhang, L., Lu, Z., Lu, F., & Bie, X. (2006). Effect of γ irradiation on quality-maintaining of fresh-cut lettuce. *Food Control, 17*(3), 225–228.

Zhang, B.-Y., Samapundo, S., Pothakos, V., de Baenst, I., Sürengil, G., Noseda, B., et al. (2013). Effect of atmospheres combining high oxygen and carbon dioxide levels on microbial spoilage and sensory quality of fresh-cut pineapple. *Postharvest Biology and Technology, 86,* 73–84.

2

Status and recent trends in fresh-cut fruits and vegetables

Sucheta*, Gisha Singla*, Kartikey Chaturvedi†, Pankaj Preet Sandhu*

**Center of Innovative and Applied Bioprocessing, Mohali, India. †National Institute of Food Technology Entrepreneurship and Management, Sonipat, India*

1 Introduction

Fruits and vegetables are a perishable commodity, which suffers extensive postharvest losses in absentia of preservation methodologies, especially in low-income countries. Various essential components are present in fruits and vegetables like vitamins, fibers, and micronutrients, which, when consumed in considerable quantity, impart health benefits (Abadias, Usall, Anguera, Solson, & Vinas, 2008). Consumption of fruits and vegetables has been recommended by various organizations such as the WHO (World health organization), FAO (Food and agriculture organization), USDA (United states Department of Agriculture), and EFSA (European food safety authority) to minimize risk of cardiovascular disease and cancer (Allende, Tomás-Barberán, & Gil, 2006; Su & Arab, 2006). Appearance, freshness, and color are the mainly used as criteria to evaluate "farm to fork" immediate quality attributes of fresh fruits and vegetables, so as to indicate quality in the supply chain (Barrett, Beaulieu, & Shewfelt, 2010; Clydesdale, 1991). Fresh fruits and vegetables form part of a well-balanced diet and are sources of carbohydrates, fiber, vitamins, minerals, and poly-phenols (Nunes, 2015). In view of the developments in processing technologies, low-cost shelf-life enhancement methodologies, rising incomes, and busy lifestyle have attracted minimal processing of fruits and vegetables in the form of fresh-cut produce. Current research efforts in the direction of sanitation, edible coatings, innovative disinfection treatments, MAP, and hurdle technologies by combinations of one or more of the above lead to possibilities in increased

Fresh-cut Fruits and Vegetables. https://doi.org/10.1016/B978-0-12-816184-5.00002-1

consumer acceptance as well as assured safety of fresh-cut produce worldwide. This chapter deals with trends in production, consumption, recent methodologies, preservation methods, and future marketing prospects of fresh-cut fruits and vegetables (FFVs).

2 Overview of FFVs worldwide

With change in lifestyle and awareness among consumers, demand for healthy, minimally processed fruits and vegetables has increased (Ramos, Miller, Brandão, Teixeira, & Silva, 2013). Minimal processing, include nonthermal techniques such as food preservation methods, and safe standards, which maintain fresh-like characteristics and increase the shelf-life from 5 to 7 days at 4° C (Allende, Tomás-Barberán, & Gil, 2006; Cliffe-Byrnes & O'Beirne, 2002). FFVs are a popular choice of meal nowadays, as no cooking or preparation is required, and are an abundant source of vitamins, minerals, and phytochemicals.

The International Fresh cut Produce Association (IFPA) coined the definition of FFVs as: fresh fruits and vegetables that are peeled and cut into usable product, having high nutrition, convenience, and flavor and, after proper packaging, offered to consumers. These fruits and vegetables are not frozen and have not undergone thermal processing. They are ready to eat or serve (RTE or RTS). Increasing consumer demand for fresh, healthy, convenient, and additive free food products are the main reasons behind market growth of these FFVs (Oliveira et al., 2015).

To access ripening stages (color charts of banana, tomato, and avocado) and to grade (size, color, and defects) color charts and visual inspection methods have also been frequently used in food industries (Nunes, 2015). Various rating scales and color charts have developed and studies have shown that there are correlations between sensory, physical, or chemical attributes of fresh fruits and vegetables (Corollaro et al., 2014; Pace, Renna, & Attolico, 2011). Examples include that firmness and color are sensory attributes of tomatoes (Ressureccion & Shewfelt, 1985), while volatile compounds are correlated with sweetness and sourness (Maul et al., 2000).

To restore freshness and shelf-life of fruits and vegetables, proper handling, preparation, and storage is required to assure the reduction of spoilage. Refrigeration also helps to maintain freshness (Tournas, 2005). So, various nonthermal methods have been employed, such as ozone based treatments, ultraviolet radiation, pulsed light, cold plasma, and novel packaging practices to maintain freshness. Effectiveness of these methods depends upon the food material, type of contamination, and process employed (Ramos et al., 2013). Type and amount of microflora depends upon type of produce, agronomic practices, geographical area, and weather conditions on fresh fruits and vegetables such as roots, leaves, bulbs, and tubers (Olaimat & Holley, 2012). Raw fruits and vegetables containing natural microflora are nonpathogenic and pathogenic microflora gets added from human, animal, and environmental sources as, during cutting, shredding, and peeling, the surface gets exposed and prone to bacteria, yeasts, and molds (Food and drug administration, 2008; Sánchez, Elizaquível, & Aznar, 2012).

Fruits and vegetables fall under the category of perishable foods that are a rich source of carbohydrates, poor in proteins, and having neutral to slightly acidic pH, which is optimum for the growth of micro-flora responsible for spoilage (Del Nobile, Conte, Cannarsi, & Sinigaglia, 2008). Processing conditions alter the physical integrity of food products and minimal processing reduces the shelf-life to 4–7 days, compared to weeks or months for the original raw material (Sanz, Gimenez, Olarte, Lomas, & Portu, 2002). A decrease in shelf-life after minimal processing is due to production of ethylene, and enzymatic and nonenzymatic browning (Ramos et al., 2013). Production of off-odor, off-flavors, and microbial proliferation, discoloration (browning, whitening, and translucency), and firmness loss results in changes in product appearance, decrease in nutritional value, and leads to microbial growth (Klein, 1987; Ruiz-Cruz et al., 2010). To different extents and paces, these physical, biological, biochemical, and microbial changes are influenced by intrinsic and extrinsic factors, leading to significant quality losses in harvest, processing, storing, and consumption (Mahdavian, Kalantari, & Ghorbanli, 2007; Shirzadeh & Kazemi, 2011).

Despite the increase in health consciousness and awareness of food safety issues among consumers, food-borne diseases related to these fruits and vegetables keep on increasing constantly (Vitale & Schillaci, 2016). Contamination can take place from air, water, soil, insect vectors, equipment, or by improper handling (Martinez-Vaz, Fink, Diez-Gonzalez, & Sadowsky, 2014). Bio-films formed on surfaces of fresh-cut produce have a negative impact as they impair surface sanitation and lead to increase in contamination (Kumar & Anand, 1998; Martinez-Vaz et al., 2014).

Prevention of food-borne diseases is an important step, which can be done by reducing the microorganism population on whole or fresh-cut produce (Corbo, Del Nobile, & Sinigaglia, 2006). Various methods such as chemical and physical methods have proved efficient in reduction of microorganisms. Chemical methods include mechanical washing with chemicals such as hypochlorite, chlorine dioxide, bromine, iodine, tri-sodium phosphate, electrolyzed water, ozone, etc., followed by rinsing with potable water. Physical methods using shear forces such as modified atmosphere packaging, active and intelligent packaging, nano-composite packaging, irradiation, high pressure processing, and ultrasound are effective in removing bacteria from plant surfaces (Parish et al., 2003). Due to efficacy, cost effectiveness ratio, and simplicity, the most commonly used method is chlorine. Lack of efficiency can lead to disinfectant inaccessibility, resulting in the growth of microorganisms (Ramos et al., 2013). According to Randhawa, Khan, Javed, & Sajid, 2015 from a consumer point of view, these FFVs help in maintaining a healthy diet, freshness, and nutritional status.

Consumption of fruits and vegetable depends upon flavor, quality, postharvest technology, and supply chain (Kyriacoua & Rouphael, 2018). Fresh-cut produce is more perishable than whole produce, as it is physically altered during processing operations. Fresh-cut fruit and vegetables reduce meal preparation time and the volume of kitchen waste produced but are fragile, as processing does not result in biological stabilization, but rather in reduction of shelf-life to a few days. During processing, breaking down of cells of FFVs leads to release of intracellular products, thus enhanced bacterial growth (Oliveira et al., 2015). Preliminary operations on these FFVs cause some mechanical and physiological damage, resulting in speeding up of chemical and enzymatic reactions. To prevent this, and to produce high-quality raw products, integrated pest management practices (IPM) have been adopted to reduce massive use of pesticides and produce selective products having low environmental risk (Arienzo, Cataldo, & Ferrara, 2013). The EC has defined the new term "maximum residue level" (MRL) for the active compounds present in pesticides to replace harmful compounds, for sustainable use of chemicals and encouraging production processes that have limited or zero utilization of pesticides (Commission Regulation EC no 396/2005; Commission Regulation EC no 128/2009).

Fresh-cut produce has been associated with growth of various pathogens such as *Escherichia coli* and *Salmonella* spp. Water washing of this fresh-cut produce removes soil exudates and associated microorganisms, which are then transferred to the water.

Therefore, pathogen cross-contamination can occur via water (Holvoet, Jacxsens, Sampers, & Uyttendaele, 2012). Washing of these FFVs using disinfectant solutions will enhance microbial removal and avoids cross contamination. Chemical method such as peracetic acid (PAA), due to its stability in presence of organic matter and nonproduction of harmful by products, has been an alternate disinfectant method (Santoro et al., 2007). FFVs suffer from evaporation of water, resulting in wilting, shriveling, flaccidness, soft texture, and loss of nutritional value (Nunes & Emond, 2007). Consequently, best postharvest practices have been required for maintaining product quality for sufficient storage time.

From the results of Nunes (2015), it has been concluded that subjective quality evaluation and physiochemical analysis are simple and economic methods to estimate changes in color, texture, water content, and specific chemical components when fresh fruits and vegetables are exposed to environmental conditions. From the studies of Ali, Nawaz, Ejaz, Haider, and Alam (2019), it has been concluded that during low temperature storage of fruits and vegetables, H_2S played a major role in delayed ripening, inhibiting senescence, inhibition of physiological disorders, reduction in color changes, and suppression of respiration oxidative stress elevation. Slow H_2S releasing pads can be considered beneficial in the global supply and transport chain of fruits and vegetables. Further research has been required to standardize optimum concentration of H_2S before its application at commercial levels.

To control microbial spoilage, modified atmosphere packaging (MAP) has been used for MPFC fruits and vegetables. Nonmicrobial spoilage has been reduced by using antibrowning agents and edible coating methods (Putnik et al., 2017). In Modified atmospheric packaging (MAP), the atmosphere inside the package has been altered by active or passive MAP by maintaining interaction between respiration rate natural material and gases transfer through packaging material of fresh fruits and vegetables. Active MAP includes transfer of gases in or out of package, or use of gas scavengers or absorbents, thereby establishing the desired mixture of gases. Passive packaging includes the use of specific packaging film, resulting in a naturally developed atmosphere due to exchange of gases through film (Lee, Arul, Lencki, & Castaigne, 1996). Oxygen, nitrogen, and carbon dioxide are commonly used gases in MAP. O_2 has been consumed and CO_2 has been generated. N_2 has been used as a filler gas for balancing and to prevent collapse (Sandhya, 2010).

3 Fresh-cut produce trends in Europe, the United States, and Asian countries

Trends in total Fruit and vegetable consumption in European countries reflect total fruit and vegetable production, as reported by FAO (2013). According to Van Haute, Sampers, Holvoet, and Uyttendaele (2013), chlorine was the most commonly used disinfectant in the food industry, due to its low price, easily availability, and facility to apply. Under certain conditions its efficiency of reducing microbial load is limited (Yaron & Romling, 2014) or inactivated due to presence of organic matter (Parish et al., 2003; Ramos et al., 2013); this results in production of unhealthy by-products, which can be carcinogenic and mutagenic, present in an indicative list of industrial emissions (IPPC, 2007/0286 (COD). Therefore, its use has been banned in some countries such as Belgium, Denmark, Germany, and the Netherlands (Bilek & Turantaş, 2013; Fallik, 2014; Ramos et al., 2013). Consequently, various greener alternatives such as bacteriocins, bacteriophages, enzymes, and phytochemicals; chemical methods such as ClO_2, electrolyzed oxidizing water, H_2O_2, ozone, organic acids, etc.; and physical methods including irradiation, filtration, US, UV light; or a combination of strategies are employed (Meireles, Giaouris, & Simões, 2016).

In Europe, the market was characterized by double-digit growth for FFVs in the early 1980s (Rabobank, 2011). Among the FFVs, 50% of total fresh-cut volume has been accounted for by packaged lettuce; the remaining 50% has been fresh-cut fruits and ready to cook food (Rabobank, 2010). In a survey in Western Europe, Euromonitor recorded growth of 19% per capita volume (2009–14) (Anastasi, 2015). Due to the convenience and freshness of FFVs, the "fresh-cut" sector plays an important role. Overall daily consumption of fruits and vegetables within the 28 EU member states, according to the consumption monitor of the European Fresh Produce Association (2014), decreased by 13.1% in 2013 as compared with average years 2000–06 (Freshfel, 2014). European governments have lunched various informational and educational initiatives concerning contemporary dietary habits and overweight related health problems, aimed at increasing public awareness about the benefits of fruits and vegetables (Gordon, McDermott, Stead, & Angus, 2006; Mazzocchi, Traill, & Shogren, 2009). An EU School Fruit Scheme (SFS) has been introduced by the agriculture ministry of the European Commission (EU) and approved, in 2008, for supporting consumption of fresh fruit and vegetables among children in school. The concept of FFVs first appeared in 1980 in Europe, and subsequently became popular. According to the definition given by the International Fresh cut Produce Association (IFPA), FFVs were just washed, cut, mixed and packed. Consumer acceptance has increased towards these FFVs as it ensures quality and safety of products (Artés, Gómez, Aguayo, Escalona, & Artés-Hernández, 2009; Baselice, Colantuoni, Lass, & Nardone, 2017). The FFVs market is growing at a larger pace than any other segment of this market, so the fresh-cut industry has reported a constant growth in terms of quantity and turnover as fresh-cut segment serves both the food service industry and retail outlets (Rosa & Rocculi, 2007). In the United States, fresh-cut fruit salad and vegetables comprised more than 10% and 35%, respectively, of their value in this segment (Lamikanra, 2002).

According to Rabobank (2010), countries like Germany and Spain have emerging markets in FFVs as compared to Italy and Netherlands having established markets. A forecast by Euromonitors 2015–20 survey data revealed strong growth of FFVs sales in Spain and Germany followed by France and Italy; while growth and consumption in southeastern European countries still shows as negligible.

According to data by ISMEA (2011), sales of fresh-cut, ready-to-use vegetables grew considerably during 2008, with volumes rising by 16.7% and sales up to 7.5%. In the European Union, the United Kingdom accounted for roughly one-third of total EU FFV consumption. The retail market recorded 5% growth in 2008 and 4% growth in 2009 (Rabobank, 2010). According to the FAO, 2010 report, the largest consumer of FFVs was identified as the United Kingdom, having 97% interviewees from this country in terms of purchasing and consumption of FFVs, followed by Spain (88.4%), Italy (84%), and Greece (70.3%). Out of the total fresh-cut produce, green salad represented 36.7% in United Kingdom, 52.7% in Greece, 87% in Italy, and 53.2% in Spain. Therefore, green salads represented about 50% of total fresh consumption, except in Italy (Baselice et al., 2017).

In Canada, the growing season is short, so most fresh fruits and vegetables need to be imported. Canada imports 88% of fruits and 41% of vegetables (Statistics Canada, 2001). According to import data from 2011, Canada imports leafy greens, soft fruits, citrus fruits, grapes, cauliflower, broccoli, onions, and carrots from the United States, and also peppers, tomatoes, avocados, cucumbers, asparagus, etc. (AAFC, 2014). The Canadian government has taken various initiatives to encourage consumers in consumption of a nutritious and healthy diet in the form of fresh fruits and vegetables.

Internationally, a number of outbreaks regarding food-borne illness have been identified and no of efforts have been made to resolve food safety problems (Berger et al., 2010; Denis, Zhang, Leroux, Trudel, & Bietlot, 2016). Examples include leafy greens and fresh herbs that are potential sources of bacterial infections (FAO/WHO, 2008). In 2006, the United States experienced an outbreak of *E. coli* infection due to bagged spinach consumption, resulting in 200 cases of food poisoning (Grant et al., 2008; Wendel et al., 2009). Then, in 2007, an outbreak of salmonella infection in fresh herbs took place that affected at least 51 individuals from England, Wales, Scotland, Denmark, the Netherlands, and the United States (Elviss et al., 2009; Pezzoli et al., 2008). In 2011–12, Salmonella infection from watermelon was reported with 63 cases of food poisoning (Byrne et al., 2014). Salmonellosis outbreak from fresh tomatoes has also been reported (Hanning, Nutt, & Ricke, 2009).

Experts were gathered in Canada in a meeting held by the Canadian Food Inspection Agency (CFIA) and Food Safety Science Committee (FSSC) in 2008 to assess and rank food-hazard combinations. CFIA initiated a food safety system in same year to modernize and strengthen Canada's food safety system (CFIA, 2013). So, based upon the above recommendations from the FAO/WHO and FSSC, along with a review of the scientific literature, and documented outbreaks of food-borne illness, fresh leafy vegetables, herbs, green onions, tomatoes were selected for targeted surveillance under FSAP, with focus on *Salmonella* spp., *E. coli, Campylobacter* spp., and *Shigella* spp. (Denis et al., 2016).

Minimally Processed Fresh Cut (MPFC) apples occupy a significant proportion in international food markets, as their consumption has been recommended, with increased daily consumption of FFVs (Nicola, Tibaldi, & Fontana, 2009). A major challenge with MPFC foods is shorter life span—of less than 2 weeks—as minimal processing results in physiological damage to fruits and induces various enzymatic and microbial changes, which lead to production of ethylene, oxidation of Polyphenols, surface browning, loss of structural integrity, and nutritive and sensorial properties (Putnik, Kovacevic, Herceg, & Levaj, 2016).

In Spain, an annual increase of 5%–6% in sales of fresh-cut, ready-to-eat fruits and vegetables has been seen. It has reached 70,600 and 74,064 tons in 2010 and 2011, respectively (Anonymous, 2013). In 2013, the Spanish market recorded sales of 77,000 ton (Anonymous, 2014).

Central American supermarket chains implemented private standards in order to cut the cost and compete with wet markets by shifting the traditional whole systems towards centralized purchases and using implicit contracts and specialized/dedicated wholesalers. Private standards implementations are good for consumers, but the tougher standards are challenging for producers in making significant investments.

China is the largest tropical fruits producing and importing country in the world (FAO, 2009). The Philippines, Thailand, Vietnam, and Malaysia export fruits to China. Banana, mango, pineapple, papaya, and dragon fruits are consumed in whole form or as juice in the Chinese market. Major outbreaks reports of foodborne illness with consumption of these tropical fruits caused by Salmonella by tropical fruits such as Saintpaul, Heidelberg, Litchfield, Typhi, Newport, Paratyphi, and Senftenburg has been mentioned by CDC (2003, 2004) and Strawn, Schneider, & Danyluk (2011).

Sales of organic food in 2011 worldwide were approximately 63 billion US dollars (Low, 2013). These organic vegetables grown using agricultural fertilizers such as animal manure may result in microbial contamination of vegetables (Goodburn & Wallace, 2013). Thus, microbial safety is an important issue in organic and conventional vegetables.

In developing countries, rapid transformation in food systems leads to high value products and modern supply chains. It has been found that demand for products was highly elastic in supermarkets and nontraditional imports. In addition, the income impact was stronger than price and supermarket penetration (Mergenthaler, Weinberger, & Qaim, 2009).

Several studies have reported the difficulties faced by students studying in foreign countries, migrants to Korea, and female immigrants, in terms of dietary acculturation and dietary changes (Cha, Bu, Kim, Kim, & Choi, 2012; Song & Kim, 2015). Studies on similarities between cultural adaption and dietary habits have been done overseas that influenced the dietary changes and health of individuals and families (Franzen & Smith, 2009; Unger et al., 2004). Results concluded that fresh fruits are available at affordable prices in South East Asia, but not in Korea. Sautéed vegetables are preferred mainly by South East Asian workers (Lee, Lee, & Lee, 2017).

Significant amounts of β-Carotene and pro-vitamin A are present in vegetables and fruits; these are absorbed and converted into Vitamin A in the human body. This pro-vitamin A from plant foods fulfills the need of vitamin A intake in developing countries. From global surveys, it has been found that 254 million preschool children and infants suffer from Vitamin A deficiency (VAD) (WHO, 2000; WHO/UNICEF, 1995). In developing countries, food based strategies are useful to combat VAD (Ruel, 2001).

Daily recommendations for consumption of fresh fruits and vegetables have been met by less than 15% of US adults (Moore, Thompson, & Demissie, 2017). To improve environment of local food market, rotating schedule of renovated buses, vans to sell the local fruits and vegetables at affordable prices, focused on areas to provide affordable and nutritious food to low income and minority neighborhoods. Obesity has become an epidemic in the United States, due to unhealthy eating habits. The US Department of Agriculture (USDA), between 2009 and 2015, invested more than $1 billion in mobile food markets. Such popular food farmers markets include Arcadia's Mobile market in Washington, DC and Fresh truck in Boston (Ylitalo et al., 2019).

To predict the interactions between food fraud, a suitable Bayesian Network (BN) methodology can be applied (Marvin et al., 2016). BN is a graphical model applying probability relationships among variables representing knowledge, reasoning under uncertainty, and conclusions (Cheng, Greiner, Kelly, Bell, & Liu, 2002). BN can be applied to hazard risk assessment, fraud detection, accident prevention, food safety, nano-materials, sample size optimization, medical image analysis domains, and so on. Due to diverse geographical and meteorological conditions, and the high frequency of notifications for fruits and vegetables to the European Rapid Alert System for Food and Feed (RASFF), fruits and vegetables that are imported to Europe from India, Turkey, and the Netherlands were the best examples of BN methodology. A total of 5610 RASFF notifications occurred in the period 2005–15; fruits and vegetables were collected from Turkey, India, and Thailand and linked to climatic, agricultural, and economic factors by means of BN. The presence of mycotoxins, foreign bodies, chemical contaminants, pesticide residues, unauthorized additives, and adulteration were observed in hazard reports (Bouzembraka & Marvina, 2019).

4 Fresh-cut processing methodologies, traceability, and challenges

4.1 Receiving, inspection, precooling, cleaning/washing, and disinfection

After harvesting, fruits are received and inspected according to their prescribed quality standard and washed with water to obtain good quality products, dependent on good quality of the raw material.

Temperature control in the postharvest chain helps to reduce the postharvest losses and maintain the fresh produce quality. Precooling of the produce is necessary after harvesting to remove the stored heat associated with the temperature of the produce. Precooling of the produce also assists in reducing the metabolic activity (respiration, ethylene production, and transpiration) of the produce and maintains the fruit quality.

Generally, pure water is used for the washing of fruits having a smooth surface. In case of fruit having a rough surface (cantaloupes) or where there may be chances of microflora on the fruit surface, pure water with the addition of disinfectant/sanitizer (chlorinated water, hydrogen peroxide, trisodium phosphate, and ozone) is used for the washing of fruits (Beuchat, 2000; Heard, 2002; Sapers and Simmons, 1998).

4.2 Sorting and grading

Sorting plays an important role in increasing the produce quality and consumer acceptability of the product. After harvesting, fruits and vegetable are sorted according to their size, color, shape, and appearance. During the process of sorting, the immature and the damaged produce are removed to maintain the quality of the product (Mishra & Gamage, 2007).

After sorting, the produce is graded into different quality standards according to the market (local and international) and undergoes removal of immature, off grade, and undersized produce. The quality standard for the produce is country specific. Traditionally, this work was done manually; later, optical methods were adopted for the grading of fruits and vegetable (Mishra & Gamage, 2007).

4.3 Peeling, coring, and cutting

Peeling and cutting is an essential processing step for the processing of FFV produce to subdivide the intact fruit and vegetables into small fractions manually or mechanically, according to the type of fruits and vegetables. Both peeling and cutting increase fruit and vegetable senescence, due to increased metabolic activity, resulting in increased respiration and ethylene production rate. The peeling and cutting of the produces makes more substrate available on the surface for microbial growth and enzymatic activity. The cutting of fruit develops wounding stress in the fruit tissue, thus accelerating the rate of spoilage in the cut produce. Therefore, the cutting of the harvested fruit and vegetables is the prime factor responsible for the rapid deterioration of fruits and vegetable as compared to the intact fruits and vegetables (Portela & Cantwell, 2001). The increased respiration rate in cut fruit and vegetables develops rapid deteriorative changes, which are linked with tissue senescence, thus reducing the shelf-life and quality of the produce. The cutting blade sharpness also affects the quality of the cut produce. Cutting with a sharp blade helps to reduce the electrolyte leakage, lignin accumulation, and also visual quality in fresh-cut produce during processing (Li et al., 2017).

During the cutting process, the phenolic metabolisms come into existence because the cutting of fruits and vegetables breaks the plasma membrane and induces/triggers the oxidative enzyme (polyphenol oxidase), resulting in tissue browning (Degl'Innocenti, Pardossi, Tognoni, & Guidi, 2007; Li et al., 2017).

4.4 Washing and sanitation

Cutting of the fresh produce release the fluids from the produce tissue and this tissue fluid must be washed to avoid undesirable biochemical reactions and microbial contamination.

The quality and shelf-life of the fresh-cut produce depends upon the sanitation and hygiene maintained during the processing. The reduce rate/level of the contamination is dependent on the application of suitable sanitizer and disinfectant. Because after cutting the produce comes in direct contact with the processing environment, the presence of microbes in the processing environment affects the quality of cut produce or even spoils the fresh-cut produce. Generally, chlorine is used as disinfectant due to its low coat and antimicrobial effectiveness. The concentration in the range of 50–200 ppm with 5 min exposure time is mostly used as hypochlorite and hypochlorous in fruit and vegetable industry (Ramos et al., 2013).

However, due to health issues, the use of chlorine in the food industry as a disinfectant is prohibited in some European countries such as Belgium, the Netherlands, and Denmark. Alternative sanitization techniques are used in the food industry to maintain the quality of products, such as ozone (O_3), chlorine dioxide (ClO_2), and calcium (Ca), to destroy the microbes, pathogen spores, and bacteria, and inhibit enzymatic browning (Meireles et al., 2016; Yaron & Romling, 2014). The excess moisture on the fresh produce due to washing must be removed before packaging to maintain the quality and safety of the produce. Therefore, the fresh produce is passed through semifluidized beds in combination with forced air to remove the excess moisture.

4.5 Packaging of FFVs

4.5.1 Modified atmospheric packaging

MAP is generally used for the packaging of fresh-cut produce by altering the composition of the gases within the package and creating a modified environment within the package. The MAP of fresh-cut produce consists of reduced oxygen and increased carbon dioxide environment to reduce the degradation process. The success of MAP is depending upon the produced respiration rate and gases transfer/permeability of the package (Zagory & Kader, 1988).

4.5.2 Active modification

An active modification in the package environment involves creating a partial vacuum within the package and replace the package internal gases with the desired mixture of gases (Scetar, Kurek, & Galic, 2010).

4.5.3 Passive modification

In passive modification, the respiration rate of the produce is responsible for the change in the internal environment of the package. Final equilibrium in the internal environment depends upon the gases permeability of the packaging material and the characteristics of the produce in the package (Scetar et al., 2010).

4.5.4 Active packaging

Active packaging involves the application of some active agents, such as carbon dioxide, ethylene, and oxygen scavengers, to replace or modify the composition of the gases within the package.

4.6 Record keeping and traceability

4.6.1 Record keeping

The information at ground level must be recorded for the complete traceability of the product even at the farm level. The traceability system at ground level includes record keeping of:

- agriculture field where the crop is grown;
- manufacturing practice;
- use of pesticides/fertilizer;
- date of harvesting;
- method of harvesting;
- method transportation; and
- temperature during transportation.

Previously, different types of code were printed on the produce package, which were used for produce traceability. The major drawback of these types of code is that only limited information was obtained from the traceability code (James & Ngarmsak, 2010). Nowadays, the bar code is an improved way to track the movement and complete information related to the produce.

The Uniform Code Council (UCC) takes a global leadership role in establishing and promoting multiindustry standards for the identification of products and electronic communication.

4.6.2 Product traceability in supply chain

Consumer awareness about their health and food safety has forced manufacturers to ensure the safety of the product by traceability during supply chain.

Product traceability is an important step to determine the real time status and exact location of the produce in the supply/distribution chain, and also refers to the recall and reconstruction of the produce from the supply chain based on the data/record maintained (James & Ngarmsak, 2010). It collects all the information from harvesting to the retail. According to the Codex Alimentarius, traceability, or product tracing, is "the ability to follow the movement of a food through specified stages of production, processing, and distribution." The products traceability has advantages for both consumer and the producers (James & Ngarmsak, 2010):

- It helps to enhance the efficiency at every level from manufacturing company to the supply chain.
- The product having quality and safety issues can be segregated from the lot/batch specific.
- Helps to recall the products during quality issues and other outbreaks related to food-borne illness, thus avoiding shipping of products having quality issues.
- Decreases the chances of product rejection, bad publicity, and lawsuits.

4.7 Challenge

FFVs have advantages over whole produce; however, the processes used to maintain the quality and the freshness of the produce face a number of challenges (James & Ngarmsak, 2010). The major challenges in the fruit cut processing industry are:

- high processing/production cost;
- high perishability of the cut produce;
- increased metabolic activity;

- more prone to quality degradation/losses;
- preservation of shelf-life; and
- microbial contamination.

5 Safety practices during processing of FFVs

Fresh fruits and vegetables are highly perishable in nature and their quality deteriorates rapidly after harvesting due to a number of physical and physiological processes. As the fruit and vegetable are a live food commodity, both quantitative and qualitative losses can take place anywhere in the supply chain after harvesting (transportation, storage, packaging, and distribution). Proper postharvest management can only provide the produce in minimum time with safe conditions to the processed market.

Mainly due to processing, FFVs have a shorter shelf-life as compared to the intact fruits. A number of processing steps involved in the processing of the FFV effect the stability and shelf-life of FFVs. The shelf-life stability of FFVs depends upon the number of unit operation (washing/disinfecting, peeling, shredding, cutting, drying, and packaging) involved during the processing, use of additives, temperature management, and good manufacturing practice. FFVs are considered ready to eat, thus it is necessary to ensure the safety of the produce with food safety programs like Good Agricultural Practices (GAPs), Good Manufacturing Practices (GMPs), and Hazard Analysis Critical Control Point (HACCP) programs (IFPA, 2001).

Immediately after the cutting of fruits, the proper washing of the fruit is an important step for the safety of the produce to remove tissue fluid, remove cell exudate, and reduce the microbial load responsible for FFV quality loss. Water washing also removes the cellular contents of FFVs. As water used for the washing is generally recirculated and reused in the washing system, it can contaminate the produce with pathogen microbes. Therefore, disinfection of water is necessary to eliminate the pathogen bacteria and other micro-flora to ensure the product safety (Allende, Selma, López-Gálvez, Villaescusa, & Gil, 2008; IFPA, 2001). The purpose of sanitization is to reduce the microbial load from the cut produce. Washing with disinfectants only reduces the microbial load from the surface of the produce. Washing has no significant effect on the pathogen microbes that are present within the produce. Thus, the quality of washing water is gaining more attention to maintain the quality of FFVs. The efficacy of the method of the sanitization method is dependent upon the method of sanitization, concentrations of sanitizer, exposure time and temperature, physiology of the microbes, and surface characteristic of the produce (Allende et al., 2008; IFPA, 2001; López-Gálvez, Allende, Selma, & Gil, 2008).

6 Newer preservation methods for enhancement of shelf-life of fresh-cut produce

Intact fruit or vegetables contain certain enzymes and phenolic compounds, which remain compartmentalized in endoplasmic reticulum until the fruit is cut. Once a fruit is bruised or cut, leading to disruption of its native cellular structure, it results in interaction of certain enzymes (polyphenol oxidases, catechol oxidases, etc.) and phenolic compounds. The fresh-cut produce thus undergoes enzymatic browning in the presence of oxygen. Several mechanisms of this reaction has been reported, all of which form complex brown polymers at the end of the reaction. In addition, secondary browning in FFVs is also associated with microbial assisted exudation of enzymes. Therefore, to prevent such deteriorative changes, fresh-cut fruits or vegetables need to be preserved through using food additives, physical methods, or packaging innovations.

6.1 Additives

Sanitizers, antioxidants, antibrowning agents, and texturizers are additives being used to preserve fresh-cut fruits or vegetables. Liquid chlorine and hypochlorite in concentrations of 50–200 ppm (contact time less than 5 min) are mostly used currently for disinfestation purposes. Several studies indicate the efficacy of chlorine utilization for apples, cantaloupe, tomatoes, lettuce, etc. However, due to regulatory restrictions, chlorine is still banned in some countries like the Netherlands, Belgium, and Switzerland for utilization in ready-to-eat food products. Citric, ascorbic, and other organic acids are known for reducing pH and surface microflora in fresh-cut fruits or vegetables and have been the leading generally recognized as safe (GRAS) antioxidants for fresh produce. They have been used for shelf-life extension of salad vegetables (Shapiro & Holder, 1960), Chinese water chestnut (Jiang, Pen, & Li, 2004), fresh-cut apples (Aguayo, Requejo-Jackman, Stanley, & Woolf, 2010), and fresh-cut mango (Chiumarell, Ferrari, Sarantópoulos, & Hubinger, 2011), etc. Additionally, ascorbic acid is widely demonstrated for its effectiveness in reducing browning through reduction of *o*-quinones, thereby retarding browning polymer formation (Rico, Martin-Diana, Barat, & Barry-Ryan, 2007).

Considering sensorial impact, different plant essential oils have recently been introduced to prevent microflora growth over surfaces of fresh-cut fruits or vegetables. These consist mainly of terpenoids and are extracted from garlic, coriander, peppermint, thyme, cinnamon, etc. The inclusion of aroma volatiles in headspace of fresh-cut commodities has been reported to enhance shelf-life significantly. Hexanal, 2-*E*-hexenal, oregano, and thyme inhibited pathogenic microflora in fresh-cut apples, lettuce, and carrots (Gutierrez, Zaldivar, & Contreras, 2009). Purslane is a herb, aqueous extract of which retained membrane integrity and prevented browning in fresh-cut potato slices at 4°C (Liu et al., 2019). Calcium lactate, an effective bio preservative on infusion with heat treatment or ozone, has been used for firmness retention in fresh-cut cantaloupe, strawberry, and lettuce (Martin-Diana et al., 2006). Calcium interacts with pectin polymers forming a cross-polymer network, thus increasing the strength of plant cell walls. 1-Methylcyclopropene also retains firmness of FFVs by blocking ethylene receptors and inhibiting extent of ripening.

6.2 Packaging innovations

The modification of atmosphere alters the gaseous exchange requirements of the fresh produce, resulting in enhancement of shelf-life by retarding ripening or reducing respiration rate. This has led to several applications and innovations in packaging, which have proved to reduce several pathogenic microflora, inhibition of enzymatic activity, and thus decay of fresh-cut produce (Oliveira et al., 2015). Earlier, modifications of O_2 (1%–10%) and CO_2 (0%–20%) was considered for shelf-life enhancement; recently attention has been given to incorporation noble gases such as Ar, He, and N_2O. In sliced apples and kiwifruits, high N_2O and Ar mixture increased the firmness, color retention, and total soluble solids content during storage (Rocculi, Romani, & Rosa, 2005).

When an inert gas is dissolved in water at optimum temperature and pressures, gas molecules are enclosed in a cage-like structure formed by water molecules. The final structure so formed is known as clathrate hydrate, which retards water movement in fruit tissues and retards enzymatic reactions. Pressurized argon and nitrogen (4 MPa) in combination with MAP (4% O_2, 94% N_2, and 2% CO_2) extended shelf-life of fresh-cut potatoes for 12 days (Shen, Zhang, Devahastin, & Guo, 2019). Likewise, MAP edible coatings are also believed to extend or preserve the shelf-life of fresh-cut produce by aiding moisture loss prevention, restricting oxygen absorption, and microbial penetration (Vargas, Pastor, Chiralt, McClements, & Gonzalez-Martinez, 2008). Previously, coatings were applied on whole fruits or vegetables, but their ease to use and eco-friendly aspects raised awareness, and they are now being utilized for fresh-cut produce.

The incorporation of certain active ingredients like essential oils, organic acids, preservatives, nanoparticles, probiotics, etc. has proved to enhance the effectiveness of edible coatings for fresh-cut produce application. Although, due to gelling property, biopolymers such as polysaccharides, lipids, and proteins are used for coating of fresh produce, chitosan edible coatings have been appreciated for their preservative effects. In addition, to alter permeability, the fresh-cut produce surface, which becomes more sensitive to microbial spoilage, is found to be better protected with chitosan application due to its polycationic nature, resulting in formation of complexes with polymers at bacterial surface (Durango, Soares, & Andrade, 2006). It has been shown to inhibit coliforms, yeast, molds, and psychotropics on fresh-cut broccoli stored at refrigerated temperatures (Moreira, Roura, & Ponce, 2011). Enrichment of edible coatings with certain bioactive compounds from tea tree, rosemary, pollen, and propolis enhanced antimicrobial activity of chitosan against *E. coli* and *L. monocytogenes* applied over fresh-cut broccoli (Alvarez, Pounce, & Moreira, 2013). The antimicrobial activity of chitosan was also observed to be enhanced with incorporation of Ag^{+} nanoparticles because of the binding of metal ion with microbial DNA, proteins, and enzymes (Cavaliere et al., 2015).

The need for diversification in terms of value addition and sensory parameters has led to application of cocoa and chocolate-based coatings on fresh-cut grapes, apples, and dried chestnut (Glicerina et al., 2019). Climacteric fruit and chilling injury sensitive salad vegetables like cucumber can be preserved with combinations of packaging, i.e., chitosan and air, and nitrogen and argon based modified atmosphere packaging. Argon based modified atmosphere packaged cucumbers retarded degradative changes and respiration better than other combinations (Olawuyi, Park, Lee, & Lee, 2018). Some recent innovations in use of edible coatings are summarized in Table 1.

Additionally, to survive the robust conditions in marketing channels, the extension of shelf-life needs more self-sufficiency in preservation methodologies. Combinations of modified atmosphere packaging with edible coatings have been attempted and have presented significant results. Fruit and vegetables puree and waste flour have recently been investigated for ripening delay and retarding respiration. Fai et al. (2016) utilized wastes of different fruits or vegetables, i.e., Selecta orange, passion fruit, watermelon, lettuce, zucchini, carrot, spinach, mint, taro, cucumber, and rocket. The solid residues obtained after fruit or vegetable processing was dried for 1 h at 90°C and then ground. The sliced and shredded carrots coated with films delayed weight loss and prevented whitening of carrot during storage. The fresh-cut vegetables, asparagus, baby corn, and oyster mushroom when packed in films bags from banana flour incorporated with chitosan reduced the growth of *Staphylococcus aureus*, along with retention of texture and freshness (Pitak & Rakshit, 2011). Olawuyi et al. (2018) coated fresh-cut cucumbers with 1% chitosan and

Table 1 Recent trends of edible films incorporated with active ingredients for FFVs preservation

Fresh-cut fruit/ vegetable	Basic edible coating polymer	Active ingredients	Shelf-life or quality effects	Reference
Chinese Yam	Xanthan gum	Star anise essential oil	Reduced weight loss, inhibited PPO activity and browning 8 times as compared to control	Zhang, Gu, Lu, Yuan, and Sun (2019)
Melon	Chitosan	Ag-chitosan	Reduced respiration rate, prevention of softening, higher retention of vitamin C	Ortiz-Duarte, Perez-Cabrera, Artes-Hernandez, and Martinez-Hernandez (2019)
Strawberry	Chitosan-CMC	Layer-by layer	Retention of aroma volatiles, little effect on TSS and acidity, lower metabolites content	Yan et al. (2019)
Melon	Carboxymethyl cellulose (CMC)	Citral	Improved quality, prevented microbial damage	Arnon-Rips, Porat, and Poverenov (2019)
Melon	Mushroom waste	Fungal chitosan	Reduced microflora, retained physiological quality	Poverenov et al. (2018)
"Bravo de Esmolfe" apple	Sodium alginate Pectin	Eugenol Citral Ascorbic acid	Significant reduction of browning, best results achieved with ascorbic acid	Guerreiro, Gago, Faleiro, Miguel, and Antunes (2015)
Apple	Alginate	Cinnamon oil Rosemary oil	Active components reduced weight loss and improved color	Chiabrando and Giacalone (2015)

subjected them to storage for 12 days at 5°C after argon-based MAP. The combination retained quality as well as prolonging shelf-life, as compared to individual treatments.

6.3 Processing methodologies trends

Fruits and vegetables subjected to washing, peeling, and cutting must follow some enzyme or microbial inactivation treatment such as blanching or heat-shock treatment. Blanching consists of heating in water at 85–100°C, depending upon fruit or vegetable. Heat-shock is a washing step at 45–70°C for less than 5 min. Heat-shock represses aggregation of phenolics, which form brown polymers. Lettuce or leafy vegetables suffer more enzymatic assisted degradative quality changes. On an industrial scale, still only chemical disinfestation is being done, followed by computerized color sorting and form fill seal packaging.

If not blanched, then the extent of cutting of vegetables determines rate of respiration. UV-C light in the range of 240–260 nm is used for surface sanitization and approved by USFDA. It prevents growth of microbes by altering their DNA or proteins. In recent years, combinations of processing methodologies like UV-C light, cold plasma, HPP, and MAP are being adopted, together with blanching treatment, to create hurdles for microbes and enzymes for more effective results. Immersion in hot water at 55°C for 30 s and subjecting to UV-C treatment followed by packaging under super atmospheric oxygen MAP resulted in controlled respiration rate during 14 days storage with minimum yeast and mold count (Maghoumi et al., 2012). Artes-Hernandez, Robles, Gomez, Tomas-Callejas, and Artes (2010) studied the effect of UV-C illumination on quality of watermelon for up to 11 days stored at 5°C. Higher doses induced higher carbon dioxide production, but overall quality was retained in lower doses treatment.

Another rising technology for surface disinfestation is photodynamic treatment (PDT), in which reactive oxygen species (ROS) are produced by photosensitizer on illumination by light. These ROS inactivate bacteria by oxidizing DNA or proteins and retarding metabolites production. Despite its antimicrobial activity, enzymatic activity also reduces without affecting quality of fresh-cut produce significantly. Tao et al. (2019) observed antimicrobial effect and shelf-life retention of fresh-cut Fuji apples on application of curcumin-PDT.

Gas-plasma is an ionized gas composed of gas particles such as ions, electrons, and free radicals in ground and at excited state. At low pressure, nonthermal plasmas are produced, which in turn generates reactive oxygen and nitrogen species, i.e., ozone, -OH radicals, and atomic oxygen. These species are able to cause DNA oxidation, lipid peroxidation, and other degradative changes. The Dielectric barrier discharged (DBD) was able to decrease browning the fresh-cut apples "Pink-lady" through inhibition of polyphenol oxidases. The processing method did not affect texture and nonsignificantly affected other quality parameters (TSS, reducing sugars, etc.) (Tappi et al., 2014). Similarly, the authors studied effects of cold plasma in fresh-cut melon and found a stress response induced in fruit tissue, which might lead to reduction of metabolic heat and altered respiratory mechanism. Although, microbial shelf-life was improved, there was limited effect on enzymatic activity, dependent on the kind of enzyme. Therefore, there are upcoming novel methodologies for processing and preservation of fresh-cut produce, which, however, need to be optimized for further commercial application.

7 Marketing aspects of fresh-cut produce worldwide

In terms of the fresh-cut produce market among European Union (EU) countries, the largest share is FFVs products available in the United Kingdom, accounting for around a third of total EU consumption. In other countries like Germany and Spain, the development in processing of fresh-cut produce is still naïve. However, the market is still emerging, despite the economic crisis of recent years. Sales of ready-to-eat vegetables in Spain showed an annual increase of 5%–6%, reaching 70,600 and 74,064 tons in 2010 and 2011, respectively (Anonymous, 2013). After this period the Spanish market stabilized, with sales of approximately 77,000 tons recorded in 2013 (Anonymous, 2014).

The market for FFV is burgeoning in the food sector with time, as the demand for healthier, convenient, and fresh food is increasing. The nutritional and sensory aspects play a major role in marketing of such products as people nowadays are more concerned about the safety of the food they eat. Another factor that has promoted and maintained fresh-cut sales is technology; however, permanent innovations are required to drive new growth in this sector. Some of the fresh-cut produce produced in developed countries include melons, cantaloupe, watermelon, mango, jackfruit, papaya, grapefruit, pineapple, fruit mixes, shredded leafy vegetables and salad mixes; vegetables for cooking, like peeled baby carrots, baby corn, broccoli and cauliflower florets, cut celery stalks, shredded cabbage, cut asparagus, cut sweet potatoes, and many more (James & Ngarmsak, 2010). Fresh-cut sales across all retail channels are $10–15 billion per year in the United States.

Fresh-cut produce has a significantly longer shelf-life than whole produce, ranging from 3 to 13 days for bagged salads and cut vegetables to 21 days for bagged apple slices and baby carrots. The product is customer friendly: many items can fit in vehicle cup holders. Fresh-cut produce is also ready to consume on the go without need of additional washing and is more recipe ready, eliminating the preparation time required to fix meals. From a retail perspective, fresh-cut produce has a longer shelf-life. The global marketing of such minimally processed fruits and vegetable in 2017 was USD 245.97 Billion and is projected to reach USD 346.05 Billion by 2022. This growth is also because of the growing number of supermarkets and hypermarkets, which are a result of the rising middle-class population and disposable income in developing economies such as China, India, and Mexico. In the near future, these products will be in more demand as people's requirement for ready to eat and use produce is increasing. These kinds of products offer great value to the growers or the processors by adding value to the product. The dominant product in this sector is fresh-cut salads, with a sale of 2.4 billion. Continued growth is forecast for this segment, as the need for convenience continues to be relevant for consumers. Fresh-cut produce offers healthier grab-and-go options for new consumption patterns, such as a renewed focus on breakfast and more snacking occasions.

In India, the market for FFV is in the initial stages, but this segment of produce is emerging as a strong growth category in the organized retail markets. Consumer lifestyles continue to get busier and, in turn, demand more convenient meal solutions. Sales of washed, chopped, or bagged fresh produce have been growing at a faster rate. Not only are prices of chopped produce rising, consumers are also adding higher-priced items to their shopping carts as people are ready to pay more for convenience. The marketing of FFVs requires the appropriate combination of technologies for extending the shelf-life of the products, maintaining the sensory and organoleptic characteristics of the original fresh product. The marketing aspects in FFVs can be categories into different areas like introduction of new products, better and innovative packaging, and by improving the shelf-life of the product.

7.1 New products developments

New products in the market can affect the sales of the product. Modification to the existing fresh-cut products can also increase the market for this produce, e.g., shredded broccoli, or microwave-ready fresh vegetables. In addition to the conventional broccoli, today consumers may select from broccoli Romanesco, purple broccoli, and brocco flower. In addition, fruits like watermelon are now being packed into different products like seedless and with seeds. Cut fruits can be packed in variations of peeled and unpeeled so that consumer can pick up the products according to their needs and demands. Thus new products are being introduced in the market in line with the needs of consumers. Introduction of new fruit mixtures with more variety and combinations of flavor can also be a new trend for marketing.

Products that are minimally processed are always favored by the consumer because of their nutritional value. In addition, products labeled as "organic" stand out from the rest, corresponding to produce that has been grown without the aid of chemicals and delivered free from preservatives, with emphasis on the purity of the product and the effectiveness of the packaging that protects it. Consequently, organic products have gained a lot of importance in recent years. Nowadays, these organic products have special sections in growing numbers of retail stores. Sales of organic packaged salad mixes, one of the fastest-growing categories, have grown at a rate of 200% over the past 3 years and show no sign of slowing down.

7.2 Better and innovative packaging

Packaging plays a vital role in the marketing of a product. Packaging has become a key component in promoting and merchandising fresh-cut produce. The presentation of the package is almost as important as the product in the package—the package must present the product well and stand out among all the other products, as consumers generally make purchases with their eyes. New packaging innovations foster the increase in the market and demand for certain produce, enabling producers and retailers to maintain quality for longer. Use of innovative packaging technology can improve the product quality and shelf-life. Innovative approaches to extending shelf-life include active modified atmospheric packages with differentially permeable films, films that incorporate antimicrobial properties, edible coatings that confer barriers properties, and the use of nontraditional gases to modify respiration. Intelligent packaging is also available in the market, which utilizes integrated sensor technologies that can indicate maturity, ripeness, respiration rate, and spoilage. Many innovative packaging solutions are appearing in the market. Some of the latest trends in packaging are steam table packaging, which help to steam the product within the packaging material. The use of these steamer bags for vegetables is just one option that could expand the market for fresh-cut products. Vegetables like broccoli, beans, and sweet potatoes are packed in such packaging materials.

More initiatives are being taken to make the packaging material antifog as, when FFVs are stored in refrigerated conditions, after a certain period of time water beads up on the front of the bag. This does not affect the quality of the produce, but a fogged package reduces the merchandising value of the product because consumers cannot judge the quality of the food contents. To create antifogging packaging material, manufacturers typically coat the material with a surfactant that chemically changes the surface tension of the water and prevents the packaging material from fogging. In addition, to solve this problem of antifogging, micro perforated bags have been introduced, containing small holes that help in venting of the gasses. Peel and re-seal technology is also being utilized as a marketing strategy for these products, as the consumer can tightly reseal the package multiple times and keep the produce fresh.

7.3 Shelf-life extension

Shelf-life of the FFVs play a vital role in the marketing of the produce. Therefore, the most important aspect is an integrated approach, where different aspects, such as raw materials, handling, processing, packaging, and distribution are being properly managed to make shelf-life extension possible. The intelligent selection of different preservation techniques, without obviating the intensity of each treatment and the sequence of application to achieve a specified outcome, is expected to have significant prospects for the future of minimally processed fruit and vegetables. Unit operations such as peeling and shredding need further development to make them more gentle. There is no sense in disturbing the quality of produce by rough treatment during processing, and then trying to limit the damage by subsequent use of preservatives. Nowadays, ultraviolet light (UV) is a preservation technique that is widely being used instead of adding preservatives to the fresh-cut produce. Modified atmospheric packaging is being utilized, which means modifying the atmosphere of the packaged material. Edible coatings are being applied in order to improve the shelf-life of fresh-cut fruits. Edible films and coatings may help to reduce the deleterious effects of the resulting from minimal processing. The application of edible coatings to deliver active substances is one of the major advances reached so far in order to increase the shelf-life of fresh-cut produce. The functionality of edible coatings is being improved by incorporating antimicrobial agents (chemical preservatives or antimicrobial compounds obtained from a natural source), antioxidants, and functional ingredients such as minerals and vitamins. In addition, techniques like nano-encapsulation are widely being used to incorporate some functional ingredients into these edible films, which will be released under specific conditions (Vargas et al., 2008).

8 Conclusion

Worldwide, there is a wide range of vegetables and fruits that could be used to elaborate and expand the product offerings in the market, which can result in expanding and increasing the sale of such products. The market for such products is on the rise owing to its tremendous scope for expansion of this industry. The consumption of such kinds of products has gone up, partly because of changes in lifestyle leading to a demand for convenience. However, it is necessary to improve preparation and preservation techniques, taking into consideration the food laws, as FFVs are highly perishable. So adequate technologies can result in keeping the product safe and maintaining its quality long enough to make the distribution of fresh-cut commodities feasible, also enhancing the market for such products.

References

AAFC Agriculture and Agri-Food Canada. (2014). *Statistical overview of the Canadian fruit industry 2013. Retrieved from (2014). http://www5.agr.gc.ca/resources/prod/doc/horticulture/fruit_report_2013-en.pdf.*

Abadias, M., Usall, J., Anguera, M., Solson, C., & Vinas, I. (2008). Microbiological quality of fresh, minimally processed fruit and vegetables, and sprouts from retail establishments. *International Journal of Food Microbiology, 123*(1–2), 121–129.

Aguayo, E., Requejo-Jackman, C., Stanley, R., & Woolf, A. (2010). Effects of calcium ascorbate treatments and storage atmosphere on antioxidant activity and quality of fresh-cut apple slices. *Postharvest Biology and Technology, 57*, 52–60.

Ali, S., Nawaz, S., Ejaz, T.-A., Haider, M. W., & Alam, H. U. J. (2019). Effects of hydrogen sulphide on postharvest physiology of fruits and vegetables: an overview. *Scientia Horticulturae, 243*, 290–299.

Allende, A., Selma, M. V., López-Gálvez, F., Villaescusa, R., & Gil, M. I. (2008). Impact of wash water quality on sensory and microbial quality, including Escherichia coli cross-contamination, of fresh-cut escarole. *Journal of Food Protection, 71*, 2514–2518.

Allende, A., Tomás-Barberán, F. A., & Gil, M. I. (2006). Minimal processing for healthy traditional foods. *Trends in Food Science and Technology, 17*(9), 513–519.

Alvarez, M. V., Pounce, A. G., & Moreira, D. R. (2013). Antimicrobial efficiency of chitosan enriched with bioactive compounds to improve the safety of fresh cut broccoli. *LWT-Food Science and Technology, 50*, 78–87.

Anastasi, A. (2015). *Euromonitor international report on fruit and vegetables on Western Europe. Retrieved from: http://blog.euromonitor.com/2015/03/fruits-and-vegetables-in-western-europe.html.*

Anonymous, El consumo de productos IV gama crece, pero a menor ritmo, 2013. Available at: www.revistamercados.com/imprimir_articulo.asp?articulo_ID%4372 (Accessed 20 January 2019).

Anonymous, IV gama: Crecimiento lento, pero firme., 2014. Available at: www.alimarket.es/noticia/150889/IV-gama-crecimiento-lento-pero-firme. (Accessed 20 January 2019).

Arienzo, M., Cataldo, D., & Ferrara, L. (2013). Pesticide residues in fresh-cut vegetables from integrated pest management by ultra-performance liquid chromatography coupled to tandem mass spectrometry. *Food Control, 31*, 108–115.

Arnon-Rips, H., Porat, R., & Poverenov, E. (2019). Enhancement of agricultural produce quality and storability using citral-based edible coatings; the valuable effect of nano-emulsification in a solid-state delivery on fresh-cut melons model. *Food Chemistry, 277*, 205–212.

Artés, F., Gómez, P., Aguayo, E., Escalona, V., & Artés-Hernández, F. (2009). Sustainable sanitation techniques for keeping quality and safety of fresh-cut plant commodities. *Postharvest Biology and Technology, 51*(3), 287–296.

Artes-Hernandez, F., Robles, P. A., Gomez, P. A., Tomas-Callejas, A., & Artes, F. (2010). Low UV-C illumination for keeping overall quality of fresh-cut watermelon. *Postharvest Biology and Technology, 55*, 114–120.

Barrett, D. M., Beaulieu, J. C., & Shewfelt, R. (2010). Color, flavor, texture, and nutritional quality of fresh-cut fruits and vegetables: Desirable levels, instrumental and sensory measurements, and the effects of processing. *Critical Reviews in Food Science and Nutrition, 50*, 369–389.

Baselice, F., Colantuoni, D. A., Lass, G., & Nardone, A. S. (2017). Trends in EU consumers' attitude towards fresh-cut fruit and vegetables. *Food Quality and Preference, 59*, 87–96.

Berger, C. N., Sodha, S. V., Shaw, R. K., Griffin, P. M., Pink, D., & Hand, P. (2010). Fresh fruit and vegetables as vehicles for the transmission of human pathogens. *Environmental Microbiology, 12*(9), 2385–2397.

Beuchat, L. R. (2000). Use of sanitizers in raw fruit and vegetable processing. In S. M. Alzamora, M. S. Tapia, & A. López-Malo (Eds.), *Minimally processed fruits and vegetables: Fundamental aspects and applications* (pp. 63–78). Gaithersburg, MD: Aspen.

Bilek, S. E., & Turantaş, F. (2013). Decontamination efficiency of high-power ultrasound in the fruit and vegetable industry, a review. *International Journal of Food Microbiology, 166*(1), 155–162.

Bouzembraka, Y., & Marvina, H. J. P. (2019). Impact of drivers of change, including climatic factors, on the occurrence of chemical food safety hazards in fruits and vegetables: a Bayesian network approach. *Food Control, 97*, 67–76.

Byrne, L., Fisher, I., Peters, T., Mather, A., Thomson, N., Rosner, B., et al. (2014). A multi-country outbreak of Salmonella Newport gastroenteritis in Europe associated with watermelon from Brazil, confirmed by whole genome sequencing: October 2011 to January 2012. *Euro Surveillance, 19*(31), 6–13.

Cavaliere, E., De Cesari, S., Landini, G., Riccobono, E., Pallecchi, L., Rossolini, G. M., et al. (2015). Highly bactericidal Ag nanoparticle films obtained by cluster beam deposition. *Nanomedicine: Nanotechnology, Biology and Medicine, 11*(6), 1417–1423.

Centers for Disease Control and Prevention (CDC) (2003). *Foodborne outbreaks due to confirmed bacterial etiologies, 2006. Available at:(2003). http://www.cdc.gov/foodborneoutbreaks/usoutb/fbo2003/fbofinal2003.pdf.*

Centers for Disease Control and Prevention (CDC) (2004). *Foodborne outbreaks due to confirmed bacterial etiologies, 2006. Available at:(2004). http://www.cdc.gov/foodborneoutbreaks/usoutb/fbo2004/fbofinal2004.pdf.*

CFIA Canadian Food Inspection Agency (2013). *ARCHIVED e food safety action plan evaluation. Retrieved from http://www.inspection.gc.ca/about-the-cfia/accountability/other-activities/audits-reviews-and-evaluations/fsap/report/eng/1385242045975/1385242106674.*

Cha, S. M., Bu, S. Y., Kim, E. J., Kim, M. H., & Choi, M. K. (2012). Study of dietary attitudes and diet management of married immigrant women in Korea according to residence period. *Journal of Korean Dietary Association, 18*, 297–307.

Cheng, J., Greiner, R., Kelly, J., Bell, D., & Liu, W. (2002). Learning Bayesian networks from data: An information-theory based approach. *Artificial Intelligence, 137* (1), 43–90.

Chiabrando, V., & Giacalone, G. (2015). Effect of essential oils incorporated into an alginate-based edible coating on fresh-cut apple quality during storage. *Quality Assurance and Safety of Crops and Foods, 7*, 251–259.

Chiumarell, M., Ferrari, C. C., Sarantópoulos, C. I. G. L., & Hubinger, M. D. (2011). Fresh cut 'Tommy Atkins' mango pre-treated with citric acid and coated with cassava (Manihot esculenta Crantz) starch or sodium alginate. *Innovative Food Science and Emerging Technologies, 12*, 381–387.

Cliffe-Byrnes, D., & O'Beirne, J. (2002). *Effects of chlorine treatment and packaging on the quality and shelf-life of modified atmosphere (MA) packaged coleslaw mix, 7th Karlsruhe nutrition congress on food safety* (pp. 707–716). Germany Elsevier Sci Ltd.

Clydesdale, F. M. (1991). Color perception and food quality. *Journal of Food Quality, 14*, 61–74.

Corbo, M. R., Del Nobile, M. A., & Sinigaglia, M. (2006). A novel approach for calculating shelf life of minimally processed vegetables. *International Journal of Food Microbiology, 106*(1), 69–73.

Corollaro, M. L., Aprea, E., Endrizzi, I., Betta, E., Demattè, M., Bergamaschi, M., et al. (2014). A combined sensory-instrumental tool for apple quality evaluation. *Postharvest Biology and Technology, 96*, 135–144.

Degl'Innocenti, E., Pardossi, A., Tognoni, F., & Guidi, L. (2007). Physiological basis of sensitivity to enzymatic browning in "lettuce", "escarole" and "rocket salad" when stored as fresh-cut products. *Food Chemistry, 104*(1), 209–215.

Del Nobile, M. A., Conte, A., Cannarsi, M., & Sinigaglia, M. (2008). Use of biodegradable films for prolonging the shelf life of minimally processed lettuce. *Journal of Food Engineering, 85*(3), 317–325.

Denis, N., Zhang, H., Leroux, A., Trudel, R., & Bietlot, H. (2016). Prevalence and trends of bacterial contamination in fresh fruits and vegetables sold at retail in Canada. *Food Control, 67*, 225–234.

do Nunes, M. C. N. (2015). Correlations between subjective quality and physicochemical attributes of fresh fruits and vegetables. *Postharvest Biology and Technology, 107*, 43–54.

Durango, A. M., Soares, N. F. F., & Andrade, N. J. (2006). Microbiological evaluation of an edible antimicrobial coating on minimally processed carrots. *Food Control, 17*(5), 336–341.

Elviss, N. C., Little, C. L., Hucklesby, L., Sagoo, S., Surman-Lee, S., & de Pinna, E. (2009). Microbiological study of fresh herbs from retail premises uncovers an international outbreak of salmonellosis. *International Journal of Food Microbiology, 134*, 83–88.

Fai, A. E. C., de Souza, M. R. A., de Barros, S. T., Bruno, N. V., Ferreira, M. S. L., & de Goncalves, E. C. B. A. (2016). Development and evaluation of biodegradable films and coatings obtained from fruit and vegetable residues applied to fresh-cut carrot. *Postharvest Biology and Technology, 112*, 194–204.

Fallik, E. (2014). Microbial quality and safety of fresh produce. In W. J. Florkowski, R. L. Shewfelt, B. Brueckner, & S. E. Prussia (Eds.), *Postharvest handling* (3rd ed., pp. 313–333). San Diego, CA: Academic Press.

FAO (2010). Retrieved from(2010). http://www.fao.org/docrep/014/i1909e/i1909e00.pdf.

FAO (2013). Retrieved from(2013). http://www.fao.org/docrep/017/i3138e/i3138e05.pdf.

FAO/WHO (2008). *Microbiological risk assessment series 14: Microbiological hazards in fresh leafy vegetables and herbs. Retrieved from(2008). ftp://ftp.fao.org/docrep/fao/011/i0452e/i0452e00.pdf.*

Food & Drug Administration, H.H.S (2008). Irradiation in the production, processing and handling of food. Final rule. *Federal Register, 73*(164), 49593–49603.

Food Agriculture Organization (FAO), 2009. Food and agriculture organization of the United Nations. FAOSTAT. http://faostat.fao.org/site/609/DesktopDefault.aspx?PageID¼609#ancor Retrieved 07/09/2013, 2013.

Franzen, L., & Smith, C. (2009). Acculturation and environmental change impacts dietary habits among adult Hmong. *Appetite, 52*, 173–183.

Freshfel European Fresh Produce Association. (2014). *Freshfel "Consumption monitor." Retrieved from http://www.freshfel.org/asp/what_we%20do/consump tion_monitor.asp.*

Glicerina, V., Tylewicz, U., Canali, G., Siroli, L., Rosa, M. D., Lanciotti, R., et al. (2019). Influence of two different cocoa-based coatings on quality characteristics of fresh-cut fruits during storage. *LWT-Food Science and Technology, 101*, 152–160.

Goodburn, C., & Wallace, C. A. (2013). The microbiological efficacy of decontamination methodologies for fresh produce: A review. *Food Control, 32*, 418–427.

Gordon, R., McDermott, L., Stead, M., & Angus, K. (2006). The effectiveness of social marketing interventions for health improvement: What's the evidence? *Public Health, 120*(12), 1133–1139.

Grant, J., Wendelboe, A. M., Wendel, A., Jepson, B., Torres, P., & Smelser, C. (2008). Spinach-associated Escherichia coli O157:H7 outbreak, Utah and New Mexico, 2006. *Emerging Infectious Diseases, 14*(10), 1633–1636.

Guerreiro, A. C., Gago, C. M. L., Faleiro, M. L., Miguel, M. G. C., & Antunes, M. D. C. (2015). Edible coatings enriched with essential oils for extending the shelf life of "Bravo de Esmolfe" fresh-cut apples. *International Journal of Food Science and Technology, 51*, 87–95.

Gutierrez, A. F. W., Zaldivar, R. J., & Contreras, S. G. (2009). Effect of various levels of digestible energy and protein in the diet on the growth of gamitana (Colossoma macropomum) Cuvier 1818. *Revista de Investigaciones Veterinarias del Perú, 20*, 178–186.

Hanning, I. B., Nutt, J. D., & Ricke, S. C. (2009). Salmonellosis outbreaks in the United States due to fresh produce: Sources and potential intervention measures. *Foodborne Pathogens and Disease, 6*(6), 635–648.

Heard, G. M. (2002). Microbiology of fresh-cut produce. In O. Lamikanra (Ed.), *Fresh-cut fruits and vegetables: Science, technology, and market* (pp. 187–248). Boca Raton, FL: CRC Press.

Holvoet, K., Jacxsens, L., Sampers, I., & Uyttendaele, M. (2012). Insight into the prevalence and distribution of microbial contamination to evaluate water management in the fresh produce industry. *Journal of Food Protection, 75*, 671–681.

IFPA, International Fresh-Cut Produce Association. (2001). In J. R. Gorny (Ed.), *Food safety guidelines for the fresh-cut produce industry* (4th ed.). Alexandria, VA. See http://www.fresh-cuts.org.

ISMEA. (2011). *Gli ortaggi di IV gamma nel rapporto competitivo con il fresco tradizionale Cesena. 7 ottobre http://www.ismea.it/flex/cm/pages/ServeBLOB.php/L/IT/IDPagina/6548.*

James, J. B., & Ngarmsak, T. (2010). *Processing of fresh-cut tropical fruits and vegetables: A technical guide.* Available from http://www.fao.org/docrep/014/i1909e/i1909e00.

Jiang, Y., Pen, L., & Li, J. (2004). Use of citric acid for shelf life and quality maintenance of fresh-cut Chinese water chestnut. *Journal of Food Engineering, 63*, 325–328.

Klein, B. P. (1987). Nutritional consequences of minimal processing of fruits and vegetables. *Journal of Food Quality, 10*, 19–193.

Kumar, C. G., & Anand, S. K. (1998). Significance of microbial bio-films in food industry: A review. *International Journal of Food Microbiology, 42*(1–2), 9–27.

Kyriacoua, M. C., & Rouphael, Y. (2018). Towards a new definition of quality for fresh fruits and vegetables. *Scientia Horticulturae, 234*, 463–469.

Lamikanra, O. (2002). Preface. In O. Lamikanra (Ed.), *Fresh-cut fruits and vegetables science, technology and market.* Boca Raton, FL: CRC Press.

Lee, L., Arul, J., Lencki, R., & Castaigne, F. (1996). A review on modified atmosphere packaging and preservation of fresh fruits and vegetables: Physiological basis and practical aspects e part 2. *Packaging Technology and Science, 9*, 1–17.

Lee, E. J., Lee, K.-R., & Lee, S.-J. (2017). Study on the change and acculturation of dietary pattern of southeast Asian workers living in South Korea. *Appetite, 117*, 203–213.

Li, X., Long, Q., Gao, F., Han, C., Jin, P., & Zheng, Y. (2017). Effect of cutting styles on quality antioxidant activity in fresh-cut pitaya fruit. *Postharvest Biology and Technology, 124*, 1–7.

Liu, X., Yang, Q., Lu, Y., Li, Y., Li, T., Zhou, B., et al. (2019). Effect of purslane (Portulaca oleracea L.) extract on anti-browning of fresh-cut potato slices during storage. *Food Chemistry, 283*, 445–453.

López-Gálvez, F., Allende, A., Selma, M. V., & Gil, M. I. (2008). Prevention of Escherichia coli cross-contamination by different commercial sanitizers during washing of fresh-cut lettuce. *Journal of Food Protection, 133*, 167–171.

Low, D. (2013). *Global organic food and beverage sales approach $US63 billion. Available at http://weedsnetwork.com/traction/permalink/WeedsNews4324. Accessed 15 May 2016.*

Maghoumi, M., Gomez, P. A., Artes-Hernandez, F., Mostofi, Y., Zamani, Z., & Artes, F. (2012). Hot water, UV-C and superatmospheric oxygen packaging as hurdle techniques for maintaining overall quality of fresh-cut pomegranate arils. *Journal of Science of Food and Agriculture, 93*, 1162–1168. https://doi.org/10.1002/jsfa.5868.

Mahdavian, K., Kalantari, M., & Ghorbanli, M. (2007). The effect of different concentrations of salicylic acid on protective enzyme activities of pepper (Capsicum annuum L.) plants. *Pakistan Journal of Biological Sciences, 10*, 3162–3165.

Martin-Diana, A. B., Rico, D., Frias, J., Henehan, G. T. M., Mulcahy, J., Barat, J. M., et al. (2006). Effect of calcium lactate and heat-shock on texture in fresh-cut lettuce during storage. *Journal of Food Engineering, 77*, 1069–1077.

Martinez-Vaz, B. M., Fink, R. C., Diez-Gonzalez, F., & Sadowsky, M. J. (2014). Enteric pathogen-plant interactions: Molecular connections leading to colonization and growth and implications for food safety. *Microbes and Environments, 29* (2), 123–135.

Marvin, J. P., Bouzembrak, Y., Janssen, E. M., van der, H. J. F.-K., van Asselt, E. D., & Kleter, G. A. (2016). Holistic approach to food safety risks: food fraud as an example. *Food Research International, 89*(1), 463–470.

Maul, F., Sargent, S. A., Sims, C. A., Baldwin, E. A., Balaban, M. O., & Huber, D. J. (2000). Tomato flavor and aroma quality as affected by storage temperature. *Journal of Food Science, 65*, 1228–1237.

Mazzocchi, M., Traill, W. B., & Shogren, J. F. (2009). *Fat economics: Nutrition, health, and economic policy.* Oxford University Press.

Meireles, A., Giaouris, E., & Simões, M. (2016). Alternative disinfection methods to chlorine for use in the fresh-cut industry. *Food Research International, 82*, 71–85.

Mergenthaler, M., Weinberger, K., & Qaim, M. (2009). The food system transformation in developing countries: A disaggregate demand analysis for fruits and vegetables in Vietnam. *Food Policy, 34*(5), 426–436.

Mishra, V. K., & Gamage, T. V. (2007). *Postharvest handling and treatments of fruits and vegetables. Handbook of food preservation* (2nd ed., pp. 49–69).

Moore, L. V., Thompson, F. E., & Demissie, Z. (2017). Percentage of youth meeting federal fruit and vegetable intake recommendations, Youth Risk Behavior Surveillance System, United States and 33 states, 2013. *Journal of the Academy of Nutrition and Dietetics, 117*, 545–553.

Moreira, M. D. R., Roura, S. I., & Ponce, A. (2011). Effectiveness of chitosan edible coatings to improve microbiological and sensory quality of fresh cut broccoli. *LWT-Food Science and Technology, 44*, 2335–2341.

Nicola, S., Tibaldi, G., & Fontana, E. (2009). Fresh-cut produce quality: Implications for a systems approach. In *Postharvest handling: A systems approach (2nd ed.)* (pp. 247–282). San Diego, CA: Academic Press.

Nunes, C. N., & Emond, J. P. (2007). Relationship between weight loss and visual quality of fruits and vegetables. In *vol. 120. Proceedings of the Florida state horticultural society* (pp. 235–245).

Olaimat, A. N., & Holley, R. A. (2012). Factors influencing the microbial safety of fresh produce: A review. *Food Microbiology, 32*(1), 1–19.

Olawuyi, I. F., Park, J. J., Lee, J. J., & Lee, W. Y. (2018). Combined effect of chitosan coating and modified atmosphere packaging on fresh-cut cucumber. *Food Science and Nutrition, 7*, 1–10.

Oliveira, M., Abadias, M., Usall, J., Torres, R., Teixido, N., & Vinas, I. (2015). Application of modified atmosphere packaging as a safety approach to fresh-cut-fruits and vegetables a review. *Trends in Food Science & Technology, 46*, 13–26.

Ortiz-Duarte, G., Perez-Cabrera, L. E., Artes-Hernandez, F., & Martinez-Hernandez, G. B. (2019). Ag-chitosan nanocomposites in edible coatings affect the quality of fresh-cut melon. *Postharvest Biology and Technology, 147*, 174–184.

Pace, M. C., Renna, F., & Attolico, G. (2011). Relationship between visual appearance and browning as evaluated by image analysis and chemical traits in fresh-cut nectarines. *Postharvest Biology and Technology, 61*, 178–183.

Parish, M. E., Beuchat, L. R., Suslow, T. V., Harris, L. J., Garrett, E. H., Farber, J. N., et al. (2003). Methods to reduce/eliminate pathogens from fresh and fresh-cut produce. *Comprehensive Reviews in Food Science and Food Safety, 2*, 161–173.

Pezzoli, L., Elson, R., Little, C. L., Yip, H., Fisher, I., & Yishai, R. (2008). Packed with Salmonella-investigation of an international outbreak of Salmonella Senftenberg infection linked to contamination of prepacked basil in 2007. *Foodborne Pathogens and Disease, 5*(5), 661–668.

Pitak, N., & Rakshit, S. K. (2011). Physical and antimicrobial properties of banana flour/chitosan biodegradable and self-sealing films used for preserving fresh-cut vegetables. *LWT-Food Science and Technology, 44*, 2310–2315.

Poverenov, E., Arnon-Rips, H., Zaitsev, Y., Bar, V., Danay, O., Horev, B., et al. (2018). Potential of chitosan from mushroom waste to enhance quality and storability of fresh-cut melons. *Food Chemistry, 268,* 233–241.

Portela, S. I., & Cantwell, M. I. (2001). Cutting blade sharpness affects appearance and other quality attributes of fresh-cut cantaloupe melon. *Journal of Food Science, 66,* 1265–1270.

Putnik, P., Kovacevic, D. B., Herceg, K., & Levaj, B. (2016). Influence of cultivar, anti-browning solutions, packaging gasses, and advanced technology on browning in fresh-cut apples during storage. *Journal of Food Process Engineering, 40,* 12400.

Putnik, P., Roohinejad, S., Greiner, R., Granato, D., Bekhit, A. D. A., & Danijela, B. K. (2017). Prediction and modelling of microbial growth in minimally processed fresh-cut apples packaged in a modified atmosphere: A review. *Food Control, 80,* 411–419.

Rabobank International (2010). Retrieved from http://s3.amazonaws.com/zanran_storage/www.agf.nl/ContentPages/682927373.pdf.

Rabobank International (2011). Retrieved from http://www.freshconvenience congress.com/resources/documents/1308561709cindyvanrijswick.pdf.

Ramos, B., Miller, F. A., Brandão, T. R. S., Teixeira, P., & Silva, C. L. M. (2013). Fresh fruits and vegetables-overview on applied methodologies to improve its quality and safety. *Innovative Food Science & Emerging Technologies, 20,* 1–15.

Randhawa, M. A., Khan, A. A., Javed, M. S., & Sajid, M. W. (2015). Green leafy vegetables: A health promoting source. In R. R. Watson (Ed.), *Handbook of fertility* (pp. 205–220). San Diego, CA: Academic Press.

Ressureccion, A. V. A., & Shewfelt, R. L. (1985). Relationships between sensory attributes and objective measurements of postharvest quality of tomatoes. *Journal of Food Science, 50,* 1242–1245.

Rico, D., Martin-Diana, A. B., Barat, J. M., & Barry-Ryan, C. (2007). Extending and measuring the quality of fresh-cut fruit and vegetables: A review. *Trends in Food Science and Technology, 18,* 373–386.

Rosa, M. D., & Rocculi, P. (2007). Prodotti di IV gamma, aspetti qualitativie tecnologici. *Agricoltura, 34,* 33–34.

Rocculi, P., Romani, S., & Rosa, M. D. (2005). Effect of MAP with argon and nitrous oxide on quality maintenance of minimally processed kiwifruit. *Postharvest Biology and Technology, 35,* 319–328.

Ruel, M. T. (2001). *Can food-based strategies help reduce vitamin A and iron deficiencies. A review of recent evidence.* Washington, DC: International Food Policy Research Institute.

Ruiz-Cruz, S., Alvarez-Parrilla, E., de la Rosa, L. A., Martinez-Gonzalez, A. I., Ornelas-Paz, J. D. J, Mendoza-Wilson, A. M., et al. (2010). Effect of different sanitizers on microbial, sensory and nutritional quality of fresh-cut jalapeno peppers. *American Journal of Agricultural and Biological Sciences, 5,* 331–341.

Sánchez, G., Elizaquível, P., & Aznar, R. (2012). A single method for recovery and concentration of enteric viruses and bacteria from fresh-cut vegetables. *International Journal of Food Microbiology, 152*(1–2), 9–13.

Sandhya. (2010). Modified atmosphere packaging of fresh produce: Current status and future needs. *Food Science and Technology, 43,* 381–392.

Santoro, D., Gehr, R., Bartrand, T. A., Liberti, L., Notarnicola, M., Dell'Erba, A., et al. (2007). Waste water disinfection by peracetic acid: assessment of models for tracking residual measurements and inactivation. *Water Environment Research, 79,* 775–787.

Sanz, S., Gimenez, M., Olarte, C., Lomas, C., & Portu, J. (2002). Effectiveness of chlorine washing disinfection and effects on the appearance of artichoke and borage. *Journal of Applied Microbiology, 93*(6), 986–993.

Sapers, G. M., & Simmons, G. F. (1998). Hydrogen peroxide disinfection of minimally processed fruits and vegetables. *Food Technology, 52*, 48–52.
Scetar, M., Kurek, M., & Galic, K. (2010). Trends in meat and meat products packaging—A review. *Croatian Journal of Food Science and Technology, 2*, 32–48.
Shapiro, J. E., & Holder, I. A. (1960). Effect of antibiotic and chemical dips on the microflora of packaged salad mix. *Applied Microbiology, 8*, 341.
Shen, X., Zhang, M., Devahastin, S., & Guo, Z. (2019). Effects of pressurized argon and nitrogen treatments in combination with modified atmosphere on quality characteristics of fresh-cut potatoes. *Postharvest Biology and Technology, 149*, 159–165.
Shirzadeh, E., & Kazemi, M. (2011). Effect of malic acid and calcium treatments on quality characteristics of apple fruits during storage. *American Journal of Plant Physiology, 6*, 176–182.
Song, F., & Kim, M. J. (2015). Acculturation, food intake and dietary behaviors of Chinese college students in Busan by residential period. *Journal of the East Asian Society of Dietary Life, 24*, 594–606.
Statistics Canada (2001). *Food consumption in Canada Part II.*
Strawn, L. K., Schneider, K. R., & Danyluk, M. D. (2011). Microbial safety of tropical fruits. *Critical Reviews in Food Science and Nutrition, 51*, 132–145.
Su, L. J., & Arab, L. (2006). Salad and raw vegetable consumption and nutritional status in the adult US population: Results from the third national health and nutrition examination survey. *Journal of the American Dietetic Association, 106*(9), 1394–1404.
Tao, R., Zhang, F., Tang, Q. J., Xu, C. S., Ni, Z. J., & Meng, X. H. (2019). Effects of curcumin-based photodynamic treatment on the storage quality of fresh-cut apples. *Food Chemistry, 274*, 415–421.
Tappi, S., Berardinelli, A., Ragni, L., Rosa, M. D., Guarnieri, A., & Rocculi, P. (2014). Atmospheric gas plasma treatment of fresh-cut apples. *Innovative Food Science and Emerging Technologies, 21*, 114–122.
Tournas, V. H. (2005). Moulds and yeasts in fresh and minimally processed vegetables and sprouts. *International Journal of Food Microbiology, 99*(1), 71–77.
Unger, J. B., Reynolds, K., Shakib, S., Spruijt-Metz, D., Sun, P., & Johnson, C. A. (2004). Acculturation, physical activity, and fast-food consumption among Asian American and Hispanic adolescents. *Journal of Community Health, 29*, 467–481.
Van Haute, S., Sampers, I., Holvoet, K., & Uyttendaele, M. (2013). Physicochemical quality and chemical safety of chlorine as a reconditioning agent and wash water disinfectant for fresh-cut lettuce washing. *Applied and Environmental Microbiology, 79*, 2850–2861.
Vargas, M., Pastor, C., Chiralt, A., McClements, D. J., & Gonzalez-Martinez, C. (2008). Recent advances in edible coatings for fresh and minimally processed fruits. *Critical Reviews in Food Science and Nutrition, 48*, 496–511.
Vitale, M., & Schillaci, D. (2016). *Food processing and foodborne illness, reference module in food science.* Elsevier.
Wendel, A. M., Johnson, D. H., Sharapov, U., Grant, J., Archer, J. R., & Monson, T. (2009). Multistate outbreak of Escherichia coli O157:H7 infection associated with consumption of packaged spinach, August-September 2006: The Wisconsin investigation. *Clinical Infectious Diseases, 48*(8), 1079–1086.
WHO (World Health Organization). (2000). *Vitamin A deficiency. Retrieved June 2000(2000). http://www.who.int/vaccines-diseases.*
WHO/UNICEF (World Health Organization/United Nations Children's Fund). (1995). *Global prevalence of vitamin A deficiency (1995). Micronutrient deficiency information system, MDIS working paper #2, WHO/NUT/95.3, Geneva.*

Yan, J., Luo, Z., Ban, Z., Lu, H., Li, D., Yang, D., et al. (2019). The effect of the layer-by-layer (LBL) edible coating on strawberry quality and metabolites during storage. *Postharvest Biology and Technology, 147*, 29–38.

Yaron, S., & Romling, U. (2014). Biofilm formation by enteric pathogens and its role in plant colonization and persistence. *Microbial Biotechnology, 7*(6), 496–516.

Ylitalo, K. R., During, C., Thomas, K., Ezell, K., Lillard, P., & Scott, J. (2019). The veggie van: Customer characteristics, fruit and vegetable consumption, and barriers to healthy eating among shoppers at a mobile farmers market in the United States. *Appetite, 133*, 279–285.

Zagory, D., & Kader, A. A. (1988). Modified atmosphere packaging of fresh produce. *Food Technology, 42*, 70–77.

Zhang, G., Gu, L., Lu, Z., Yuan, C., & Sun, Y. (2019). Browning control of fresh-cut Chinese yam by edible coatings enriched with an inclusion complex containing star anise essential oil. *RSC Advances, 9*, 5002–5008.

Further reading

Abadias, M., Alegre, I., Oliveira, M., Altisent, R., & Viñas, I. (2012). Growth potential of Escherichia coli O157:H7 on fresh-cut fruits (melon and pineapple) and vegetables (carrot and escarole) stored under different conditions. *Food Control, 27*, 37–44.

Allende, J. L., McEvoy, Y. L., Artes, F., & Wang, C. Y. (2006). Effectiveness of two-sided UV-C treatments in inhibiting natural microflora and extending the shelf-life of minimally processed 'red oak leaf' lettuce. *Food Microbiology, 23* (3), 241–249.

Ayala-Zavala, J. F., Gonzalez-Aguilar, G. A., & Toro-Sanchez, L. D. (2009). Enhancing safety and aroma appealing of fresh-cut fruits and vegetables using the antimicrobial and aroma power of essential oils. *Journal of Food Science, 74*(7), R84–R91.

Kader, A. A., & Cantwell, M. (2006). *Produce quality rating scales and color charts. Postharvest horticultural series no. 23.* Postharvest Technology Research and Information Center, University of California Davis.

3

Enzymatic browning and its amelioration in fresh-cut tropical fruits

Sarana Rose Sommano*, Usawadee Chanasut†, Wilawan Kumpoun‡

**Plant Bioactive Compound laboratory (BAC), Department of Plant and Soil Sciences, Faculty of Agriculture, Chiang Mai University, Chiang Mai, Thailand. †Department of Biology, Faculty of Science, Chiang Mai University, Chiang Mai, Thailand. ‡Science and Technology Research Institute, Chiang Mai University, Chiang Mai, Thailand*

1 Introduction

The trend for minimal processing of tropical horticultural products including fruit and vegetables has increased in recent years, stimulated largely by the demands of consumers for fresh, healthy, convenient, and additive-free foods that are safe and nutritious (James & Ngarmsak, 2010). There is also a growing interest in international markets for exotic flavors, which has prompted the international trade of fresh-cut products. Fresh-cut tropical fruits on the market today include watermelon, mangoes, mangosteen, rambutan, jackfruit, pomelo, papaya, durian, grapefruit, pineapples, and fruit mixes (Falah, Nadine, & Suryandono, 2015; James & Ngarmsak, 2010; Ketsa & Paull, 2011; Voon, Hamid, Rusul, Osman, & Quek, 2006). Minimally processed fruits have a relatively short shelf-life due to tissue disruption and wound ethylene, which increase the respiration and transpiration and, therefore, accelerate ripening and senescence (Paull & Chen, 1997).

Tropical fruits are physiologically processed, including being washed, peeled, sliced, chopped, and then packed, which make products susceptible to enzymatic browning. Browning of these lightly processed fruits is the result of enzymatic or nonenzymatic oxidation of phenolic compounds present in plant cells (Friedman, 1996; Martinez & Whitaker, 1995). Enzymatic browning is of one of the major concerns in the fresh-cut industry as

Fresh-cut Fruits and Vegetables. https://doi.org/10.1016/B978-0-12-816184-5.00003-3

it directly affects the commercial value of the products (Ma, Zhang, Bhandari, & Gao, 2017). In horticultural produce for instance, enzymatic browning is a sign of deterioration, and is associated with diminished quality and sensory properties such as undesirable changes in flavor, color, texture, and nutritional value (Whitaker & Lee, 1995), and consequently decreases the marketability of the produce. Nonetheless, certain types of browning can be desirable in products such as cocoa, coffee, and fermented tea leaves, in which their quality is enhanced by developing distinctive flavors and fragrances (Yoruk & Marshall, 2003).

In both cases, however, the browning mechanism is explained by the oxidation of plant phenolics by browning enzymes (Marshall, Kim, & Wei, 2000). Within plant cells, phenolic substrates are essentially contained within vacuoles bounded by a tonoplast—a lipoprotein membrane. At the early stages of senescence, which can be accelerated by external stresses, plant cells create reactive oxygen species (ROS) that include superoxide radicals (O_2^-), hydrogen peroxide (H_2O_2), hydroxyl radicals (•OH), and singlet oxygen ($^1O_2$) (Ke & Sun, 2004). When such ROS react with membrane lipoproteins, it results in cell membrane damage, subsequently releasing the secluded phenolic substrates. The final browning reaction occurs when these phenolic substrates come into contact with the intracellular enzymes, polyphenol oxidase (PPO), and polyphenol peroxidise (POD) (Toivonen & Brummell, 2008). Plants, however, also have a self-defense mechanism against browning by producing natural antioxidants (e.g., ascorbic acid) that removes free radicals and ultimately inhibits the browning reaction (Sommano, 2015). The current chapter describes and highlights the influential factors underlying the mechanisms of enzymatic browning in tropical fresh-cut fruits and treatments that could minimize the browning and also increasing the marketability of the products.

2 Minimal processing of tropical fruits

The consumer preference for fresh-cut fruits is not only that they have to look appealing, but must have aroma, taste, texture, and visual appeal similar to the fresh produce. Thus, these products must undergo only minimal processing to maintain the freshness quality and also with the low risk of harm from chemicals and food-borne pathogens (Yu, Neal, & Sirsat, 2018). Fresh-cut processing involves washing, disinfection, peeling, trimming, and deseeding fresh produce, and slicing or cutting to the preferred shape and size (Fig. 1). The products may undergo

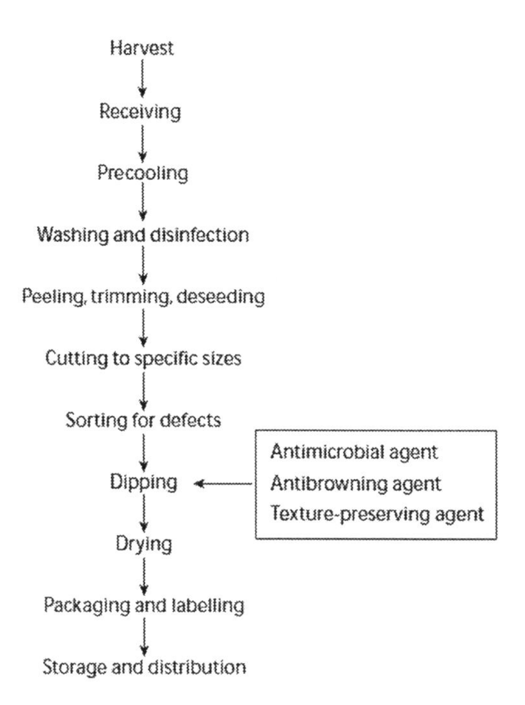

Fig. 1 Minimal processing of fresh-cut tropical fruits. From James, J. B., Ngarmsak, T. (2010). *Processing of fresh-cut tropical fruits and vegetables: A technical guide.* Bangkok: Food and Agriculture Organization of the United Nations Regional Office (FAO) for Asia and the Pacific.

treatments that prolong shelf-life including antibrowning, antimicrobial, or preservative prior to packaging (James & Ngarmsak, 2010).

3 Sanitation of fresh-cut fruits

Many plant pathogens infect fruits prior to harvesting and these pathogens may remain latent and result in decay during fruit minimal processing, handling, and storage. Contamination caused by farmers or workers during harvesting through equipment, transport vehicles, wash water, and human hygiene can be important contributors to postharvest loss. For fruits to undergo minimal processing, whole fruit sanitation plays a role in maintaining the best quality and safety of the cut fruits. It is recommended that the sanitizer systems (i.e., peroxyacetic acid (200 ppm) on whole fruit followed by peroxyacetic acid (50 ppm) or acidified $NaClO_2$ (200 ppm) on the fresh-cuts) can reduce the contaminations (Narciso & Plotto, 2005).

3.1 Styles of cutting

3.1.1 Papaya

The commercial ripening stage for fresh-cut papaya is usually when fruits are 60%–70% yellow surface color. Knife and cutting surface are sanitized with 150 μL/L free chlorine. The whole fruit is peeled, halved longitudinally, and seeds removed. As shown in Fig. 2, the flesh is cut into preferred pieces (~7 × 5 × 3 cm) (Karakurt & Huber, 2003). Shelf-life of papaya pieces is approximately 2 days at refrigerated temperature (4–13°C), with the rate being faster at higher temperature. Desiccation is a major problem with cut papaya pieces, which can be reduced by wrapping with plastic (Paull & Chen, 1997).

3.1.2 Mango

Fruit at the commercial ripening stage, i.e., color break, shoulders developed, and/or shoulder bloom, are chosen (Beaulieu & Lea, 2003). Clean fruit are peeled thoroughly with sharp paring knifes (2–3 mm into the subepidermal tissue) to remove all visible veins and browning. Stem scar ends are also removed completely. Peeled fruit is submerged in clean water for up to 5 min. The fruit is cut in a filet-like fashion, following the flat side of the seeds, and then cut into 2 × 2 cm cubes (Beaulieu & Lea, 2003; Plotto, Narciso,

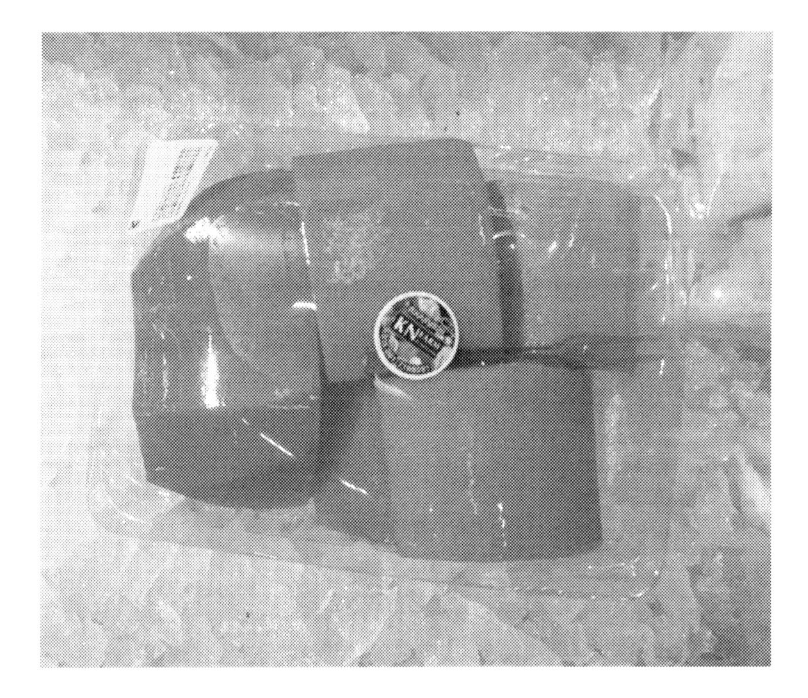

Fig. 2 Example of cutting style and fresh-cut packaging of papaya.

Fig. 3 Example of cutting style and fresh-cut packaging of ripe mango.

Rattanapanone, & Baldwin, 2010; Rattanapanone, Lee, Wu, & Watada, 2001). In another style, the fresh is sliced into a bite-size pieces and packed on tray with clear plastic wrap (Fig. 3).

3.1.3 Rose apple

The mature fruits are harvested carefully at ~45 days after full bloom. After washing, fresh-cut rose apple is cut into four pieces, with the core and ends of each fruit removed (Mola, Uthairatanakij, Srilaong, Aiamla-or, & Jitareerat, 2016).

3.1.4 Watermelon

Whole-undamaged watermelons are cleaned and dipped in 100 μL/L sodium hypochlorite (pH 6.5). Fruit is sliced latitudinally, cut simultaneously from the center of the fruit into 2.5–4 cm wide rings (Guo et al., 2011; Saftner, Luo, McEvoy, Abbott, & Vinyard, 2007). Each ring is then processed into six equally sized wedge-shaped slices. The wedges are placed in rigid polypropylene trays and sealed with plastic film (Saftner et al., 2007). In some cases, fruit is halved and wrapped with plastic for display (Fig. 4).

3.1.5 Jack fruit

After cleaning in acidified hypochlorite, whole jackfruit fruit is cut manually with sanitized stainless steel knives. The peel and nonedible latex part are removed and the edible part is cut in

Fig. 4 Example of cutting style and fresh-cut packaging of water melon.

Fig. 5 Example of cutting style and fresh-cut packaging of jackfruit.

small slices of size $4 \times 3 \times 2$ cm. Then the slices may be dipped in chilled chlorinated water (30 ppm) for 5 min for sanitization as described in Saxena, Bawa, and Raju (2009) and Rana, Pradhan, and Mishra (2018) (Fig. 5).

3.1.6 Pineapple

Undamaged or nondefective fruit at the maturity stage (maturity index 4 or firmness of 54–58 N and soluble solids content of 12–13° Brix) of a commercial (weighing about 1.3–1.5 kg) size is selected. After removing crown, the fruit is cleaned with 100 μL/L sodium hypochlorite (pH 6.5). Peels are manually removed with a sharp knife, the fruit halved, and sliced transversely (2 cm thick), or diced into 2.5 cm triangular chunks (Yeoh & Ali, 2017). Pineapple slices (average weight 100 g) are placed in a 250 mL polystyrene plastic tray covered with a clear plastic lid (González-Aguilar, Ruiz-Cruz, Cruz-Valenzuela, Rodrı´guez-Félix, & Wang, 2004). In many tropical countries, the cut fruits are kept into food-grade plastic bag (Fig. 6).

Fig. 6 Example of cutting style and fresh-cut packaging of pineapple.

3.1.7 Mangosteen

The mature fruit has a dark purple skin that is firm and soft, and white flesh. Burst latex vessels leave a yellow dried mark on the fruit skin that should be scraped off and the fruit then washed with a soft brush (Ketsa & Paull, 2011). Fruit skin is cut open and the entire white flesh is removed from the fruit. The flesh is then immediately placed in chilled chlorinated water (30 ppm) for 5 min. Thereafter, the arils are air-dried and approximately 200 g are packed in polypropylene (PP) trays with lid (Ayudhya, Sophanodora, Pisuchpen, Pisuchpen, & Phongpaichit, 2014).

3.1.8 Durian

In Thailand, the optimum ripe stage of durian is usually 3–7 days after harvest, depending on the varieties (Charoenkiatkul, Thiyajai, & Judprasong, 2016). After cleaning, durian husk is carefully cut along the rind by knife. Durian pods, usually firm textured and creamy colored, are removed. Selected pods of similar size (~300–350 g) are placed in a plastic box with lid or on a Styrofoam tray and wrapped with cling wrap. Fresh-cut durian is stored at ambient temperature or refrigerated conditions (0–4°C) (Voon et al., 2006) (Fig. 7).

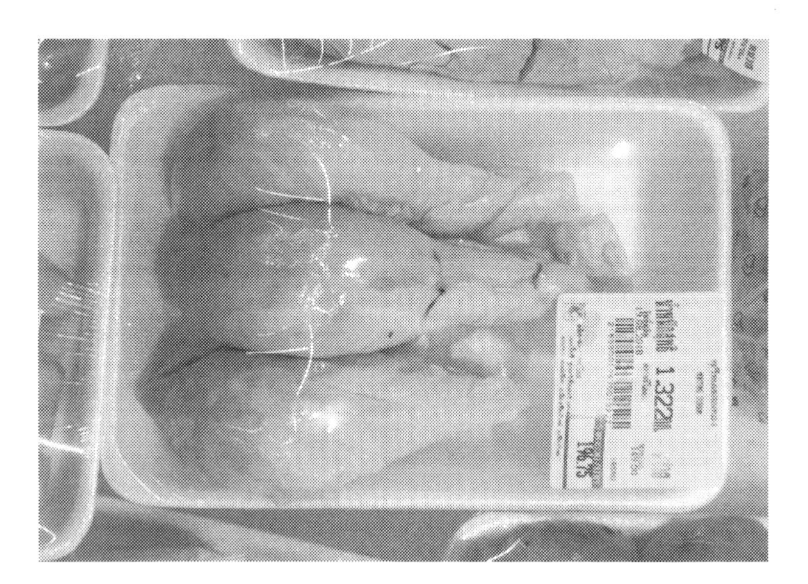

Fig. 7 Example of cutting style and fresh-cut packaging of durian.

4 Stresses affecting enzymatic browning of fresh-cut tropical fruits

Plants and their harvested parts, including fresh-cut commodities, function optimally within a limited range of conditions; beyond these optimum conditions, they are subject to stress (Kays & Paull, 2004). Stress is an external factor that stimulates the alteration of metabolic processes. In the postharvest industry, stress applied to plant produce after they have been harvested can be beneficial. For instance, heat stress can be used to accelerate the ripening process in some tropical climacteric fruits, i.e., fruit that ripen in association with increased ethylene production and a rise in cellular respiration (Alexander & Grierson, 2002). Mangoes, for example, can ripen by induced heating when they undergo a smoke pit ripening method (Kaswija, Peter, & Leonard, 2006). However, ripening put aside, external stress applied to plant tissues may also contribute to adverse effects by causing postharvest injuries or disorders in the plant produce (Kays & Paull, 2004). We here review a list of abiotic and biotic stressors and their effects on the harvested tropical fruits and their fresh-cuts.

4.1 Abiotic stresses

4.1.1 Temperature stress

The respiration rate in plants has been shown to rise following an increase in temperature. In lychee fruit, heating at 60°C for 10 min leads to an irreversible degradation of intracellular components of the lychee pericarp (Underhill & Critchley, 1993). In tropical countries, due to relatively high ambient temperatures throughout the year, produce harvested from plants are often stored in cooling chambers to preserve them from heat stress, a procedure that may also affect temperature-sensitive produce. Indeed, detrimental chilling injury (CI) occurs via the development of internal tissue browning when pineapple fruits are exposed to low temperatures, at 10°C (Nukuntornprakit, Chanjirakul, van Doorn, & Siriphanich, 2015; Weerahewa & Adikaram, 2005). There is the evidence of CI in fresh-cut mango stored at 5°C versus 12°C (Dea, Brecht, Nunes, & Baldwin, 2010).

The study by Kays and Paull (2004) showed that cold temperatures can change the physical properties of cellular lipoprotein membranes, which in turn leads to a number of secondary effects, such as the loss of membrane integrity, a leakage of secluded electrolytes, to significant changes in enzyme activity, and to a permanent loss of subcellular compartmentation. It is also advised that the development of chilling injury symptoms was due to reactive oxygen species metabolism and inadequate antioxidant response (Nukuntornprakit et al., 2015). CI may also encourage many undesirable postharvest symptoms including softening or other textural changes, skin/peel darkening, loss of pigments, and increase in CO_2 production (Hodges & Toivonen, 2008).

4.1.2 White light exposure

A nonthermal emerging Pulsed Light (PL) that can produce intense and short time pulses of broad spectrum "white light" (ultraviolet to the near infrared region) is currently used to microorganism decontaminate in the surfaces of fresh-cut fruits (Salinas-Roca, Soliva-Fortuny, Welti-Chanes, & Martín-Belloso, 2016). The UV spectrums (ranging from 200 to 280 nm) are the most active ranges for microbial inactivation (Gómez-López, Ragaert, Debevere, & Devlieghere, 2007). However, undesirable color can be seen as a consequence of exposure to this visible light, due to a decompartmentalization process allowing browning substances such as phenolic compounds and carotenoids to come into contact with oxidative enzymes like PPO and POD (Charles, Vidal, Olive, Filgueiras, & Sallanon, 2013; de Sousa et al., 2017; Gómez et al., 2012).

4.1.3 Dehydration

Browning and cracking in the pericarp of lychee were found as fruit dehydrated (Jiang, Duan, Joyce, Zhang, & Li, 2004; Joas, Caro, Ducamp, & Reynes, 2005). It was determined that the dehydration of the pericarp introduced micro-cracking at points of weakness where squamous collenchyma disrupts the sclerenchyma continuity (Underhill & Simons, 1993). Underhill and Critchley (1994) suggested that desiccation or moisture loss from the pericarp tends to increase the pericarp pH, therefore conveying the changes of the pericarp pigment (anthocyanin) to the colorless form, thus browning pigment through enzymatic reaction has become more obvious. This, however, is not the case in other sapindus species such as longan (Sommano, Ittipunya, & Sruamsiri, 2009; Sommano, Kanphet, Siritana, & Ittipunya, 2011)

4.1.4 Ethylene stress

Ethylene (C_2H_4) is a growth-regulating plant hormone that is associated with deterioration in harvested produce (Martinez-Romero et al., 2007). This plant hormone affects the cell metabolism by increasing respiration and affects the cellular physiology by increasing membrane permeability, thereby altering cellular compartmentation (Kays & Paull, 2004). Fruits are categorized into climacteric and nonclimacteric based on their ethylene production and respiration. In climacteric fruits like mango, ethylene production rises after the fruit is harvested at full maturity until it reaches its climacteric peak and then gradually declines (Tharanathan, Yashoda, & Prabha, 2006). In contrast, nonclimacteric fruits like longan (Jiang, Yao, Lichter, & Li, 2003; Jiang, Zhang, Joyce, & Ketsa, 2002) or lychee (Jiang, 2000) produce only a small amount of ethylene after harvest. Nonetheless, there is evidence of an increase in ethylene production that is associated with decay in both classes of fruits (Martinez-Romero et al., 2007).

4.1.5 Mechanical stress

Mechanical stress can cause physical injuries in horticultural produce at any time during harvesting, packing, or transporting (Kays & Paull, 2004). This stress brings about cellular changes in vegetables and fruits. Martinez-Romero, Valero, Serrano, Martinez-Sanchez, and Riquelme (1999) and Serrano, Martinez-Romero, Castillo, Guillen, and Valero (2004) determined that mechanical damage causes changes in the levels of polyphenol oxidase (PPO), ethylene, endogenous polyamines (Pas, putrescine (Put), spermidine (Spd), and spermine (Spm)) and abscisic acid (ABA). PPO and ethylene are involved in the aging and damage reaction of fruits and vegetables. Accumulation of PA may be a biochemical measure against mechanical stress and an increase in PA accumulation is found in fruits that have undergone mechanical stress (Serrano et al., 2004). Increased ABA levels are also found in stored fruits and are significantly higher in mechanically damaged fruits (Martinez-Romero et al., 1999). In carrots, physical damage manifest as splitting is mainly caused during mechanical harvesting. This damage is evidently worse in the rainy season due to the high turgor mediated brittleness of the carrots (Galindo, Herppich, Gekas, & Sjoholm, 2004). Bruising in tomato (Thiagu, Chand, & Ramana, 1993) and plum (Serrano et al., 2004) also occurs during the bulk handling and transport process.

Oleocellosis is a result of mechanical damage to the skin of citrus fruits during harvest and handling (Hall, Scott, Wild, & Tugwell, 1989) as well as plants containing essential oil (Sommano et al., 2012; Sommano, 2015). In this disorder, the essential oil released because of mechanical damage kills the surrounding cells and results in dark brown patches and rind collapse (Hall et al., 1989; Knight, Klieber, & Sedgley, 2001, 2002). Browning in peach is associated with damage in epidermal cells due to skin abrasion (Cheng & Crisosto, 1994, 1995).

Minimal processing such as washing, scrubbing, peeling, trimming, cutting, and shredding causes mechanical injury to plant tissues, therefore eliciting physiological and biochemical changes of fresh-cut products including accelerating the respiration rate and ethylene production, which lead to rapid loss of water, firmness, aroma, and flavor (Mola et al., 2016). Moreover, through these processes, cut tissues are also exposed easily to air and the release of the endogenous enzymes to contact with their substrates is liberated (Garcia & Barrett, 2002). Cutting and slicing processes of rose apple are said to be the main cause of its browning (Mola et al., 2016).

4.2 Biotic stress

Biotic stress such as microbial infection during poor sanitizing process can cause visible damage in fresh-cut products such as mango (Salinas-Roca et al., 2016), rose apple (Mola et al., 2016), and guava (Lima, Pires, Maciel, & Oliveira, 2010). *Listeria* sp. is a serious contaminant of fresh-cut fruits as it is able to survive in a wide range of pH and temperature conditions (Salinas-Roca et al., 2016). It has been found that an increase in PPO activity and resultant browning follows attacks by pests, bacteria, or fungi in plants (Taranto et al., 2017). The rapid discoloration is probably due to the solubilization or activation of the latent PPO, which is normally particulate, or to its de novo synthesis (Mayer & Harel, 1979). PPO may play an important role in disease resistance due to quinone formation as an intermediate of PPO oxidation activity and which is toxic to fungus (Thipyapong, Stout, & Attajarusit, 2007).

5 Browning pathway

The substrates of browning are *o*-dihydroxyphenols with oxidizable OH groups (Martinez & Whitaker, 1995). The oxidizable OH groups are those phenolic OHs that are adjacent, or orth, to each other. Within the plant cell, phenolic substrates are contained largely within the vacuole where they are bounded by the lipoprotein tonoplast membrane. Fig. 8 illustrates the location in the cell of the phenolic compounds and the browning enzymes PPO and POD. They are also abundant in the cell wall. Once physical or aging stress occurs and the deteriorative processes (i.e., wounding responses or senescence) are initiated, the plant cell

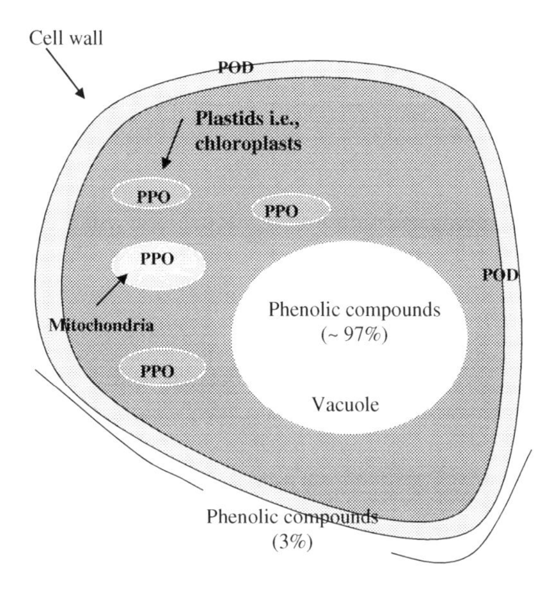

Fig. 8 Structures with in an individual plant cell and indications of the location of the phenolic substrates and of the browning enzymes, PPO (polyphenol oxidase) and POD (peroxidases). Modified from Toivonen, P. M. A., Brummell, D. A. (2008). Biochemical bases of appearance and texture changes in fresh-cut fruit and vegetables. *Postharvest Biology and Technology, 48*, 1–14.

creates reactive oxygen species (ROS) that then react with the lipoprotein membrane and result in a failure of cell compartmentation and the release of phenol substrates (Toivonen & Brummell, 2008).

Tambussi, Bartoli, Guiamet, Beltrano, and Araus (2004) described the process of free radical generation causing cell damage in response to chilling stress as where free radicals or reactive oxygen species (ROS) in the thylakoid membrane were formed through the univalent reduction of oxygen (O_2) to create superoxide at the donor side of photosystem I (PSI) and/or through energy transfer from excited chlorophyll molecules to O_2 and resulting in singlet oxygen. Bartoli et al. (1999) observed similarly that water stress may actuate the formation of superoxide radicals and also hydrogen peroxide, which react directly with the cell membranes. After the loss of cell compartmentation due to the reaction of free radicals with the cellular membranes, the released substrates can interact with oxidative enzymes like PPO and lead to the browning reaction. During senesce and upon protolysis and in becoming more concentrated, cellular solution may alter in pH, which could activate the enzymes associated with membrane breakdown. Study of cell membrane during storage of fresh-cut pear, also described that cell membrane degraded disrupting cell compartmentalization of fresh-cut pears, therefore leading to browning (Li, Zhang, & Ge, 2017).

The reactions of enzymatic browning involves the enzymatic oxidation of phenols to *o*-quinone and the condensation or polymerization of *o*-quinones to form polymers (Yoruk & Marshall, 2003). The *o*-quinones themselves are dark in color. Nonetheless, brown colored pigments are predominantly caused by nonenzymatic secondary reactions of the *o*-quinones by condensation to form the complex polymer compound of melanin (Whitaker & Lee, 1995; Yoruk & Marshall, 2003). The characteristic activity of polyphenol oxidases comprises two basic catalytic reactions (Marshall et al., 2000; Mayer & Harel, 1979). The first involves hydroxylation of the *o*-position adjacent to an existing hydroxyl group of the phenolic substrate (monophenol oxidase activity). The second is oxidation of diphenols to *o*-benzoquinones (diphenol oxidase activity).

6 Browning substrates

Phenolic compounds are characteristically present in higher plants. They act as potential antioxidants in relatively small amounts, whereas at high concentrations they can behave as oxidative substrates (Robards, Prenzler, Tucker, Swatsitang, & Glover, 1999). As being aromatic bearing ring with the hydroxyl groups and several other substituents, the phenolic compounds are of good candidates for enzymatic browning (Cheynier, 2005; Marshall et al., 2000). Plant polyphenols can be categorized into nonflavonoids and flavonoids depending on the complexity of the molecule (Cheynier, 2005).

Anthocyanins are plant pigments representing blue, red, and purple colors in fruits, vegetables, and flowers. These pigments can be found in almost every part of plants. Within the anthocyanin group, they differ from each other in the number of hydroxyl groups, the nature and number of sugars attached to the molecule, the position of this attachment, and in other molecular groups attached to sugar molecules. Pelargonidin, cyanidin, peonidin, delphinidin, petunidin, and malvidin are the six most common anthocyanins in higher plants (Kong, Chia, Goh, Chia, & Brouillard, 2003).

PPO is associated with the degradation of anthocyanins in many kinds of fruits (Jiang, 2000). Browning in the lychee pericarp has been attributed to rapid degradation of anthocyanin. As PPO has low affinity for anthocyanin due to a sugar moiety that may cause a steric hindrance (Jiang, 2000; Underhill & Critchley, 1994), browning of the lychee peel is possibly related to anthocyanase hydrolyzation. The result is that major substrates are accessible by PPO or POD (Jiang et al., 2004). Anthocyanin discoloration ascribable to pH alteration enhances the pericarp browning appearance (Underhill & Critchley, 1994).

6.1 Browning in fresh-cut guava

Fig. 9 illustrates the occurrence of browning discoloration in fresh guava slices. The slices were stored at ambient temperature (30°C) and refrigerated conditions (5°C) for 20h. they did not evidently change in terms of visual color whether they were stored in open conditions or in a nonair tight polypropylene box. Nasution, Ye, and Hamzah (2015) Advised that guava contained ascorbic acid that may act as a natural antibrowning agent.

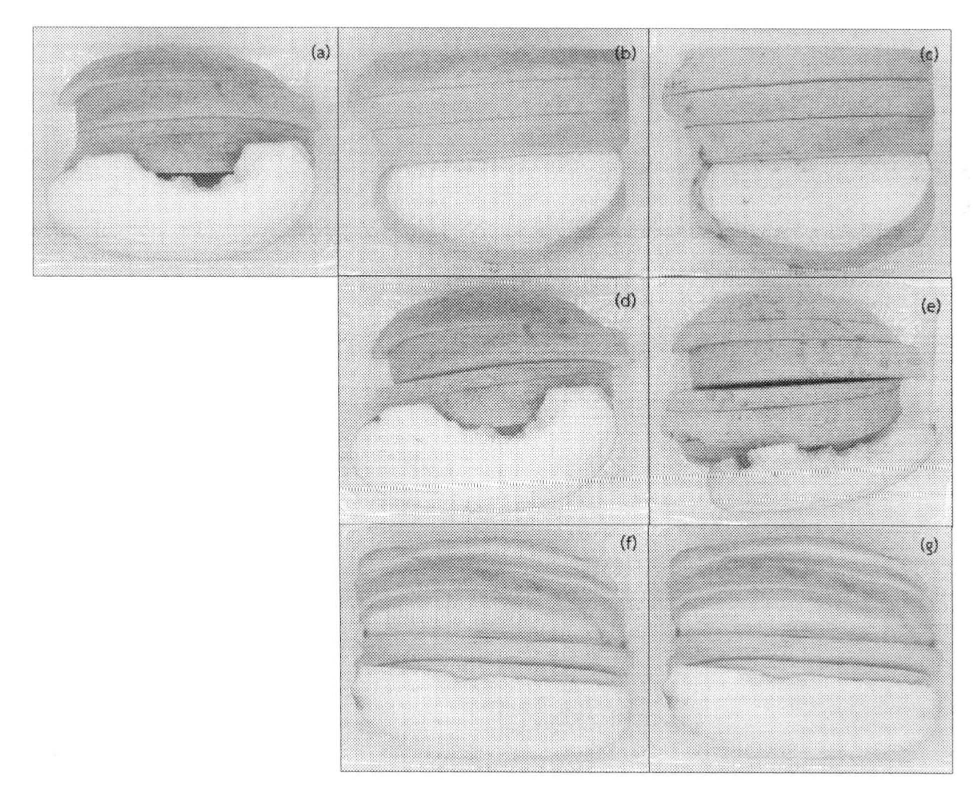

Fig. 9 Browning in fresh guava after minimal processing (A) right after cutting, (B–C) leaving the slices at 30°C for 5 and 20h. After cutting, the slices were placed in a polypropylene box and left at 30°C for 5 and 20h (D–E), and at 5°C for 5 and 20h (F–G).

This antioxidant reduces quinones generated by the oxidation of polyphenols by PPO back to phenolic substrates. This might be the reason why browning was less evident in fresh-cut guava. However, the ascorbic acid could be oxidized to dehydroascorbic acid after storing, thereafter leading to the accumulation of *o*-quinones and browning.

6.2 Browning in fresh-cut mango

Browning in fresh-cut mango usually occurs through mechanical destruction of the membrane system at the cut surface. The wound accelerates ethylene production that degrades other cell membranes, disrupting and destroying the tissue (Brecht, 1993). This loss of cellular compartmentalization increase chances of tissue exposure to oxygen as well as providing greater contact between the oxidative enzymes (de Sousa et al., 2017). Browning is more obvious in fresh-cut raw mango stored at ambient temperature as compared to ripe mango in the same conditions (Figs. 9 and 10) possibly due to the active PPO and the availability of phenolic substrates (Thomas & Janave, 1973; Vithana, Singh, &

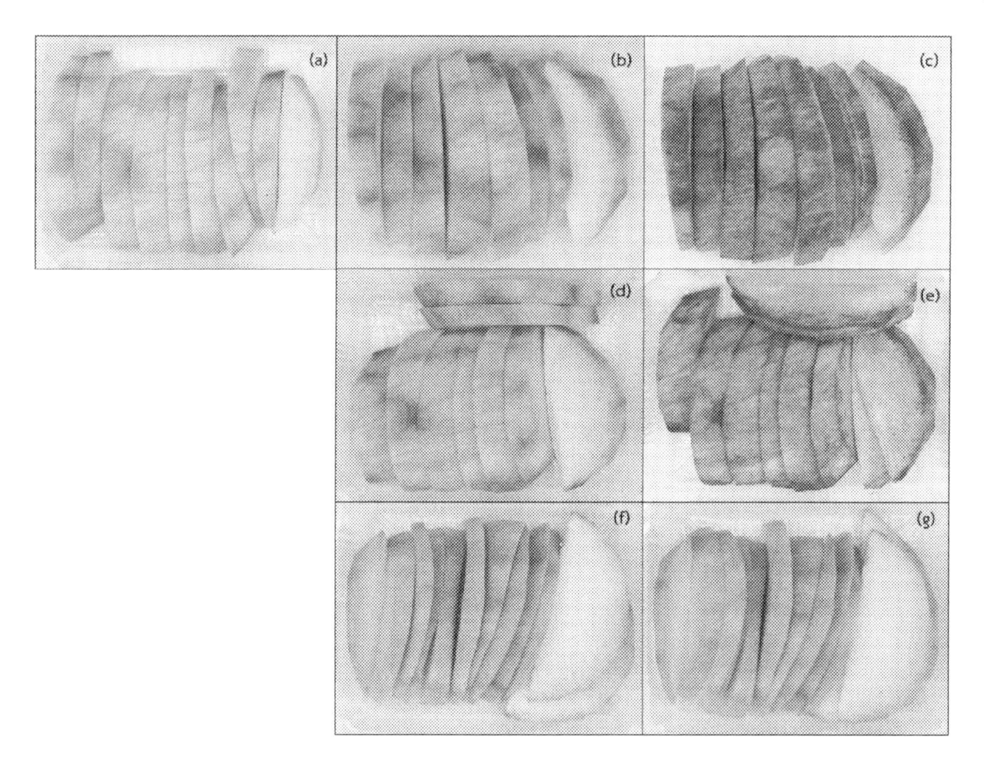

Fig. 10 Browning in fresh raw mango (cv. Keaw) after minimal processing (A) right after cutting, (B–C) leaving the slices at 30°C for 5 and 20h. After cutting, the slices were placed in a polypropylene cbox and left at 30°C for 5 and 20h (D–E), and at 5°C for 5 and 20h (F–G).

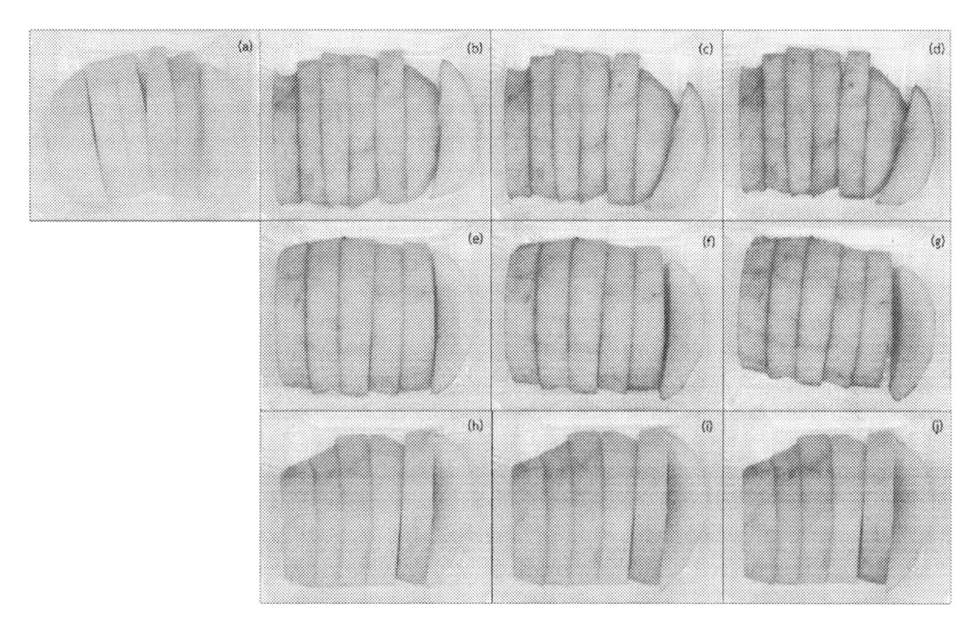

Fig. 11 Browning in fresh ripe mango (cv. Keaw) after minimal processing (A) right after cutting, (B–D) leaving the slices at 30°C for 5, 10, and 20 h. After cut, the slices were placed in polypropylene container and left at 30°C for 5, 10, and 20 h (E–G) and at 5°C for 5, 10, and 20 h (H–J).

Johnson, 2018). Maintaining the fresh-cuts at low temperature could evidently retain the freshness appearance in both raw and ripe types (plates F–G in Figs. 10 and 11).

7 Antibrowning treatments

The approaches to controlling enzymatic browning of fresh-cut tropical fruits can be applied through the use of both physical and chemical methods. The chemical treatments utilize compounds that could inhibit the active enzymes and remove the substrates (viz., phenolics and oxygen), while the physical approaches may include reduction of temperature, the use of modified atmosphere packaging or edible coating (Garcia & Barrett, 2002).

7.1 Chemical treatments

The most commonly used chemical treatments used in tropical fresh-cut products are those that can retard or inhibit enzymatic reactions. Examples are as follows:

- **Vitamin C** (ascorbic acid) is the antioxidation agent, and is able to alter quinones reversibly to phenolic substrates (Hasan, Manzocco, Morozova, Nicoli, & Scampicchio, 2017; Soliva-Fortuny & Martı´n-Belloso, 2003).

- **Acidulants** (i.e., 1% citric acid) are used to adjust the pH of the products to the lower range, which is not preferable for browning enzymes to be active (optimum pH of PPO ~6–7) (Capotorto, Amodio, Diaz, de Chiara, & Colelli, 2018).
- **Aromatic carboxylic acids** can also inhibit PPO and they include;
  - *Benzoic acid and derivatives of cinnamic acid* that are able to react with the enzymes and block the reaction between the enzymes and the substrates (Ding, Alborzi, Bastarrachea, & Tikekar, 2018; Sharma & Rao, 2015).
  - *Sodium metabisulfite or other sulfate compounds* function as chelating agents that complex with copper at the enzyme active site, therefore inhibiting enzymatic reaction.
  - *Cysteine* is a reducing agent that causes the reduction of colorless *o*-quinones resulting from enzymatic reaction with PPO to reverse back to *o*-diphenols (Garcia & Barrett, 2002).
- **Sodium chloride (NaCl)** is a weak inhibitor, thus an elevated amount is required to minimize browning incidence in fresh-cut products (Mola et al., 2016).

7.2 Physical treatments

- **Reducing oxygen availability** as the substrate of enzymatic reactions (apart from phenolics) can be done by immersing the products in water, brine, or syrup to diminish diffusion of oxygen. However, exposing the fresh-cut in the solutions may result in the loss of solute and imbibition of the solution as well as microbial growth during storage (Garcia & Barrett, 2002). To extend shelf-life, these types of minimally processed products usually require pasteurization, which alters texture and flavor.
  - *Dense Phase Carbon Dioxide* (*DPCD*): Recently, DPCD was applied at low temperature to a solid-liquid product consisting of fresh-cut fruit in syrup as an alternative to the conventional thermal processing. This technique can inactivate PPO while maintaining other qualities of the products such as pH, color, and total acidity (Ferrentino & Spilimbergo, 2017).
  - *Edible coating*: Fresh-cut produce has a much larger cut surface through the process of peeling, cutting, and slicing. Edible coating is also used to reduce the chance of the fresh-cut product coming into contact with oxygen, especially on the surface of the product. The protective layer is also able to help protect the product from water loss, color change, reduce respiration, improve texture, and retain aromatic characteristics, and also reduce microbial growth (Garcia & Barrett, 2002; Yousuf, Qadri, & Srivastava, 2018).

– *Modified Atmosphere Packaging (MAP):* MAP adjust ideal gas (CO_2 and O_2) compositions that could reduce the respiration of the fresh-cut thereby, prolonging the storage quality of the products. Browning could be minimized by limiting the O_2 condition in the package (Waghmare & Annapure, 2013; Xing et al., 2010).

- **Temperature management** before or after minimal-processing is crucial to reduce the metabolic processes that affect the quality of the products during storage. Besides the increasing of respiration, high temperature coincides with the increase in related enzyme activities and the browning degree of the fresh-cuts (Min et al., 2017). Therefore, it is recommended that fresh-cut products should be kept at a temperature just above the freezing point; however, it should be chosen carefully to avoid damage such as chilling injury in highly sensitive produce such as tropical fruits (Garcia & Barrett, 2002). There are also benefits of heat treatment in the inhibition of browning due to enzymatic inactivation (Li et al., 2018; Zhang, Li, Wang, Li, & Zong, 2017).

Table 1 sets out antibrowning treatments and their applications in some tropical fresh-cut fruits.

Table 1 Antibrowning treatments and applications of high potential tropical fresh-cut fruits

Fresh-cut tropical fruits	Treatments	Method of application	Quality assessments	References
I. Chemical treatments				
Fresh mango slices	2% (*w/v*) malic acid	Dipping in solution	Reduce visual browning	Salinas-Roca et al. (2016)
II. Physical treatments				
Fresh-cut pineapple	Edible coatings (ECs) based on chitosan (CH), pullulan (PU), linseed (LM), nopal cactus (NM), and aloe mucilage (AM)	layer-by-layer technique by dipping pineapple pieces in the ECs then packing into polyethylene terephthalate containers and storage for 18 days at 4°C	Coated samples reduce darkening of pineapple pieces	Treviño-Garza, García, Heredia, Alanís-Guzmán, and Arévalo-Niño (2017)
Peeled guava cube	*Aloe vera* coating gel with additives	Dipping in the coating solution for 5 min,	Less change in lightness and	Nasution et al. (2015)

Continued

Table 1 Antibrowning treatments and applications of high potential tropical fresh-cut fruits—Cont'd

Fresh-cut tropical fruits	Treatments	Method of application	Quality assessments	References
		drained and dried at 30°C for 20 min	yellowness of the color	
Fresh-cut melon and papaya	Controlling temperature	Storing fresh-cut fruits in controlled room temperature in a showcase with air temperature varied in the range of 14–16°C	Maintaining the color and extend storage life for 3–4 days	Falah et al. (2015)
III. Combine treatments				
Fresh-cut rose apple	Temperature management, chemical treatment and edible coating	Hydro cooling rose apple fruits before dipping in 6% calcium ascorbate and 0.2% chitosan coating solution and storage at 5°C	Delay browning for up to 72 h	Worakeeratikul, Srilaong, Uthairatanakij, and Jitareerat (2007)
Fresh-cut rose apple	Chemical treatments and temperature management	200 mg/L NaCl and in 200 mg/L NaCl combined with 20 g/L $CaCl_2$ and 20 g/L CaAs (calcium ascorbate) for 3 min, then subjected to low storage temperature at 4 ± 2°C	Reduce browning of fresh-cut rose apple without any significant effects on firmness	Mola et al. (2016)
Fresh mango slices	Combination treatments of pulsed light (PL), alginate coating (ALC), and malic acid dipping (MA)	PL (20 pulses at fluence of 0.4 J cm^{-2}/pulse), 2 min dipping in ALC (2% *w/v*) and MA (2% w/v) solutions	Reduce browning and microbial counts	Salinas-Roca et al. (2016)
Fresh mango cubes	Application of alginate edible coating in conjunction with antibrowning agents (ascorbic and citric acid) with low temperature control 4°C were evaluated	Dipping the fruit into alginate solution for 2 min; the excess of coating material was allowed to drip off for 1 min before submerging for 2 min in the calcium chloride solution containing antibrowning agents	Maintain visual color (higher L* and Hue) and improves the antioxidant potential of fresh-cut mango	Robles-Sánchez, Rojas-Graü, Odriozola-Serrano, González-Aguilar, and Martin-Belloso (2013)

8 Conclusions

Enzymatic browning discoloration is a crucial issue in the postharvest industry as it affects quality attributes including appearance and nutritional value, and therefore limits the marketability of the produce. The browning mechanism involves the generating of radical oxygen species when harvested produce experiences physical stresses (e.g., heat and cold temperatures). As a result, cells lose their compartmentation, which brings about the release of enzymes (e.g., polyphenol oxidase and peroxidase) from subcellular structures (e.g., plastids) and phenolic compounds (i.e., browning substrates) from the vacuole. The reaction of these compounds is thereby initiated and browning occurs. However, plants do have their own antioxidant chemicals and mechanisms to inhibit the browning mechanism by scavenging free radicals produced by extrinsic stresses. This review has highlighted the mechanism of enzymatic browning in tropical fresh-cut produce.

References

Alexander, L., & Grierson, D. (2002). Ethylene biosynthesis and action in tomato: A model for climacteric fruit ripening. *Journal of Experimental Botany, 53*, 2039–2055.

Ayudhya, C. P. N., Sophanodora, P., Pisuchpen, S., Pisuchpen, S., & Phongpaichit, S. (2014). Influence of sealing film lid on the quality of packaged fresh-cut mangosteen. *Chiang Mai University Journal of Natural Sciences, 13*(1), 529–540.

Bartoli, C. G., Simontacchi, M., Tambussi, E., Beltrano, J., Montaldi, E., & Puntarulo, S. (1999). Drought and watering-dependent oxidative stress: Effect on antioxidant content in *Triticum aestivum L.* leaves. *Journal of Experimental Botany, 50*, 375–383.

Beaulieu, J. C., & Lea, J. M. (2003). Volatile and quality changes in fresh-cut mangos prepared from firm-ripe and soft-ripe fruit, stored in clamshell containers and passive MAP. *Postharvest Biology and Technology, 30*(1), 15–28.

Brecht, J. K. (1993). Physiology of lightly processed fruits and vegetables. *HortScience, 28*(5), 472.

Capotorto, I., Amodio, M. L., Diaz, M. T. B., de Chiara, M. L. V., & Colelli, G. (2018). Effect of anti-browning solutions on quality of fresh-cut fennel during storage. *Postharvest Biology and Technology, 137*, 21–30.

Charles, F., Vidal, V., Olive, F., Filgueiras, H., & Sallanon, H. (2013). Pulsed light treatment as new method to maintain physical and nutritional quality of fresh-cut mangoes. *Innovative Food Science & Emerging Technologies, 18*, 190–195.

Charoenkiatkul, S., Thiyajai, P., & Judprasong, K. (2016). Nutrients and bioactive compounds in popular and indigenous durian (*Durio zibethinus* murr.). *Food Chemistry, 193*, 181–186.

Cheng, G. W., & Crisosto, C. H. (1994). Development of dark skin colouration on peach and nectarine fruit in response to exogenous contaminations. *Journal of the American Society for Horticultural Science, 119*, 529–533.

Cheng, G. W. W., & Crisosto, C. H. (1995). Browning potential, phenolic composition, and polyphenol oxidase activity of buffer extracts of peach and nectarine skin tissue. *Journal of the American Society for Horticultural Science, 120*, 835–838.

Cheynier, V. (2005). Polyphenols in foods are more complex than often thought. *American Journal of Clinical Nutrition, 81*, 223S–229S.

de Sousa, A. E. D., Fonseca, K. S., da Silva Gomes, W. K., Monteiro da Silva, A. P., de Oliveira Silva, E., & Puschmann, R. (2017). Control of browning of minimally processed mangoes subjected to ultraviolet radiation pulses. *Journal of Food Science and Technology, 54*(1), 253–259.

Dea, S., Brecht, J. K., Nunes, M. C. N., & Baldwin, E. A. (2010). Occurrence of chilling injury in fresh-cut 'Kent' mangoes. *Postharvest Biology and Technology, 57*(1), 61–71.

Ding, Q., Alborzi, S., Bastarrachea, L. J., & Tikekar, R. V. (2018). Novel sanitization approach based on synergistic action of UV-A light and benzoic acid: Inactivation mechanism and a potential application in washing fresh produce. *Food Microbiology, 72*, 39–54.

Falah, M. A. F., Nadine, M. D., & Suryandono, A. (2015). Effects of storage conditions on quality and shelf-life of fresh-cut melon (*Cucumis Melo* L.) and papaya (*Carica Papaya* L.). *Procedia Food Science, 3*, 313–322.

Ferrentino, G., & Spilimbergo, S. (2017). Non-thermal pasteurization of apples in syrup with dense phase carbon dioxide. *Journal of Food Engineering, 207*, 18–23.

Friedman, M. (1996). Food browning and its prevention: An overview. *Journal of Agricultural and Food Chemistry, 44*, 631–653.

Galindo, F. G., Herppich, W., Gekas, V., & Sjoholm, I. (2004). Factors affecting quality and postharvest properties of vegetables: Integration of water relations and metabolism. *Critical Reviews in Food Science and Nutrition, 44*, 139–154.

Garcia, E., & Barrett, D. M. (2002). *Preservative treatments for fresh-cut fruits and vegetables*. Davis, CA: Department of Food Science and Technology, University of California.

Gómez, P. L., García-Loredo, A., Nieto, A., Salvatori, D. M., Guerrero, S., & Alzamora, S. M. (2012). Effect of pulsed light combined with an antibrowning pretreatment on quality of fresh cut apple. *Innovative Food Science & Emerging Technologies, 16*, 102–112.

Gómez-López, V. M., Ragaert, P., Debevere, J., & Devlieghere, F. (2007). Pulsed light for food decontamination: A review. *Trends in Food Science & Technology, 18*(9), 464–473.

González-Aguilar, G. A., Ruiz-Cruz, S., Cruz-Valenzuela, R., Rodrı´guez-Félix, A., & Wang, C. Y. (2004). Physiological and quality changes of fresh-cut pineapple treated with antibrowning agents. *LWT-Food Science and Technology, 37*(3), 369–376.

Guo, Q., Cheng, L., Wang, J., Che, F., Zhang, P., & Wu, B. (2011). Quality characteristics of fresh-cut 'Hami' melon treated with 1-methylcyclopropene. *African Journal of Biotechnology, 10*(79), 18200–18209.

Hall, E. G., Scott, K. J., Wild, B. L., & Tugwell, B. L. (1989). Citrus. In W. B. McGlasson, B. B. Beattie, & N. L. Wade (Eds.), *Temperate fruit: Posthavest diseases of horticultural produce*. East Melbourne, VIC: CSIRO.

Hasan, S. M. K., Manzocco, L., Morozova, K., Nicoli, M. C., & Scampicchio, M. (2017). Effects of ascorbic acid and light on reactions in fresh-cut apples by microcalorimetry. *Thermochimica Acta, 649*, 63–68.

Hodges, D. M., & Toivonen, P. M. A. (2008). Quality of fresh-cut fruits and vegetables as affected by exposure to abiotic stress. *Postharvest Biology and Technology, 48*(2), 155–162.

James, J. B., & Ngarmsak, T. (2010). *Processing of fresh-cut tropical fruits and vegetables: A technical guide.* Bangkok: Food and Agriculture Organization of the United Nations Regional Office (FAO) for Asia and the Pacific.

Jiang, Y. M. (2000). Role of anthocyanins, polyphenol oxidase and phenols in lychee pericarp browning. *Journal of the Science of Food and Agriculture, 80,* 305–310.

Jiang, Y. M., Duan, X. W., Joyce, D., Zhang, Z. Q., & Li, J. R. (2004). Advances in understanding of enzymatic browning in harvested litchi fruit. *Food Chemistry, 88,* 443–446.

Jiang, Y., Yao, L., Lichter, A., & Li, J. (2003). Postharvest biology and technology of litchi fruit. *Food Agriculture and Environment, 1,* 76–81.

Jiang, Y. M., Zhang, Z. Q., Joyce, D. C., & Ketsa, S. (2002). Postharvest biology and handling of longan fruit (*Dimocarpus longan* Lour.). *Postharvest Biology and Technology, 26,* 241–252.

Joas, J., Caro, Y., Ducamp, M. N., & Reynes, M. (2005). Postharvest control of pericarp browning of litchi fruit (*Litchi chinensis* Sonn cv Kwaï Mi) by treatment with chitosan and organic acids: I. Effect of pH and pericarp dehydration. *Postharvest Biology and Technology, 38*(2), 128–136.

Karakurt, Y., & Huber, D. J. (2003). Activities of several membrane and cell-wall hydrolases, ethylene biosynthetic enzymes, and cell wall polyuronide degradation during low-temperature storage of intact and fresh-cut papaya (*Carica papaya*) fruit. *Postharvest Biology and Technology, 28*(2), 219–229.

Kaswija, M., Peter, M., & Leonard, F. (2006). Sensory attributes, microbial quality and aroma profiles of off vine ripened mango (*Mangifera indica* L.) fruit. *African Journal of Biotechnology, 5,* 201–205.

Kays, S. J., & Paull, R. E. (2004). *Postharvest biology.* Athens, GA: Exon Press.

Ke, D. S., & Sun, G. C. (2004). The effect of reactive oxygen species on ethylene production induced by osmotic stress in etiolated Mungbean seedling. *Plant Growth Regulation, 44,* 199–206.

Ketsa, S., & Paull, R. E. (2011). 1—Mangosteen (*Garcinia mangostana* L.). In E. M. Yahia (Ed.), *Postharvest biology and technology of tropical and subtropical fruits* (pp. 1–32e). Cambridge: Woodhead Publishing.

Knight, T. G., Klieber, A., & Sedgley, M. (2001). The relationship between oil gland and fruit development in Washington Navel orange (*Citrus sinensis* L. Osbeck). *Annals of Botany, 88,* 1039–1047.

Knight, T. G., Klieber, A., & Sedgley, M. (2002). Structural basis of the rind disorder oleocellosis in Washington navel orange (*Citrus sinensis* L. Osbeck). *Annals of Botany, 90,* 765–773.

Kong, J. M., Chia, L. S., Goh, N. K., Chia, T. F., & Brouillard, R. (2003). Analysis and biological activities of anthocyanins. *Phytochemistry, 64,* 923–933.

Li, S., Li, X., He, X., Liu, Z., Yi, Y., Wang, H., & Lamikanra, O. (2018). Effect of mild heat treatment on shelf life of fresh lotus root. *LWT-Food Science and Technology, 90,* 83–89.

Li, Z., Zhang, Y., & Ge, H. (2017). The membrane may be an important factor in browning of fresh-cut pear. *Food Chemistry, 230,* 265–270.

Lima, M. S., Pires, E. M. F., Maciel, M. I. S., & Oliveira, V. A. (2010). Quality of minimally processed guava with different types of cut, sanification and packing. *Food Science and Technology, 30,* 79–87.

Ma, L., Zhang, M., Bhandari, B., & Gao, Z. (2017). Recent developments in novel shelf life extension technologies of fresh-cut fruits and vegetables. *Trends in Food Science & Technology, 64,* 23–38.

Marshall, M. R., Kim, J., & Wei, C. (2000). *Enzymatic browning in fruits, vegetables and seafoods. In Agriculture and consumer protection department*: Food and

Agriculture Oganisation (FAO) of the United Nation. *http://www.fao.org/ag/ags/agsi/ENZYMEFINAL/Enzymatic%20Browning.html.*

Martinez, M. V., & Whitaker, J. R. (1995). The biochemistry and control of enzymatic browning. *Trends in Food Science and Technology, 6*, 195–200.

Martinez-Romero, D., Bailen, G., Serrano, M., Guillen, F., Valverde, J. M., Zapata, P., Castillo, S., & Valero, D. (2007). Tools to maintain postharvest fruit and vegetable quality through the inhibition of ethylene action: A review. *Critical Reviews in Food Science and Nutrition, 47*, 543–560.

Martinez-Romero, D., Valero, D., Serrano, M., Martinez-Sanchez, F., & Riquelme, F. (1999). Effects of post-harvest putrescine and calcium treatments on reducing mechanical damage and polyamines and abscisic acid levels during lemon storage. *Journal of the Science of Food and Agriculture, 79*, 1589–1595.

Mayer, A. M., & Harel, E. (1979). Polyphenol oxidase in plants. *Phytochemistry, 18*, 193–215.

Min, T., Xie, J., Zheng, M., Yi, Y., Hou, W., Wang, L., Ai, Y., & Wang, H. (2017). The effect of different temperatures on browning incidence and phenol compound metabolism in fresh-cut lotus (*Nelumbo nucifera* G.) root. *Postharvest Biology and Technology, 123*, 69–76.

Mola, S., Uthairatanakij, A., Srilaong, V., Aiamla-or, S., & Jitareerat, P. (2016). Impacts of sodium chlorite combined with calcium chloride, and calcium ascorbate on microbial population, browning, and quality of fresh-cut rose apple. *Agriculture and Natural Resources, 50*(5), 331–337.

Narciso, J., & Plotto, A. (2005). A comparison of sanitation systems for fresh-cut mango. *HortTechnology, 15*(4), 837–842.

Nasution, Z., Ye, J. N. W., & Hamzah, Y. (2015). Characteristics of fresh-cut guava coated with aloe vera gel as affected by different additives. *Kasetsart Journal (Natural Science), 49*, 1–11.

Nukuntornprakit, O.-A., Chanjirakul, K., van Doorn, W. G., & Siriphanich, J. (2015). Chilling injury in pineapple fruit: Fatty acid composition and antioxidant metabolism. *Postharvest Biology and Technology, 99*, 20–26.

Paull, R. E., & Chen, W. (1997). Minimal processing of papaya (*Carica papaya* L.) and the physiology of halved fruit. *Postharvest Biology and Technology, 12*(1), 93–99.

Plotto, A., Narciso, J. A., Rattanapanone, N., & Baldwin, E. A. (2010). Surface treatments and coatings to maintain fresh-cut mango quality in storage. *Journal of the Science of Food and Agriculture, 90*(13), 2333–2341.

Rana, S. S., Pradhan, R. C., & Mishra, S. (2018). Optimization of chemical treatment on fresh cut tender jackfruit slices for prevention of browning by using response surface methodology. *International Food Research Journal, 25*(1), 196–203.

Rattanapanone, N., Lee, Y., Wu, T., & Watada, A. E. (2001). Quality and microbial changes of fresh-cut mango cubes held in controlled atmosphere. *HortScience, 36*(6), 1091–1095.

Robards, K., Prenzler, P. D., Tucker, G., Swatsitang, P., & Glover, W. (1999). Phenolic compounds and their role in oxidative processes in fruits. *Food Chemistry, 66*, 401–436.

Robles-Sánchez, R. M., Rojas-Graü, M. A., Odriozola-Serrano, I., González-Aguilar, G., & Martin-Belloso, O. (2013). Influence of alginate-based edible coating as carrier of antibrowning agents on bioactive compounds and antioxidant activity in fresh-cut Kent mangoes. *LWT-Food Science and Technology, 50*(1), 240–246.

Saftner, R., Luo, Y., McEvoy, J., Abbott, J. A., & Vinyard, B. (2007). Quality characteristics of fresh-cut watermelon slices from non-treated and 1-methylcyclopropene-and/or ethylene-treated whole fruit. *Postharvest Biology and Technology, 44*(1), 71–79.

Salinas-Roca, B., Soliva-Fortuny, R., Welti-Chanes, J., & Martín-Belloso, O. (2016). Combined effect of pulsed light, edible coating and malic acid dipping to improve fresh-cut mango safety and quality. *Food Control, 66*, 190–197.

Saxena, A., Bawa, A. S., & Raju, P. S. (2009). Phytochemical changes in fresh-cut jackfruit (*Artocarpus heterophyllus* L.) bulbs during modified atmosphere storage. *Food Chemistry, 115*(4), 1443–1449.

Serrano, M., Martinez-Romero, D., Castillo, S., Guillen, F., & Valero, D. (2004). Role of calcium and heat treatments in alleviating physiological changes induced by mechanical damage in plum. *Postharvest Biology and Technology, 34*, 155–167.

Sharma, S., & Rao, T. V. R. (2015). Xanthan gum based edible coating enriched with cinnamic acid prevents browning and extends the shelf-life of fresh-cut pears. *LWT-Food Science and Technology, 62*(1, Pt. 2), 791–800.

Soliva-Fortuny, R. C., & Martı´n-Belloso, O. (2003). New advances in extending the shelf-life of fresh-cut fruits: A review. *Trends in Food Science & Technology, 14*(9), 341–353.

Sommano, S. (2015). Physiological and biochemical changes during heat stress induced browning of detached *Backhousia myrtifolia* (Cinnamon Myrtle) tissues. *Tropical Plant Biology, 8*(1-2), 31–39.

Sommano, S., Ittipunya, P., & Sruamsiri, P. (2009). Deterioration model for the assessment of longan senescence and decay. *Chiang Mai University Journal of Natural Sciences, 8*, 229–237.

Sommano, S., Joyce, D., Joyce, P., Ratnayake, K., Max, J. F. J., & d'Arcy, B. (2012). Phytotoxicity by essential oil may play a role in postharvest browning disorder of cinnamon myrtle (*Backhousia myrtifolia*) tissues. In *Proceedings of tropentag 2012—Resilience of agricultural systems against crises, 19-21 September, Gottingen, Germany.*

Sommano, S., Kanphet, N., Siritana, D., & Ittipunya, P. (2011). Correlation between browning index and browning parameters during the senesence of longan peel. *International Journal of Fruit Science. 11*(2), 197–205.

Tambussi, E. A., Bartoli, C. G., Guiamet, J. J., Beltrano, J., & Araus, J. L. (2004). Oxidative stress and photodamage at low temperatures in soybean (*Glycine max* L. Merr.) leaves. *Plant Science, 167*, 19–26.

Taranto, F., Pasqualone, A., Mangin, G., Tripodi, P., Miazzi, M. M., Pavan, S., & Montemurro, C. (2017). Polyphenol oxidases in crops: Biochemical, physiological and genetic aspects. *International Journal of Molecular Sciences, 18*(2), 377.

Tharanathan, R. N., Yashoda, H. M., & Prabha, T. N. (2006). Mango (*Mangifera indica* L.), "The king of fruits"—An overview. *Food Reviews International, 22*, 95–123.

Thiagu, R., Chand, N., & Ramana, K. V. R. (1993). Evolution of mechanical characteristics of tomatoes of 2 varieties during ripening. *Journal of the Science of Food and Agriculture, 62*, 175–183.

Thipyapong, P., Stout, M. J., & Attajarusit, J. (2007). Functional analysis of polyphenol oxidases by antisense/sense technology. *Molecules, 12*, 1569–1595.

Thomas, P., & Janave, M. T. (1973). Polyphenol oxidase activity and browning of mango fruits induced by gamma irradiation. *Journal of Food Science, 38*(7), 1149–1152.

Toivonen, P. M. A., & Brummell, D. A. (2008). Biochemical bases of appearance and texture changes in fresh-cut fruit and vegetables. *Postharvest Biology and Technology, 48*, 1–14.

Treviño-Garza, M. Z., García, S., Heredia, N., Alanís-Guzmán, M. G., & Arévalo-Niño, K. (2017). Layer-by-layer edible coatings based on mucilages, pullulan and chitosan and its effect on quality and preservation of fresh-cut pineapple (*Ananas comosus*). *Postharvest Biology and Technology, 128*, 63–75.

Underhill, S. J. R., & Critchley, C. (1993). Lychee pericarp browning caused by heat injury. *HortScience, 28*, 721–722.

Underhill, S., & Critchley, C. (1994). Anthocyanin decolorisation and its role in lychee pericarp browning. *Australian Journal of Experimental Agriculture, 34*, 115–122.

Underhill, S. J. R., & Simons, D. H. (1993). Lychee (*Litchi-chinensis* sonn) pericarb dessication and the importance of postharvest microcracking. *Scientia Horticulturae, 54*, 287–294.

Vithana, M. D. K., Singh, Z., & Johnson, S. K. (2018). Cold storage temperatures and durations affect the concentrations of lupeol, mangiferin, phenolic acids and other health-promoting compounds in the pulp and peel of ripe mango fruit. *Postharvest Biology and Technology, 139*, 91–98.

Voon, Y. Y., Hamid, N. S. A., Rusul, G., Osman, A., & Quek, S. Y. (2006). Physicochemical, microbial and sensory changes of minimally processed durian (*Durio zibethinus* cv. D24) during storage at 4 and 28°C. *Postharvest Biology and Technology, 42*(2), 168–175.

Waghmare, R. B., & Annapure, U. S. (2013). Combined effect of chemical treatment and/or modified atmosphere packaging (MAP) on quality of fresh-cut papaya. *Postharvest Biology and Technology, 85*, 147–153.

Weerahewa, D., & Adikaram, N. K. B. (2005). Heat-induced tolerance to internal browning of pineapple (*Ananas comosus* cv. 'Mauritius') under cold storage. *The Journal of Horticultural Science and Biotechnology, 80*, 503–509.

Whitaker, J. R., & Lee, C. Y. (1995). *Enzymatic browning and its prevention*. Washington, DC: American Chemical Society.

Worakeeratikul, W., Srilaong, V., Uthairatanakij, A., & Jitareerat, P. (2007). Effect of whey protein concentrate on quality and biochemical changes in fresh-cut rose apple. *Acta Horticulturae, 746*, 435–442.

Xing, Y., Li, X., Xu, Q., Jiang, Y., Yun, J., & Li, W. (2010). Effects of chitosan-based coating and modified atmosphere packaging (MAP) on browning and shelf life of fresh-cut lotus root (*Nelumbo nucifera* Gaerth). *Innovative Food Science & Emerging Technologies, 11*(4), 684–689.

Yeoh, W. K., & Ali, A. (2017). Ultrasound treatment on phenolic metabolism and antioxidant capacity of fresh-cut pineapple during cold storage. *Food Chemistry, 216*, 247–253.

Yoruk, R., & Marshall, M. R. (2003). Physicochemical properties and function of plant polyphenol oxidase: A review. *Journal of Food Biochemistry, 27*, 361–422.

Yousuf, B., Qadri, O. S., & Srivastava, A. K. (2018). Recent developments in shelf-life extension of fresh-cut fruits and vegetables by application of different edible coatings: A review. *LWT-Food Science and Technology, 89*, 198–209.

Yu, H., Neal, J. A., & Sirsat, S. A. (2018). Consumers' food safety risk perceptions and willingness to pay for fresh-cut produce with lower risk of foodborne illness. *Food Control, 86*, 83–89.

Zhang, L., Li, S., Wang, A., Li, J., & Zong, W. (2017). Mild heat treatment inhibits the browning of fresh-cut Agaricus bisporus during cold storage. *LWT-Food Science and Technology, 82*, 104–112.

4

Fresh-cut products: Processing operations and equipments

Aamir Hussain Dar*, Omar Bashir[†], Shafat Khan*, Aseeya Wahid[‡], Hilal A Makroo*

*Department of Food Technology, Islamic University of Science and Technology, Awantipora, India. [†]Division of Food Science and Technology, Sher-e-Kashmir University of Agricultural Sciences and Technology, Kashmir, India. [‡]Department of Processing and Food Engineering, Punjab Agricultural University, Ludhiana, India

1 Introduction

The International Fresh-Cut Produce Association (IFPA) defines fresh-cut fruit and vegetable products (FFVP) as products which are peeled, cut into a 100% usable form, and packaged to offer the consumers fresh products having high flavor and nutrition (Allende, Tomas-Barberan, & Gil, 2006; Olivas & Barbosa-Canovas, 2005). The significance of fresh-cut foods is determined by their quality attributes, i.e., freshness, retention of vital nutrients, convenience, and sensory attributes, along with enhancement of shelf-life (Bhagwat, 2006; Ragaert, Verbeke, Devlieghere, & Debevere, 2004). FFVs are prepared in a way that they don't require any additional preparation for their consumption and hence the retention of nutritional values and organoleptic characteristics is made possible to a maximum extent.

Artes-Hernandez et al. (Artes, Conesa, Hernandez, & Gil, 1999) referred to FFVP as food products that are prepared by operations such as peeling, shredding, cutting, trimming, and sanitizing, which are then stored in refrigerated conditions, and packed in semipermeable films. Ingredients and nutrient content in FFV are similar to that of the whole product, along with a benefit of having low cost and a brief period of preparation (Chien, Sheu, & Yang, 2007). The industry of fresh-cut fruits and vegetable products is growing rapidly as compared to the other industries related to fruits and vegetables; the reasons being the service, production, and global market access. The United States is the main producer in this sector followed by the United Kingdom and

Fresh-cut Fruits and Vegetables. https://doi.org/10.1016/B978-0-12-816184-5.00004-5

France (Ahvenainen, 1996). FFVPs sold presently in the global market include: leafy greens/Spanish (washed and cut), lettuce (cleaned, chopped, shredded), celery (sticks), cabbage (shredded), broccoli and cauliflower (florets), potatoes, carrots (baby, sticks, shredded), and various other tubers and roots (peeled and sliced), onions (whole peeled, sliced, and diced), garlic (fresh peeled and sliced) cucumber (sliced, diced), mushrooms (sliced), and sliced pepper and tomato (Allende et al., 2006).

The shelf-life of fruit and vegetable products are extended with the methods of food processing but the enhancement of fresh-cut product's shelf-life is reduced, resulting in the produce being highly perishable (Ahvenainen, 1996). The biological change brought by the food processing methods may cause discoloration of cut-surface, flavor loss, rapid softening, decay, shrinkage, increased rate of vitamin loss, and shorter shelf-life of the produce. The textural and flavor changes during the processing might be the result of interaction between intracellular and intercellular enzymes and water activity (Corbo et al., 2009). Fresh-cut processing causes stress on the cell tissues of vegetables, which in turn results in the phytochemical accumulation in the tissues and hence induces loss in the enzymes of secondary metabolic pathways due to their reduced activity. Moreover, fresh-cut fruit and vegetable processing results in breakdown of cell structure along with the release of various types of intracellular products like oxidizing enzymes, thereby causing a rapid decay and enzymatic browning of the product (Allende et al., 2006). Numerous factors, as reported by Gorny, Cifuentes, Hess-Pierce, and Kader (2000) and Solomos (1997), may alter the overall quality of fresh-cut products, like appearance (O'Beirne & Francis, 2003; Olivas & Barbosa-Canovas, 2005). Appearance, along with shape, size, color, and form are factors that significantly affect the consumer acceptability of fresh-cut products. Preharvest factors also influence the overall acceptability of the product. Nutrients such as antioxidant vitamins, α-tocopherol (vitamin E), beta-carotene, and vitamin C are available in fruits and vegetables in large amounts. Research has suggested that fruit and vegetable consumption on a daily basis can reduces risk of circulatory diseases, cancer, and numerous inflammatory conditions (Allende, Selma, Lopez-Galvez, Villaescusa, & Gil, 2008; Gil, Selma, Lopez-Galvez, & Allende, 2009). Healthy lifestyle today includes consumption of fruit and vegetables, with the promotion of health through the increased supply of antioxidants and other phytochemicals.

It is notable that the handling of organic products advances a quicker physiological damage, biochemical changes, and microbial invasion of the products that might cause degradation of the texture, color, and flavor, even by application of the simplest processing operations (O'Beirne & Francis, 2003). Before packaging for utilization, fresh-cut fruits are exposed to a number of unit operations, such as washing/sanitizing, cutting or slicing, peeling, dicing, and so on. During each step of processing, packaging, and storage, there is always a possibility of quality and nutritional loss of the product. A significant attention should be given to the mechanical operations that are considered extremely critical to cause shelf-life enhancement of the product, if implemented in a proper way. Mishandling and improper steps in processing cause the rupture of numerous cells and discharge of intracellular products. In the meantime, the surface of the product is subjected to air and various microbes like yeasts, bacteria, and molds. Moreover, cutting means increased injured tissues leading to increase in respiration rate, rapid disintegration, and microbial multiplication.

The damage caused by mechanical processing to the product tissues has a positive relation with the maturity of the treated product; research on the issue demonstrates that the riper the fruit, the more liable is the product to injury during processing (Gorny et al., 2000). The ideal phase of preparing the fruits so as to reduce the damage caused by cutting fluctuates significantly, depending upon the cultivar, species, and pre and postharvest conditions (Solomos, 1997). Fruits such as cactus pear, apples, and kiwis need stripping/peeling. A few techniques of peeling are accessible; however, peeling is achieved either mechanically, by a high-pressure steam peeler, or chemically, on an industrial scale (Ahvenainen, 1996). Such an operation has to be as delicate as possible; the basic method is manual peeling using knives. Abrasive, coarse peeling increases the chances of microbial invasion over peeling by hand. Hence, if peeling is done mechanically, it should resemble the peeling done manually.

Other methodologies can disrupt the cell walls of the produce, thereby increasing the degradation. The most important operation in the processing line is washing, which affects the overall quality attributes and the shelf stability of the fresh-cut products. Whole fruit washing usually removes pesticidal deposits, debris of plants, and any possible foreign contaminants. A second step of washing has to be done after peeling or cutting in order to eliminate tissue fluids and microbes. The microbiological and sensory attributes of washing water should be safe and must have a temperature below 5°C. Washing carefully extends the shelf-life of fresh-cut produce by a few days. Fruit undergoes additional handling procedures by dipping the produce in a sanitizing solution that contains chemicals like chlorine or other antibacterial. This step usually reduces the microbial load on the surface of the food (Allende et al., 2006).

Recently, Gil et al. (2009) reviewed problems that were related to the sanitation issues of fresh-cut products and disinfection. Most of the accessible literature indicated that washing that is performed with or without the aid of disinfectants reduces the microbial load on the surface of the product by 2–3 log units (Allende et al., 2008; Gomez-Lopez, Devlieghere, Ragaert, & Debevere, 2007). It was also concluded that the total bacterial counts after the storage were similar when the produce was washed with tap water or when it was sanitized (Allende et al., 2008). The sanitizers and cleansing agents also reduce the microbial counts of the produce. It is now justified that antimicrobials when used appropriately with good water quality minimize the potential risks of microbial contamination (FDA Food and Drug Administration, 2010a). A technique is desirable that is capable to sanitize the water for processing and the product, which would allow the maximum recycling ratio and hence will reduce environmental impact with less wastewater (Olmez & Kretzschmar, 2009).

2 Transport, storage, and precooling

It is necessary to ensure that fresh, good quality, safe produce is being supplied to the processors. If contamination of raw material exists at high levels, then it is not possible that further processing will decrease it to safer levels. Processors must be sure that they accept produce from the areas where they are produced in accordance to good agricultural practices. Also, it should be ensured that during transportation of produce to the factory, the produce is not further contaminated prior to processing. Maintenance of hygiene during storage and transport is necessary, and these should be enhanced by local facilities with additional measures and working practices. Some vegetables and fruits are exposed to precooling processes right after harvest in order to get a final product of a quality acceptable to the consumer. Cooling of product delays the unavoidable deterioration of quality and also extends its shelf-life. The temperature of the produce can be reduced by hydrocooling, which is a direct method of cooling of the produce by using chilled water, by direct or indirect methods of immersion, or by spraying cold water on the surface of the product. There are different durations of spraying water on the produce depending upon the size of the produce (for example asparagus and carrots require a duration less than 10 min and produce size larger than 3 in., and diameter of 7.6 cm requires 30–60 min).

For produce, which loses water quickly, vacuum coolers are more applicable (i.e., for, leafy greens, lettuce, and cauliflower). Vacuum coolers operate by decreasing the boiling temperature of water, thus creating a process of evaporative cooling. A vacuum chamber is used for the cooling process. Loss of water by the vacuum coolers is one of the disadvantages (2%–4% of moisture loss from most produce). However, a large amount of energy is needed by the coolers, which is quite expensive. Forced air cooling is a widely used system. Chilled air is discharged at high velocities over the product surface to remove the field heat, which can be damaging. It is extremely significant to take into account the kind of produce to be cooled while choosing the operation of cooling.

3 Trimming and the removal of waste

Trimming provides a proper ingredient proportion in salad preparations. The trimming table is equipped with final proportions of every dish, taking into consideration the particular process output. The unwanted components of the plant, generally green leaves and core areas, are manually removed. This type of operation can be made more efficient by utilizing sharp edged knives (Olmez & Kretzschmar, 2009). Trimming could also be mechanized. Injury to the plant structure may lead to the discharge of various enzymes and their respective substrates that are present in various cell sections. The disruption of cell microstructures causes biochemical spoilage like off-flavor, textural losses, and enzymatic browning. Unloading and dumping should be operated by use of systems that reduce tissue injuries and contamination. The initial core or trimming operation is appropriate to get rid of; core of lettuce, loose leaves, and stems.

Training and hygiene of workers is also important, and operators should wear acceptable clothes so as to meet the quality standards systematically. Many fresh processors recently installed optical and mechanical systems for inspection so as to remove foreign matter from the product. Waste removal can be challenging under food processing environments. This intensive labor process depends upon the teamwork of the staff. Wastage includes rinds, peels, tops, cores, tails, and whole rejected items that are generated during various unit operations in the plant. Waste removal systems vary from a simple solution to rather advanced ones such as flumes, belts, vacuums, and augers. Every plant of fresh fruit and vegetable must get all the undesirables to the waste truck from the trim line. Belts are simple to changeover and flexible to remove, however there might be risks at the dropping points and sanitation problems might be an issue. Moreover, the waste carried by the belt conveyors is collected in strategically located bins that arc carried by lift trucks.

Flume-carried and water-evacuated waste must be dewatered before collecting in a trailer for disposal. The solids must be separated from water. The waste water must be treated and discharged after application of various water treatment methods. The law and regulations related to this subject are getting restrictive, raising the bar on necessarily having a better hygiene design in the plant. The vacuum system for further disposal and transport of waste does not utilize water as a carrier medium. The pneumatic conveyance system contains a blower package that pressurizes a stainless steel pipe set at a lower place beneath the area of trimming. The pipeline is fed via variety of rotary airlocks that direct the waste material into the system. The waste material is then transported to a receiver hose placed on top of a waste truck for removal. The utilization of lesser amounts of water and chemicals has made this system very acceptable in recent years (Olmez & Kretzschmar, 2009).

4 Cutting, shredding, and size reduction

Numerous size reduction or shredding machines are being used depending upon the produce and application. Machines are made of better-quality carbon stainless steel of 316 grade or enhanced, while a systematic and better directed program of cleaning should be brought up to enable efficient cleaning operations. Although there is a high chance of contamination by metal due to the breakage of blades, a metal detector must be placed at a proper place to avoid such risks.

5 Washing

Washing is a crucial part of processing of any product, particularly that of the raw and processed fresh produce. The process of washing should be designed precisely, measured, and then utilized into an appropriate procedure. The washing operation must eliminate the adhering dirt and other foreign unwanted materials, decrease the chemical and microbial load on the produce, and lower the final product temperature for enhancement of shelf-life.

Usually there are three washing stages in which three different tanks are used in the optimum washing systems. The first tank is intended to remove usual field soil and other debris. A flotation washing system, in which air is blown at high velocities into the tank through spare pipe located at a distance of about 10–12 in. under the water surface is preferred as a solution for products that float. This generation of a hot tub effect initiates a tumbling movement as the produce is metered forward. Any adhering incidental dirt and debris get released and washed off and an appropriate design incorporates a system that eliminates the floating remains. Debris and dirt that sinks at the bottom of the tank needs to be removed at regular intervals. If antibacterial chemicals are used for treating water, it is directed that an automated chemical observing and dosing machine be used for optimum management of the method. These machines are accessible for chemicals like chlorine however, they might not be offered for different chemicals.

Numerous chemically based treatment methods are applied to water, like softening methods, decontamination, and flocculation. These methods/treatments are applied to kill microbes in water and on the surface of product that comes into contact with the water. The application of an antibacterial will significantly enhance the safety of the operation. A new advanced system for washing utilizes a closed pipe flume concept technology and not employing a centrifugal pump for the product movement. In this system, the contact time of the produce with the sanitizing water is precisely controlled, hence assuring full submersion along with proper treatment time. The product flow remains laminar and it separates and exposes the product to the treatment while not making any mechanical damage to it. As a result, delicate products are effectively washed, thereby maintaining a good product quality. This design safeguards the discharge of the sanitizer within the production area, hence increasing plant safety.

6 Dewatering and drying

Excess of water from the washed produce should be separated while removing from the wash tank. This can be attained by draining for a fixed time duration or by dewatering, spin drying, or using fluidized drying tunnels. Different shapes and capacities of spin dryers are available. The centrifugation cycle usually starts with the soft loading of soft leaves and acceleration of the product to the discharge end. The dewatering system can either be used as a single unit or a first stage of a continuous air drying system. The system is used more effectively for dewatering the shredded, sliced, and diced food products. However, the surface moisture content is adjusted by this system on the unwashed leafy surfaces and to regain the expensive liquids used by the fresh-cut fruit processing industries. The unique design of the dewatering systems make it ideal for handling sensitive products like sliced mushrooms and tomatoes, fruits and other fresh-cut products. Due to the high efficiency of dewatering, the process of drying could be omitted, hence reducing the mechanical damage to the products. All kinds of leafy vegetables can be dewatered by the air drying system, making the use of this developed technology over many years. Using natural technologies and safeguarding the product integrity, traditional and safe method are used in fluidized systems to remove water from the leafy vegetables. Recent studies have made its access possible by making the maintenance and cleaning operations quite simpler. It has a potential to examine the effectiveness of this system, by sampling which allows the products to be dried in an atmosphere that is free from any external contamination, hence significantly increasing the efficiency of the drying systems.

7 Packing

Packaging methods and materials that are utilized during the process should be reliable for their expected purpose. Usually, hygiene is the main concern in the production of packaging but the packaging material must be marked as food grade and its use as specified must be ensured. This will ensure that there is no migration of the chemicals from the packaging material into the product. Various formats for packing are available, from a standard polyethylene or plastic bag to a rigid plastic container, vacuum packing, boxes, buckets, tubs, and so on. In all the cases, the type of packaging should be acceptable for its application and use. Processors need to keep in mind that the product is still alive and, therefore, for some products that have a high rate of respiration, a high permeability medium packaging should be used. The rate at which several packaging materials are permeable to oxygen and carbon dioxide have to be selected as packaging materials.

8 Fresh-cut fruit

Fresh-cut fruits have recently emerged in the market and it is predicted that the fresh-cut industries will show tremendous growth in coming years. For ensuring quality and safety of fresh-cut fruit, state-of-the-art production facilities are designed.

Regular testing for any microbial contamination is performed. On receiving a batch of fruit, operators must apply good agricultural practices and carry out quality control checks of random samples at regular intervals. The quality control checks include inspection of fruit visually, so as to ensure that there is no damage or bruising of the fruits. In addition to this, color charts should be used by the operators that can help in assessing the maturity of the fruit. All the quality control checks need to be in accordance to the legal standards.

The delivery of the fruits is generally recorded on the computerized system, so as to keep records for further tracking and monitoring procedures. Large-sized fruits are processed to obtain high outputs or yields, and it is easier to prepare larger fruits than tiny fruits, hence reducing the handling and process time. Low temperatures, usually 4–6°C are preferred for the storage of fruits; however, the storage of such products is short, as they are dispatched soon after their manufacture. Suppliers should essentially prewash all the raw materials to minimize the contamination risk. It is necessary to clean and sanitize the fruit once it is cut. Usually, fresh-cut fruit pieces must be washed once they are cut, with cold chlorinated water having chlorine concentrations in the range of 50–200 ppm with a pH of <7 or equal to 7. This helps in the extension of shelf-life of the product by decreasing microbial count, removing intracellular juices at cut surfaces that might cause browning and inhibit the enzymatic reactions causing browning of fruits.

It is usually impossible to achieve a general approach for microbial safety because of the varieties in raw materials and processing operations. Steam and hot water are the systems used widely and in both cases, the external skin of the fruits like pineapple, melons, and oranges is brought to high temperatures (100°C) so as to destroy bacteria and other pathogenic microorganisms. A hot water system is preferable over the steam method as it penetrates throughout all the areas of fruits. In order to slough off the external fruit, the whole fruit is placed inside a machine manually and the working orientation offered by the blades of the machine that remove the ends of the fruits that are self-centering type to suit each fruit's dimensions.

Fruits are cut longitudinally into halves initially and the different configurations of the machine used to produce peeled whole fruits and slices of various shapes. Pineapples are washed as whole and washed carefully for the removal of contaminants and adhering unwanted material. The outer skin of the pineapple is removed by automated machines or manually. To remove microbes and any sort of foreign matter from the surface of pineapples, they are dipped in chemically treated water. Pineapples are peeled, quartered, and removal of hard inner cores is done and the peeled halves are diced. Blast chillers are used to store pineapple until its dispatch. To peel the delicate fruits, steam peeling is done and finally nitrogen cooling is done that imparts numerous advantages over other systems. As discussed earlier, controlled-atmosphere conditions are used for the process to take place, which inhibits oxidation-based chemical reactions that cause the spoilage of the product. For the reduction of numbers of bacteria present on the product, food irradiation is recognized as beneficial for the fresh-cut fruits. There are difficulties encountered in the processing of fresh-cut fruits so the requirement of new and sophisticated level of operation and techniques has to be taken into consideration. It is recommended for the manufacturers, researchers, and the industries to work together in the future so as to overcome the hindrances that obstruct the delivery of fresh-cut fruits.

9 Fresh-cut fruits preservation

For the reduction in bacterial populations in fruits, several chemical compounds have been and are still being used widely, during pre- and postcutting operations or before processing (Gil et al., 2009). Primarily, chlorine-based chemicals are used like liquid chlorine, ClO_2, and ClO^- (hypochlorite) mainly at 50–200 ppm levels of free chlorine and with exposure time of <5 min (Gómez-López, Rajkovic, Ragaert, Smigic, & Devlieghere, 2009). Traditionally, for decontamination of fresh produce, chlorinated water treatment has been used, but the usage of chlorine in ready-to-use products is forbidden in some European countries including Switzerland, Germany, Belgium, and the Netherlands due to its potential toxicity (Olmez & Kretzschmar, 2009). In addition, it is known that carcinogenic alogenated by-products are formed when chlorine reacts with natural organic matter (Gil et al., 2009). In fact, larger quantities of wastewater production with very high levels of BOD (biological oxygen demand) is associated with the employment of chlorine (Gil et al., 2009). Investigations have shown that the chemicals are incapable of inactivating microorganisms completely on fresh produce (Abadias, Usall, Oliveira, Alegre, & Vinas, 2008). Furthermore, chlorine-based dipping treatment by calcium decreases the microbial count ar d causes shelf-life enhancement of the fruit; particularly for delicate fruit, such as fresh-cut apple (Anino, Salvatori, & Alzamora, 2006) and cantaloupes (Lamikanra & Watson, 2007), calcium lactate is widely used. Dipping fresh-cut honeydew melon and cantaloupe in a solution of H_2O_2 is also documented for microbial population reduction (Rico, Martin-Diana, Barat, & Barry-Ryan, 2007).

Sensitive individuals are allergic to these chemical food preservatives, thus increasing the interest in and demand for natural antimicrobial compounds among consumers (Roller, 2003). Studies of data suggest that essential oils can be a good option in comparison to chemicals. For instance, methyl jasmonate (MJ), a natural compound found in volatile oil of Jasminum and various other varieties of plant, is established for shelf-life enhancement of fruits such as mangoes, guavas, and strawberries for whole as well as fresh-cut (Gonzalez-Aguilar, Tiznado-Hernandez, & Wang, 2006). Further valuable evidence is given by the observations made by Roller and Seedhar (2002), who suggested that microbial spoilage in fresh-cut melon and kiwifruit was delayed by carvacrol and cinnamic acid. Fruits were peeled, deseeded, sliced (melon), or cut into wedges (kiwifruit), and dipped in a solution containing 1 mM of cinnamic acid or 1, 5, 10, or 15 mM of carvacrol acid for 1 min. Immersing of fresh-cut kiwifruit in carvacrol solutions at 5–15 mM lowered total active counts from 6.6 to less than 2 log CFU g^{-1} at 4°C stored 21 days. Similarly, treatment with 1 mM of cinnamic acid or carvacrol decreased viable counts on both the fresh-cut products, increasing shelf-life greatly (Roller & Seedhar, 2002).

A recent study by D'Amato, Sinigaglia, and Corbo (2010) proposed the ability of a few natural compounds to extend the shelf-life of salads made up of fruits. In particular, investigation was done to observe the impact of chitosan, pineapple juice, and honey on the growth of psychrotrophic bacteria, lactic acid bacteria, mesophilic bacteria, and yeasts. Granny Smith apples composed the fruit-based salads. The first crop cactus pear fruits (Gialla) and table grapes (Regina), cut in identical pieces, were kept in polypropylene cups containing different solutions (1:1), which were stored at 4, 8, and 12°C. The significant antibacterial activity on the psychrotrophic and mesophilic bacteria was shown by honey. The antimicrobial effect of chitosan influenced the growth of all microbial groups taken into account, particularly, in refrigerated storage conditions. Between the various "natural" approaches, ascorbic and citric acid were often proposed for reduction in the microbial count. The antimicrobial action of these chemicals is because of the rupture of membrane transport and permeability, decrease in pH in the environment, anion absorption, or a decrease in internal cellular pH by the dissociation of hydrogen ions from the acid (Soliva-Fortuny & Martin-Belloso, 2003). Moreover, citric acid has been recognized to have effect in decreasing superficial pH of cut fruit such as apple, orange, peach, kiwifruit, apricot, bananas, and avocado (Soliva-Fortuny & Martin-Belloso, 2003). Another option for obtaining sources of natural preservative is from waste products obtained from different processing industries. For instance, whey permeates (WP) is a cheese industry by-product having potential as a sanitizing agent, as it contains lesser pH and has lactic acid, thermo-resistant bacteriocins, and other small bio-active peptides (Martin-Diana et al., 2006).

10 Novel technologies for the shelf-life extension of fresh-cut fruits and vegetables

10.1 Modified atmosphere packaging

Modified atmosphere packaging (MAP) has emerged as an effective technique for the shelf-life enhancement of fresh-cut produce Sandhya, 2010. MAP refers to the alteration in composition of gases surrounding the product or inactive gas replacement before the containers are sealed or packaged (Mangaraj, Goswami, & Mahajan, 2009). Frequently used gases for MAP are O_2, CO_2, and N_2. The applications of this packaging system have a history of about 90 years. In the 1920s, a modified atmosphere of reduced O_2 and increased CO_2 was used for the shelf-life extension of apples (Phillips, 1996). Nowadays, an extensive variety of fresh-cut fruits and vegetables accessible in the market have been developed by using MAP technology (Gorny, Hess-Pierce, Cifuentes, & Kader, 2002). The increased combination of carbon dioxide and oxygen is recommended to provide conventional improved MAP in fresh-cut vegetables. Shelf stability of sliced carrots was seen to be extended when stored under 50% O_2 and 30% CO_2 for 2–3 days, in comparison to the control stored in air, showing better or same quality than those kept under 1% O_2 and 10% CO_2 at 8°C after 8–12 days (Amanatidou, Slump, Gorris, & Smid, 2000). However, poor product quality was seen in oxygen levels above 70% when combined with 10%–30% CO_2. Zhang, Samapundo, Pothakos, Sürengil, and Devlieghere (2013) identified that MAP showed effective inhibition effect on the yeast growth and volatile organic compounds production in fresh-cut pineapple in 50% O_2 and 50% CO_2, and suggested that, for the shelf-life enhancement of fresh-cut pineapple, high O_2 and high CO_2 MAP was an efficient method. Similar results were seen in fresh-cut honeydew melon (Zhang et al., 2013).

10.2 Pressurized inert gases

Inert gases such as neon (Ne), xenon (Xe), argon (Ar) krypton (Kr), and nitrogen (N_2) form clathrate hydrate, i.e., ice-like crystal, when dissolved under higher pressure in water (Ma, Zhang, Bhandari, & Gao, 2017). The gas molecules are trapped between water molecules by physical bonding through van der Waals forces, forming cage like structures. Under certain pressures above 0°C temperatures, visible size Clathrate crystals can be developed easily (Reid & Fennema, 2008). In recent years, preservation of some of the fresh-cut fruits and vegetables has been successfully done by pressurized inert gases (Artés, Gómez, Aguayo, Escalona, & Artés-Hernández, 2009). Zhang, Zhan, Wang, and Tang (2008) found that fresh-cut green asparagus spears' shelf-life could be enhanced at 4°C from 3–5 to 12 days by treating with argon (Ar) and xenon (Xe) gas mixtures under pressure of 1.1 MPa (Ar and Xe at 2:9 (v:v) in partial pressure) for 24 h (Zhang et al., 2008). Meng, Zhang, and Adhikari (2012) found that at 4 MPa, pressurized argon treatment for 1 h on fresh-cut green peppers could decrease loss of water along with water mobility, chlorophyll, and ascorbic acid, the growth of molds and yeast, and maintain the cell integrity by inhibiting the MDA production, catalase (CAT), and POD activities. Shelf-life extension of 12 days stored at 4°C was shown by the results, which was 4 days longer when compared to control (untreated) (Meng et al., 2012).

10.3 Electron beam irradiation

Food irradiation is an innovative technology after pasteurization. Irradiation brings about minimum alteration in the color, taste, nutrients, and other attributes of food quality. Food irradiation primarily refers to the treatment of cobalt-60 radioisotopes exposure for the shelf-life extension and safety purposes of food (Farkas & Mohácsi-Farkas, 2011). Researches have shown that the application of low dose irradiation is considered as safe for the treatment of food; although conventional use of food irradiation has raised several questions due to public misunderstanding because of potential risks of the radiations to cause cancers (Kong et al., 2014). Electron beam irradiation (EBI), for the generation of ionization radiation, does not require a radioactive isotope as gamma radiation does. Electron beams are generated from machines that accelerate electrons to the speed of light at 0.15–10 MeV high energy level in a vacuum environment (Mami, Peyvast, Ziaie, Ghasemnezhad, & Salmanpour, 2014). Commercial electricity is the source of energy for irradiation and the generator can be switched on and off easily.

10.4 Pulsed light

Pulsed light (PL) is a type of novel nonthermal technology with a great potential for microorganism inactivation via short-duration and high-power pulses to the food surface or packaging substances. Inert gas (mostly xenon) and a flash lamp is involved, which produces a wide-spectrum white light (from ultraviolet to near infrared wavelength). Pulsed light microbial inactivation mechanism is generally attributed to the effect caused by photo-chemicals to the cell and DNA of viruses, bacteria, and other pathogens, which prevents replication of cells (Heinrich, Zunabovic, Bergmair, Kneifel, & Jäger, 2015).Very short treatment times of microbial reduction, lack of residual compounds, the lower energy cost, and its immense flexibility are the major advantages of pulsed light technology. This technology has been extensively used for the decontamination of eggs, meat, liquid foods, fruit juices, and milk (Farkas & Mohácsi-Farkas, 2011).

10.5 Ultraviolet light

Ultraviolet light (UV) is a nonionizing radiation within a range of 100–400 nm, and is classified into three types: UV-A ranging from 315 to 400 nm, UV-B from 280 to 315 nm, and UV-C from 100 to 280 nm (González-Aguilar, Ayala-Zavala, Olivas, de la Rosa, & Álvarez-Parrilla, 2010). At present, the highest germicidal action is presented at 478 nm UV-C irradiation. Direct damage of DNA of living organisms is a well-known mechanism for inactivation of microbes by UV. DNA photoproducts, such as cyclobutane, pyrimidine 6–4 pyrimidone, and pyrimidine dimer formations are induced by UV in DNA, which inhibits replication and transcription, leading to mutagenesis and cell death (Gayán, Condón, & Álvarez, 2014). The main advantage of UV is the wide spectrum of usage to eliminate various types of microorganisms, with reduced cost and the desired manipulation. However, insufficient penetration has limited the application in the food field. UV-C is particularly used for the decontamination of surface of fresh-cut fruits and vegetables due to enzymatic and microbial spoilage taking place mainly on the cut surface (Gayán et al., 2014).

10.6 Cold plasma

Cold plasma (CP) is another emerging nonthermal technology with the ability for preservation of foods. Plasma is treated as the fourth state of matter, it is the quasineutral ionized gas comprising of particles such as free electrons, photons, negative ions, excited or nonexcited atoms, and molecules (Fernández, Shearer, Wilson, & Thompson, 2012). Apparently, oxygen, nitrogen, air, or mixtures of noble gases (argon, helium, and neon) are often used in cold plasma. Plasma is created by utilizing energy in various forms including lasers, microwaves, electricity, alternating current, direct current, magnetic field, and radiofrequency (Niemira, 2014). Energy is dissipated in invisible form and UV light when the active particles react with each other in the recombination process. The active particles release energy stored in the target virus or bacteria when it reacts with the substrate of food (Niemira, 2014). Although the actual microbial inactivation mechanism by CP is yet to be understood completely, the primary mechanisms attributed currently are:

(1) direct chemical interaction of cell membranes with reactive species and charged particles;
(2) cellular damage of membranes by UV; and
(3) breakage of DNA strands by UV (Niemira, 2014).

The source of specific energy and mixtures of gas proposition supplied to foods not only depends on density of chemical composition and temperature generated by plasma, but also on the pH, water activity, fat and protein content, and type and texture of the food (Lacombe et al., 2015). Cold plasma CP can be created at atmospheric pressure and room temperature, without increasing temperature, and treats food at low temperatures during the process. The reduced applications of pressure ($p < 1013$ mbar) or low pressure allow food processing with no undesired phase transition ($p < 10$ mbar).

10.7 Acidic electrolyzed water

Acidic electrolyzed water (AEW), also known as functional water, was developed in Japan. AEW is manufactured by treating a diluted salt solution (mainly potassium chloride and sodium chloride) through an electrolytic cell, having cathode and anode separated through a membrane. Various ion and compounds of chlorine such as Ocl, HOCl, and Cl_2 gas are produced in the anode side of the electrolytic cell (Gil, Gómez-López, Hung, & Allende, 2015). Acidic electrolyzed water has the potential to inactivate microorganisms with minimum adverse effects on human beings and the nutritional and organoleptic quality of food. Therefore, AEW are being used as an alternative to decontamination by chlorine and as a new disinfectant for fresh-cut fruits and vegetables to ensure safety and extension of shelf-life (Park, Alexander, Taylor, Costa, & Kang, 2009).

10.8 Nanotechnology

Nanotechnology is an emerging technology for shelf-life enhancement of food. It involves the development and research of materials in the length range of 1–100 nm (Mihindukulasuriya & Lim, 2014). The physical and chemical properties are modified significantly by the nano-sized particles by improving their mechanical durability, thermo chemical stability, and electrical conductivity. Silver nanoparticles are known as auspicious antimicrobial material having antibacterial properties against various numbers of microorganisms. Inside the cell, nanoparticles act as a source for silver ions. Moreover, silver ions can alter or kill microorganisms by a sequence of loses, initiation of enzymes of antioxidant, DNA damage, protein binding depletion of antioxidant molecules (e.g., glutathione), and changes in the cell wall structure and the nuclear membrane. (Mastromatteo et al., 2015) Previous studies on quasinanoscale silver particles (101–109 nm) suggested effectual results for vegetable juice preservation. The reports showed application of silver nanoparticles-PVP coating was helpful for maintaining green asparagus quality and shelf-life enhancement of 25 days at 2°C and 20 days at 10°C, respectively. A hybrid material in combination with MAP was developed by Fernández, Picouet, and Lloret (2010) for fresh-cut melon preservation, stored for 10 days at 4°C. Results showed reduced senescence, lower yeast counts, and attainment of a juicier appearance after a storing period of 10 days, enhancing fresh-cut melon shelf-life by 5 days in comparison to control ones (Fernández et al., 2010).

10.9 Ozone

As an alternative to chlorine sanitizer, ozone is attracting more interest, as chlorination of water causes formation of carcinogenic chlorinated compounds. The USFDA approved ozone as a direct contact food-sanitizing agent after ozone achieved the status of Generally Recognized as Safe (GRAS) in 1997. Ozone bonds quickly by nucleic material, components of envelope, and intracellular enzymes, spore coats or viral capsids of the microorganisms. Decomposition of ozone also occurs quickly. Ozone decays spontaneously during water treatment by mechanisms generating hydroxyl free radicals (•OH), which are the major reactive agents of oxidation and are extremely efficient for the inhibition of virus and bacteria. Ozone oxidizes the organic chemicals into harmless elements, such as splitting of cyanide and ammonia into water and nitrogen. Oxygen is the primary byproduct after oxidation in all the chemical reactions (Krasaekoopt & Bhandari, 2010).

10.10 Bacteriophages

Bacteriophages are used as novel, effective bio-preservative and environmental-friendly techniques for control of food quality. Bacteriophages multiply and infect the host cells of bacteria effectively (Meireles, Giaouris, & Simões, 2016). Therefore, they are not harmful to plants, animals, and humans.

11 Conclusion

There is a growing demand for FFV because of their less processing operation involvement and high nutritional characteristics. Consumers should be concerned about the conditions of processing and storage as it affects the nutritional quality and other safety parameters of the produce that can change during storage period because of various enzymatic and biochemical reactions. Since the conventional methods of processing food enhance the shelf stability along with the freshness of fruits and vegetables, fresh-cut processing of fruits and vegetables show the products to be highly perishable and unacceptable to the consumers. Advanced food preservation techniques, retention of quality parameters, and consumer attraction of fresh-cut products are significantly necessary to satisfy consumers' growing demands. The industry of fresh-cut is predicted to expand greatly during the coming years, for which the shelf-life enhancing novel technologies can be put into use.

Extending the shelf-life enhancement of fresh-cut products but not affecting the nutritional qualities and sensory attributes might be accomplished by combining various useful techniques. Future research should be done for enhancing nutritional and organoleptic quality retention of fresh-cut produce with the help of nonconventional technology combinations, as the application of some nonconventional technologies is still at the infancy phase. Moreover, food safety parameters are recognized as better than convenience oriented ones. Hence, the conceivable toxicity and combined assessment of risk of various nonconventional techniques needs to be investigated and additional development in terms of legislation, efficacy, cost-effectiveness ratio, and suitable manipulation of these nonconventional techniques must also be emphasized in the works done in future.

References

Abadias, M., Usall, J., Oliveira, M., Alegre, I., & Vinas, I. (2008). Efficacy of neutral electrolyzed water (NEW) for reducing microbial contamination on minimally-processed vegetables. *International Journal of Food Microbiology, 123*, 151–158.

Ahvenainen, R. (1996). New approaches in improving the shelf life of minimally processed fruit and vegetables. *Trends in Food Science and Technology, 7*, 179–187.

Allende, A., Selma, M. V., Lopez-Galvez, F., Villaescusa, R., & Gil, M. I. (2008). Role of commercial sanitizers and washing systems on epiphytic microorganisms and sensory quality of fresh-cut escarole and lettuce. *Postharvest Biology and Technology, 49*, 155–163.

Allende, A., Tomas-Barberan, F. A., & Gil, M. I. (2006). Minimal processing for healthy traditional foods. *Trends in Food Science and Technology, 17*, 513–519.

Amanatidou, A., Slump, R. A., Gorris, L. G. M., & Smid, E. J. (2000). High oxygen and high carbon dioxide modified atmospheres for shelf-life extension of minimally processed carrots. *Journal of Food Science, 65*(1), 61–66.

Anino, S. V., Salvatori, D. M., & Alzamora, S. M. (2006). Changes in calcium level and mechanical properties of apple tissue due to impregnation with calcium salts. *Food Research International, 39*, 154–164.

Artes, F., Conesa, M. A., Hernandez, S., & Gil, M. I. (1999). Keeping quality of fresh-cut tomato. *Postharvest Biology and Technology, 17*(3), 153–162.

Artés, F., Gómez, P., Aguayo, E., Escalona, V., & Artés-Hernández, F. (2009). Sustainable sanitation techniques for keeping quality and safety of fresh-cut 1087 plant commodities. *Postharvest Biology and Technology, 51*, 287–296.

Bhagwat, A. A. (2006). Microbiological safety of fresh-cut produce: Where are we now? In K. R. Matthews (Ed.), *Microbiology of fresh produce* (pp. 121–165). Washington, DC: ASM Press.

Chien, P. J., Sheu, F., & Yang, F. H. (2007). Effects of edible chitosan coating on quality and shelf life of sliced mango fruit. *Journal of Food Engineering, 78*, 225–229.

Corbo, M. R., Bevilacqua, A., Campaniello, D., D'Amato, D., Speranza, B., & Sinigaglia, M. (2009). Prolonging microbial shelf life of foods through the use of natural compounds and non-thermal approaches- a review. *International Journal of Food Science and Technology, 44*, 223–241.

D'Amato, D., Sinigaglia, M., & Corbo, M. R. (2010). Use of chitosan, honey and pineapple juice as filling liquids for increasing the microbiological shelf life of a fruit-based salad. *International Journal of Food Science and Technology, 45*, 1033–1041.

Farkas, J., & Mohácsi-Farkas, C. (2011). History and future of food irradiation. *Trends in Food Science & Technology, 22*, 121–126.

FDA Food and Drug Administration. (2010a). *Guidance for industry: Guide to minimize microbial food safety hazards of fresh-cut fruits and vegetables. Available at: http://www.fda.gov/food/guidancecomplianceregulatoryinformation/guidancedocuments/produceandplanproducts/ucm064458. Accessed 26 May 2010.*

Fernández, A., Picouet, P., & Lloret, E. (2010). Cellulose-silver nanoparticle hybrid 1174 materials to control spoilage-related microflora in absorbent pads located in 1175 trays of fresh-cut melon. *International Journal of Food Microbiology, 142* (1176), 222–228.

Fernández, A., Shearer, N., Wilson, D. R., & Thompson, A. (2012). Effect of microbial loading on the efficiency of cold atmospheric gas plasma inactivation of Salmonella enterica serovar typhimurium. *International Journal of Food Microbiology, 152*, 175–180.

Gayán, E., Condón, S., & Álvarez, I. (2014). Biological aspects in food preservation 1210 by ultraviolet light: A review. *Food and Bioprocess Technology, 7*, 1–20.

Gil, M., Gómez-López, V., Hung, Y.-C., & Allende, A. (2015). Potential of electrolyzed water as an alternative disinfectant agent in the fresh-cut industry. *Food and Bioprocess Technology, 8*, 1336–1348.

Gil, M. I., Selma, M. V., Lopez-Galvez, F., & Allende, A. (2009). Fresh-cut product sanitation and wash water disinfection: Problems and solutions. *International Journal of Food Microbiology, 134*, 37–45.

Gomez-Lopez, V. M., Devlieghere, F., Ragaert, P., & Debevere, J. (2007). Shelf-life extension of minimally processed carrots by gaseous chlorine dioxide. *International Journal of Food Microbiology, 116*, 221–227.

González-Aguilar, G., Ayala-Zavala, J. F., Olivas, G. I., de la Rosa, L. A., & Álvarez-Parrilla, E. (2010). Preserving quality of fresh-cut products using safe technologies. *Journal für Verbraucherschutz und Lebensmittelsicherheit, 5*(1217), 65–72.

Gonzalez-Aguilar, G. A., Tiznado-Hernandez, M., & Wang, C. Y. (2006). Physiological and biochemical responses of horticultural products to methyl jasmonate. *Stewart Postharvest Review, 2*, 1–9.

Gorny, J. R., Cifuentes, R. A., Hess-Pierce, B. H., & Kader, A. A. (2000). Quality changes in fresh-cut pear slices as affected by cultivar, ripeness stage, fruit size and storage regime. *Journal of Food Science, 65*, 541–544.

Gorny, J. R., Hess-Pierce, B., Cifuentes, R. A., & Kader, A. A. (2002). Quality changes in fresh-cut pear slices as affected by controlled atmospheres and chemical preservatives. *Postharvest Biology and Technology, 24*(3), 271–278.

Gómez-López, V. M., Rajkovic, A., Ragaert, P., Smigic, N., & Devlieghere, F. (2009). Chlorine dioxide for minimally processed produce preservation: A review. *Trends in Food Science and Technology, 20*, 17–26.

Heinrich, V., Zunabovic, M., Bergmair, J., Kneifel, W., & Jäger, H. (2015). Post-packaging application of pulsed light for microbial decontamination of solid foods: A review. *Innovative Food Science & Emerging Technologies, 30*(1246), 145–156.

Kong, Q., Wu, A., Qi, W., Qi, R., Carter, J. M., Rasooly, R., et al. (2014). Effects of electron-beam irradiation on blueberries inoculated with Escherichia coli and their nutritional quality and shelf life. *Postharvest Biology and Technology, 95*, 28–35.

Krasaekoopt, W., & Bhandari, B. (2010). Fresh-cut vegetables. In *Handbook of vegetables and vegetable processing* (pp. 219–242). Wiley-Blackwell.

Lacombe, A., Niemira, B. A., Gurtler, J. B., Fan, X., Sites, J., Boyd, G., et al. (2015). Atmospheric cold plasma inactivation of aerobic microorganisms on blueberries and effects on quality attributes. *Food Microbiology, 46*, 479–484.

Lamikanra, O., & Watson, M. A. (2007). Mild heat and calcium treatment effects on fresh-cut cantaloupe melon during storage. *Food Chemistry, 102*, 1383–1388.

Ma, L., Zhang, M., Bhandari, B., & Gao, Z. (2017). Recent developments in novel shelf life extension technologies of fresh-cut fruits and vegetables. *Trends in Food Science & Technology, 64*, 23–38.

Mami, Y., Peyvast, G., Ziaie, F., Ghasemnezhad, M., & Salmanpour, V. (2014). Improvement of shelf life and postharvest quality of white button mushroom by electron beam irradiation. *Journal of Food Processing and Preservation, 38*, 1673–1681.

Mangaraj, S., Goswami, T. K., & Mahajan, P. V. (2009). Applications of plastic films for modified atmosphere packaging of fruits and vegetables: A review. *Food Engineering Reviews, 1*(2), 133.

Martin-Diana, A. B., Rico, D., Mulcahy, J., Frias, J., Henehan, G. T. M., & Barry-Ryan, C. (2006). Whey permeate as bio-preservative for shelf-life maintenance of fresh-cut vegetables. *Innovative Food Science and Technologies, 7*, 112–123.

Mastromatteo, M., Conte, A., Lucera, A., Saccotelli, M. A., Buonocore, G. G., Zambrini, A. V., et al. (2015). Packaging solutions to prolong the shelf life of Fiordilatte cheese: Bio-based nanocomposite coating and modified atmosphere packaging. *LWT-Food Science and Technology, 60*, 230–237.

Meireles, A., Giaouris, E., & Simões, M. (2016). Alternative disinfection methods to chlorine for use in the fresh-cut industry. *Food Research International, 82*, 71–85.

Meng, X., Zhang, M., & Adhikari, B. (2012). Extending shelf-life of fresh-cut green peppers using pressurized argon treatment. *Postharvest Biology and Technology, 71*, 13–20.

Mihindukulasuriya, S. D. F., & Lim, L. T. (2014). Nanotechnology development in food packaging: A review. *Trends in Food Science & Technology, 40*, 149–167.

Niemira, B. A. (2014). Decontamination of foods by cold plasma. In *Emerging technologies for food processing* (2nd ed., pp. 327–333). San Diego, CA: Academic Press.

O'Beirne, D., & Francis, G. A. (2003). Reducing the pathogen risk in MAP-prepared produce. In R. Ahvenainen (Ed.), *Novel food packaging techniques* (pp. 231–286). Cambridge: Woodhead Publishing Limited.

Olivas, G. I., & Barbosa-Canovas, G. V. (2005). Edible coatings for fresh-cut fruits. *Critical Reviews in Food Science and Nutrition, 45*, 657–663.

Olmez, H., & Kretzschmar, U. (2009). Potential alternative disinfection methods for organic fresh-cut industry for minimizing water consumption and environmental impact. *LWT-Food Science and Technology, 42*, 686–693.

Park, E.-J., Alexander, E., Taylor, G. A., Costa, R., & Kang, D.-H. (2009). The decontaminative effects of acidic electrolyzed water for Escherichia coli O157:H7, Salmonella typhimurium, and listeria monocytogenes on green onions and tomatoes with differing organic demands. *Food Microbiology, 26*, 386–390.

Phillips, C. A. (1996). Modified atmosphere packaging and its effects on the microbiological quality and safety of produce. *International Journal of Food Science & Technology, 31*(6), 463–479.

Ragaert, P., Verbeke, W., Devlieghere, F., & Debevere, J. (2004). Consumer perception and choice of minimally processed vegetables and packaged fruits. *Food Quality and Preference, 15*, 259–270.

Reid, D. S., & Fennema, O. R. (2008). *Water and ice. Fennema's food chemistry*: (4th ed., pp. 17–82). CRC Press.

Rico, D., Martin-Diana, A. B., Barat, J. M., & Barry-Ryan, C. (2007). Extending and measuring the quality of fresh-cut fruit and vegetables: A review. *Trends in Food Science and Technology, 18*, 373–386.

Roller, S. (Ed.), (2003). *Natural antimicrobials for the minimal processing of foods.* Cambridge: Woodhead Publishing Limited.

Roller, S., & Seedhar, P. (2002). Carvacrol and cinnamic acid inhibit microbial growth in fresh-cut melon and kiwifruit at 4° and 8°C. *Letters in Applied Microbiology, 35*, 390–394.

Sandhya. (2010). Modified atmosphere packaging of fresh produce: Current status and future needs. *LWT-Food Science and Technology, 43*, 381–392.

Soliva-Fortuny, R. C., & Martin-Belloso, O. (2003). New advances in extending the shelf life of fresh-cut fruits: A review. *Trends in Food Science and Technology, 14*, 341–353.

Solomos, T. (1997). Principles underlying modified atmosphere packaging. In R. C. Wiley (Ed.), *Minimally processed refrigerated fruits & vegetables* (pp. 183–225). New York: Chapman and Hall.

Zhang, B. Y., Samapundo, S., Pothakos, V., Sürengil, G., & Devlieghere, F. (2013). Effect of high oxygen and high carbon dioxide atmosphere packaging on the microbial spoilage and shelf-life of fresh-cut honeydew melon. *International Journal of Food Microbiology, 166*(3), 378–390.

Zhang, M., Zhan, Z. G., Wang, S. J., & Tang, J. M. (2008). Extending the shelf-life of asparagus spears with a compressed mix of argon and xenon gases. *LWT-Food Science and Technology, 41*, 686–691.

Further reading

Beuchat, L. R. (2002). Ecological factors influencing survival and growth of human pathogens on raw fruits and vegetables. *Microbes and Infection, 4*, 413–423.

Campaniello, D., Bevilacqua, A., Sinigaglia, M., & Corbo, M. R. (2008). Chitosan: Antimicrobial activity and potential applications for preserving minimally processed strawberries. *Food Microbiology, 25*, 992–1000.

Conway, W. S., Leverentz, B., Saftner, R. A., Janisiewicz, W. J., Sams, C. E., & Leblanc, E. (2000). Survival and growth of listeria monocytogenes on fresh-cut apples slices and its interaction with Glomerella cingulata and Penicillium expansum. *Plant Disease, 84*, 177–181.

Dingman, D. W. (2000). Growth of Escherichia coli O157:H7 in bruised apples (Malus domestica) tissue as influenced by cultivar, date of harvest and source. *Applied and Environmental Microbiology, 66*, 1077–1083.

FDA Food and Drug Administration. (2010). *Outbreaks associated with fresh produce: Incidence, growth, and survival of pathogens in fresh and fresh-cut produce.* Available at: http://www.fda.gov/Food/ScienceResearch/ResearchAreas/SafePracticesforFoodProcesses/ucm091270.htm. Accessed 26 May 2010.

Gonzalez-Aguilar, G. A., Ayala-Zavala, J. F., Olivas, G. I., de la Rosa, L. A., & Alvarez-Parrilla, E. (2010). Preserving quality of fresh-cut products using safe technologies. *Journal für Verbraucherschutz und Lebensmittelsicherheit, 5*, 65–72.

ICMSF (1998). *Microbial ecology of food commodities. Microorganisms in foods.* London: Blackie Academic & Professional.

Jay, J. M. (1992). Spoilage of fruits and vegetables. In *Modern food microbiology* (4th ed., pp. 187–198). New York: Chapman and Hall.

Kalia, A., & Gupta, R. P. (2006). Fruit microbiology. In Y. H. Hui, J. Barta, M. P. Cano, T. Gusek, J. S. Sidhu, & N. K. Sinha (Eds.), *Handbook of fruits and fruit processing* (pp. 3–28). Oxford: Blackwell publishing.

Lamikanra, O. (2002). Preface. In O. Lamikanra (Ed.), *Fresh-cut fruits and vegetables. Science, technology and market.* Boca Raton, FL: CRC Press.

Lanciotti, R., Gianotti, A., Patrignani, F., Belletti, N., Guerzoni, M. E., & Gardini, F. (2004). Use of natural aroma compounds to improve shelf life and safety of minimally processed fruits. *Trends in Food Science and Technology, 15*, 201–208.

Raybaudi-Massilia, R. M., Mosqueda-Melgar, J., Sobrino-Lopez, A., Soliva-Fortuny, R., & Martin-Belloso, O. (2007). Shelf-life extension of fresh-cut Fuji apples at different ripeness stages using natural substances. *Postharvest Biology and Technology, 45*, 265–275.

Riordan, D. C. R., Sapers, G. M., & Annous, B. A. (2000). The survival of Escherichia coli O157:H7 in the presence of Penicillium expansum and Glomerella cingulata in wounds on apple surfaces. *Journal of Food Protection, 63*, 1637–1642.

Saftner, R. A., Baj, J., Abbott, J. A., & Lee, Y. S. (2003). Sanitary dips with calcium propionate, calcium chloride, or calcium amino acid chelates maintain quality and shelf stability of fresh-cut honeydew chunks. *Postharvest Biology and Technology, 29*, 257–269.

Wade, W. I. N., & Beuchat, L. R. (2003). Proteolytic fungi isolated from decayed and damaged raw tomatoes and implications associated with changes in pericarp pH favourable for survival and growth of foodborne pathogens. *Journal of Food Protection, 66*, 911–917.

5

Sanitizers for fresh-cut fruits and vegetables

Shruti Sethi, Swarajya Laxmi Nayak, Alka Joshi, Ram Roshan Sharma
Division of Food Science and Postharvest Technology, ICAR-Indian Agricultural Research Institute, New Delhi, India

1 Introduction

Modern consumers' desire for healthy, natural food along with good sensory attributes has resulted in an upsurge in demand for and consumption of fresh-cut fruits and vegetables. Supply expansion of fresh-cut industry has taken place in the last few decades as a result of the booming of fast food chains where such commodities are in demand. However, such products are prone to pathogen outbreak as during preparation injured cells release intracellular fluids that provide a conducive environment for microbial growth. Thus, preservation and shelf-life extension have been major concerns for fresh-cut produce since their introduction to market in the 1940s (Rico, Martín-Diana, Frías, Henehan, & Barry-Ryan, 2006). Contamination of food with microbes or their toxins can cause unpleasant symptoms, ranging from gastrointestinal disturbances to long term health issues or even death. Hence, proper sanitization and handling of the fresh-cut commodities is crucial to achieving a safe global supply of these products with the desired sensory attributes.

Although simple washing with water removes the cellular components leached out during cutting and reduces the microbial load to some extent, use of sanitizing agents brings the microbial levels low enough so as to make the fresh-cut produce fit for consumption and as per the food safety guidelines. Thus, use of sanitizers plays a pivotal role in this regard. There are a growing number of sanitizing compounds used to reduce microbial load in fresh-cut produce, including chlorine dioxide, hydrogen peroxide, organic acids, peroxyacetic acid, ozone, electrolyzed water, and radiation. The chemical disinfectants are mostly added in

Fresh-cut Fruits and Vegetables. https://doi.org/10.1016/B978-0-12-816184-5.00005-7

wash water, to provide a barrier to microorganisms that may contaminate the minimally processed produce (FDA, 2000). Ozone and electrolyzed water are produced by passing the electric current directly through wash water and exposing the produce directly to the sanitizer action. Although studies have revealed that none of the abovementioned sanitizers are able to completely eliminate pathogens from these minimally processed produce, the use of the sanitizing agents can bring down the microbial levels to acceptable limits.

2 Factors affecting efficacy of sanitizers

Water used for fresh-cut fruits and vegetables should be ideally at a level of 5–10 L/kg of the product. Owing to the cost involved with the quantities of water used for washing the produce, sometimes the wash water is reused. This results in an increase in the soluble/organic matter of the wash water that results in reduction in the efficiency of the sanitizers added. However, other factors also determine the effectiveness of the sanitizers as discussed below.

2.1 Temperature

Temperature plays a pivotal role in the survival rate of pathogenic microflora (Gawande & Bhagwat, 2002). A cooler water temperature is more effective at sanitizing surfaces. Although water temperature of 0°C is ideal for getting better results, if water is at cooler temperature in comparison to the product temperature it may result in vacuum development and infiltration of bacteria along with the water into the interior of produce via open and damaged skin surface and stem tissue, thus causing contamination (Beuchat, 2000). In a study conducted by Zhuang, Beuchat, and Anjulo (1995), a temperature differential of 15°C resulted in suction of *Salmonella montevideo* into the core of tomatoes.

2.2 pH

The pH of wash water is an important factor for the reduction and inactivation of microorganisms. Each sanitizer has a particular pH range within which it is most effective. The lethal effects of chlorine (in the form of hypochlorous acid, HOCl) are much greater at pH range of 6.0–7.5 (Sapers, 2003). At pH below 6.0, chlorine gas maybe formed, which may be deleterious for personnel. On the other hand, at pH above 8.0, the hypochlorous acid splits to form hydrogen ions (H^+) and hypochlorite (OCl^-) ions. The hypochlorite ion formed has weak bactericidal effects. Ozone is generally not affected in pH range of 6–8, but its decomposition increases with high pH (>8.0). The pH of the solution may be adjusted by the addition of organic or inorganic acids.

2.3 Microbial load on food surface

The location of microorganisms on the produce surface affects their inactivation by disinfectants (Cherry, 1999). Seo and Frank (1999) reported that bacteria tend to concentrate in pores, indentations, or irregular surfaces. After cutting the fresh produce, the exposed surface increases the surface area for microbial attachment, which in turn enhances their survival. Moreover, the exposed cut surface area releases solubles into the wash water, thereby decreasing sanitizer effectiveness (Rodgers, Cash, Siddiq, & Ryser, 2004). According to Yu, Newman, Archbold, and Hamilton-Kemp (2001), the relative ineffectiveness of many sanitizers on strawberries is partially due to surface roughness, which provides an ideal site for bacteria to attach and form biofilms.

2.4 Oxidation-reduction potential

Oxidation reduction potential (ORP) is a measure of the oxidizing properties and chemical efficacy of the sanitizer in water, which is determined using a sensor. The value of this potential varies with the type of sanitizer. Suslow (2004) recommended a value of 600–650 mV for chlorine treatments. The ORP value provides the operator with a rapid and single value assessment of the wash water disinfection potential, which can prompt operators to better control the treatment by adjusting the pH of the sanitizer.

3 Sanitizing agents for minimally processed fruits and vegetables

As discussed previously, several sanitizing agents are employed for reducing the microbial population in fresh-cut fruits and vegetables. The following section discusses the various applications of these agents.

3.1 Chemical treatments

3.1.1 Chlorine based

Amongst all, chlorine is the most widely used disinfectant for fresh fruits and vegetables. Chlorine may be used as liquid or in gaseous form. Liquid chlorine in the form of sodium (NaOCl) or calcium hypochlorites ($Ca(OCl)_2$) is the most commonly used for disinfecting purposes. The recommended level of chlorine as disinfectant is 50–200 ppm at a pH below 8.0 (WHO, 1998). Chlorine is very reactive and combines with any oxidizable substrate to form secondary compounds, such as trihalomethanes (IFPA, 2001). For this reason, its use for treating fresh-cut fruits and vegetables is banned in several European countries, including Switzerland, Germany, Belgium, and the Netherlands (Artés & Allende, 2005). The inhibitory action of chlorine depends on pH and temperature of the wash water, contact time, presence of organic matter in the wash water, and exposure to light, air, or metals. If the pH of the water falls below 4.0, then chlorine gas may be formed, which is a potential health hazard for humans (IFPA, 2001). Table 1 presents an overview of application of free chlorine in varied minimally processed fruits and vegetables.

Chlorine dioxide (ClO_2) is approved for use as a sanitizer on uncut produce followed by potable water rinse (CFR, 2005). It is a yellow to red gas, having 2.5 times more oxidizing potential than chlorine gas (Suslow, 1997). A maximum of 200 ppm ClO_2 is allowed for sanitizing of processing equipment and 3 ppm is allowable for contact with whole produce. Treatment of produce with ClO_2 must be followed by a potable water rinse or blanching, cooking, or canning. Chlorine dioxide works best at neutral pH and does not react with nitrogen-containing compounds or ammonia to form dangerous chloramines, as does chlorine (White, 1972; WHO, 1998). Table 2 describes the beneficial use of chlorine dioxide (ClO_2) in various cut fruits and vegetables.

Acidified sodium chlorite (a mixture of a GRAS acid and $NaOCl_2$), another form in which foods can be exposed to chlorine is highly reactive and has lethal effect on a broad spectrum of microorganisms at pH range of 2.5–3.2. It has been approved for use on raw fruits and vegetables, as either a dip or a spray, in the range of 0.5–1.2 g/L followed by a potable water rinse (21CFR173.325). A commercial product of acidified sodium chlorite known as Sanova (Ecolab) is available that is a combination of citric acid and sodium chlorite and has about 2.5 times greater oxidizing capacity than hypochlorous acid (HOCl) (Inatsu, Bari, Kawasaki, Isshiki, & Kawamoto, 2005).

Table 1 Application of chlorine in minimally processed fruits and vegetables

Disinfecting treatment	Product	Effect	Reference
200 ppm of free chlorine for 40 min	Fresh-cut cantaloupes	Absence of *Salmonella* after 21 days of storage	Alicea, Annous, Mendez, Burke, and Orellana (2018)
Free chlorine of different concentrations	Fresh cut romaine lettuce, iceberg lettuce and cabbage	No surviving aerobic bacteria above 20 mg/L chlorine	Luo et al. (2018)
25 ppm of free chlorine	Fresh-cut romaine lettuce	1 log PFU/g reduction of coliphage MS2	Wengert, Aw, Ryser, and Rose (2017)
Free chlorine concentration varying from 1 to 3 mg/L	Spinach	Minimum residual level of chlorine effectively controlled *E. coli* O157:H7 under industrial conditions	Gómez-López, Lannoo, Gil, and Allende (2014)
200 mg/L free chlorine for 60 s	Iceberg lettuce	Lower initial count of total aerobic bacteria (*Enterobacteriaceae* and *Pseudomonades*)	Baur, Klaiber, Wei, Hammes, and Carle (2005)
200 ppm free chlorine	Shredded iceberg lettuce	2 log reduction of microbial load	Baur, Klaiber, Hammes, and Carle (2004)
200 ppm of free chlorine for 5 min	Iceberg lettuce leaves	1.75 log reduction in *E. coli* O157:H7 populations	Lang, Harris, and Beuchat (2004)
200 ppm of free chlorine for 5 min	Shredded lettuce	1.06 log reduction of *Listeria monocytogenes*	Burnett, Iturriaga, Escartin, Pettigrew, and Beuchat (2004)
100 ppm chlorine solution for 5 min	Ready-to-use vegetables (lettuce, dry coleslaw)	Significant lowering of *L. innocua* and *E. coli* populations	Francis and O'Beirne (2002)
Chlorination at 200 µg/mL	Lettuce	5.4 log reduction in *E. coli* O157:H7	Foley et al. (2002)

Use of electrolyzed water (EW) is a relatively new concept for food sanitation. EW is produced by passing a diluted salt solution—mostly sodium chloride—through an electrolytic cell, within which the anode and cathode are separated by a septum. Hypochlorous acid (HClO) is present in the range of 10–60 ppm in

Table 2 Application of chlorine dioxide on minimally processed fruits and vegetables

Disinfecting treatment	Product	Effect	Reference
3 mg/L for 1 min	Fresh-cut lollo rossa lettuce	Absence of *E. coli*	Banach, Overbeek, Groot, Zouwen, and Fels-Klerx (2018)
3.5 ppm	Grape tomatoes	3.08 log cfu/g reduction of *Escherichia coli* and 2.85 log cfu/g *Alternaria alternata* populations after 14 days of storage	Sun et al. (2017)
60 mg/L for 10 min	Fresh-cut coriander	Reduction of *Staphylococcus, Brevibacterium, Pseudomonas* and *Acinetobacter*	Jiang et al. (2017)
Aqueous chlorine dioxide 100 mg/L for 20 min	Fresh-cut asparagus lettuce	Extension of shelf-life to 14 days and reduced activities of polyphenol oxidase (PPO) and peroxidase (POD) enzymes	Chen, Zhu, Zhang, Niu, and Du (2010)
Aqueous chlorine dioxide 100 mg/L treatment for 10 min	Fresh-cut lotus root (FLR)	Inhibition of PPO activity and maintenance of high OVQ scores during 10 day storage	Du, Fu, and Wang (2009)
Gaseous chlorine dioxide 4.1 mg/L	Fresh-cut lettuce, cabbage, and carrot	1.53–1.58 log cfu/g reduction for fresh-cut lettuce, 3.13–4.42 log cfu/g reduction for fresh-cut cabbage and 5.15–5.88 log cfu/g reduction for fresh-cut carrots	Sy, Murray, Harrison, and Beuchat (2005)
5 ppm for 5 min	Lettuce	~5 log reduction of *E. coli O157:H7* and *L. monocytogenes*	Rodgers et al. (2004)
1.24 ppm for 30 min at 22°C and 90%–95% RH	Surface injured green peppers	6.45 log reduction of *E. coli O157:H7*	Han, Sherman, Linton, Nielson, and Nelson (2000)

electrolyzed acidic water, which is generated from the reaction of Cl_2 (from NaCl) and H_2O in anode site. This results in acid water with pH 2–3 (strong electrolyzed water). Electrolyzed weak acidic water at pH 5–6.5 is generated by electrolysis of NaCl solution without a separating membrane. It has a high oxidation potential between 1000 and 1150 mV (Sapers, 2003). EW shows antimicrobial activity against a broad spectrum of microorganisms and eliminates most common types of viruses, bacteria, fungi, and spores in a short time (usually within 5–20 s) in food products and food processing surfaces (Ding et al., 2015; Hao, Li, Wan, & Liu, 2015). Various factors such as current, water flow rate, electrolyte, salt concentration, electrode materials, storage conditions, water hardness, and water temperature have been reported to affect the physicochemical properties of EW and have been thought to be responsible for the sanitization effect of EW. The application of EW is a sustainable and green concept and has several advantages over traditional sanitizing systems including cost effectiveness, ease of application, effective disinfection, on-the-spot production, and safety for human beings and the environment. The EPA has approved EO water (CFR, 2005) for washing raw foods that are to be consumed without processing. It is environment friendly, as it only uses water mixed with sodium chloride. Since, the solution rapidly loses its antimicrobial activity, it is produced in-situ and utilized simultaneously. Table 3 summarizes the application of EW on fresh-cut fruits and vegetables.

Disadvantages associated with EW include the high initial investment, rapid loss of antimicrobial effect of EW due to reduction in chlorine concentration, and breathing discomfort and irritation of hands of the working personnel during chlorine gas production.

Table 3 Application of EO on minimally processed foods

Sanitizing treatment	Used on	Effect	Reference
Acidic electrolyzed water (AEW)	Fresh-cut red cabbage	3.67 log cfu/g reduction of artificially inoculated *S. typhimurium DT104*	Chen et al. (2018)
Neutral electrolyzed water containing 4000 mg/L of free chlorine	Fresh-cut chicory and lettuce leaves	Reduction of microbial loads of mesophilic bacteria and *Fnterobacteriaceae*	Pinto, Ippolito, and Baruzzi (2015)
EW containing 120 ppm free chlorine	Minimally processed lettuce	Reduced microbial load	Rico et al. (2008)
Alkaline EW followed by acidic EW for 5 min at 20°C	Cut lettuce	~1.8 log reduction of *E. coli* O157:H7 and *Salmonella*	Koseki, Yoshida, Isobe, and Itoh (2004)
Alkaline EW followed by acidic EW for 1 min	Cabbage	1.5, 1.5, 1.5, and 1.0 log reduction of total aerobic bacteria, coliform bacteria, *B. cereus*, and psychrotrophic bacteria, respectively	Koseki and Itoh (2001)

3.1.2 Hydrogen peroxide

Hydrogen peroxide (H_2O_2) is a well-known oxidizing agent, directly toxic to pathogens. It is a strong oxidizer, having 1.8 oxidizing potential—just below that of ozone. It is both bacteriostatic and bactericidal owing to its ability to generate cytotoxic hydroxyl radicals (Olmez & Kretzschmar, 2009; Parish et al., 2003). Commercially available 3% hydrogen peroxide is a stable and effective disinfectant when used on inanimate surfaces (Simons & Sanguansri, 1997). It is GRAS for some food applications (21CFR184.1366) but has not yet been approved as an antimicrobial wash-agent for produce. The sporicidal activity of H_2O_2 coupled with rapid disintegration makes it a desirable sterilant for use on food contact surfaces and packaging materials in aseptic filling operations. However, it requires repeated washing to remove the residual effects after processing. It is still preferred by processors due to its beneficial effects, as listed in Table 4.

The activity of H_2O_2 is best under pH 1–4. The recommended level of H_2O_2 is 2 ppm, while levels above 5 ppm are considered fatal.

Table 4 Application of H_2O_2 as a sanitizing agent

Sanitizer type	Used on	Effect	Reference
20 g/L solution for 1 min	Table grapes	Negligible total aerobic microbial and fungal count till the end of the storage	Ergun and Dogan (2018)
2% solution for 2 min	Fresh-cut cluster beans	Reduced microbial load	Waghmare and Annapure (2017)
3% H_2O_2 at 22°C	Fresh-cut cantaloupes	~4-log reduction of bacterial pathogens	Ukuku, Mukhopadhyay, Geveke, Olanya, and Niemira (2016)
10% vaporized H_2O_2 treatment for 10 min	Organic lettuce	3.15, 3.12 and 2.95 log10 cfu/g reduction of *Escherichia coli* O157:H7, *Salmonella typhimurium*, and *Listeria monocytogenes*	Back, Ha, and Kang (2014)
1% and 5% solutions	Red bell peppers and watercress	5% H_2O_2 provide highest reductions of microbial load	Alexandre, Brandao, and Silva (2012)
2.5% and 5% solutions	Cantaloupe and honeydew melon	4.65 and 3.13 log_{10} cfu/g reduced microbial load on cantaloupe and honeydew melon, respectively	Ukuku (2004)

Another technology, known as lactoperoxidase (LPS) technology, combines H_2O_2 with sodium thiocyanate, which generates hypothiocyanite ($OSCN^-$) in the presence of peroxidase. The commercially available product, Catallix has been approved for use on fresh-cuts (Allende, Selma, López-Gálvez, Villaescusa, & Gil, 2008) for reducing viable populations.

3.1.3 Organic acids

Organic acids are naturally present in foods and are commonly used as preservatives in the food industry (Lianou, Koutsoumanis, & Sofos, 2012). Many organic acids have GRAS status and are FDA and EC (European Commission) approved. They do not produce toxic or carcinogenic compounds after interaction with organic molecules (Lianou et al., 2012). Their mode of action towards reduction of microbial load is based on the acidification, osmotic stress, disruption of proton motive force, and inhibition of synthetic processes (Carpenter & Broadbent, 2009). Their use is majorly in those commodities that have pH >4.5 since, below this pH, most microbes are not viable.

Commercial formulations such as Citrox (Citrox Limited, Middlesbrough, United Kingdom) and Purac (PURAC Bioquímica, Spain) containing phenolic compounds and lactic acid, respectively, as the active ingredient are available for sanitizing treatments. As per manufacturer's recommendations, application of 5 mL/L for 5 min in the case of Citrox and 20 mL/L for 3 min for Purac is sufficient for achieving desirable results.

Peroxyacetic acid (CH_3CO_3H), an organic acid formulation, is approved for addition to wash water (21CFR173.315) for disinfection purposes followed by a mandatory water rinse of the produce. It is a colorless mixture of a peroxy compound, hydrogen peroxide, and acetic acid that readily decomposes into acetic acid, water, and oxygen when mixed in water. It is a strong oxidizing agent and is active in a broad pH range.

The organic acids are used largely on fresh-cut fruits and vegetables as microbial retardant, as summarized in Table 5.

3.2 Physical treatments

The safety of fresh-cut produce is a concern because of foodborne illness arising from consumption of these commodities that might have been contaminated with enteric pathogens. To attenuate such problems, a number of sanitizing methods are adopted by people. Since chemical methods may result in residues in the

Table 5 Organic acids as a sanitizer on minimally processed fruits and vegetables

Disinfecting treatment	Product	Microbial reduction	Reference
Aqueous solutions of ethanol (25%) and ascorbic acid (1%)	Fresh-cut Chinese yam	Inhibition of microbial growth and delay in browning	Gao, Zhu, and Luo (2018)
Citric (5 g/L) and lactic (5 mL/L) acid solutions	Fresh-cut iceberg lettuce	Effectively reduced *Enterobacteriaceae* population 2.2 log10 cfu/g	Akbas and Ölmez (2007)
80 ppm of peroxyacetic acid at 3–4°C for 15 s	Shredded lettuce and romaine lettuce pieces	~1 log reduction of *L. monocytogenes*	Beuchat, Adler, and Lang (2004)
2% acetic acid for 15 min	Lettuce leaves	3.37 and >2.25 log reduction in aerobic mesophilic and total coliform populations, respectively	Nascimento, Catanozi, and Silva (2003)

treated commodity, some physical treatments such as ozone, irradiation, and UV treatment may alternatively be used to disinfect the produce.

3.2.1 Ozone

Ozone (O_3) forms by the rearrangement of atoms when oxygen molecules are subjected to high-voltage electric discharge (Khadre, Yousef, & Kim, 2001). Ozone has 1.5 times the oxidizing potential of chlorine and 3000 times the potential of hypochlorous acid (EPRI, 1997). It is a blue colored gas at ambient temperature and decomposes to oxygen within a few seconds (Khadre et al., 2001). Ozone can react with contaminants directly as molecular ozone (O_3) or indirectly as ozone-derived free radicals such as OH^- and H_2O_2, that cause inactivation of microorganisms. Alternation of proteins, unsaturated lipids and respiratory enzymes and nucleic acids in the cytoplasm and/or proteins and peptidoglycan are the modes of action of O_3 (Koseki, Yoshida, Isobe, & Itoh, 2001). In 2001, the FDA approved the use of ozone on as an antimicrobial agent for the treatment, storage, and processing of foods in gas and aqueous phase in direct contact with foods, including raw and minimally processed fruits and vegetables (FDA, 2000). This gave a boost to the application of ozone in the industry, although it has the negative effect of causing

Table 6 Use of ozone on fresh-cut fruits and vegetables

Disinfecting treatment	Used on	Effect	Reference
2 mg/L of gaseous ozone	Minimally processed rocket	Inhibition of the spoilage bacteria and yeasts and extension of shelf-life	Gutiérrez, Chaves, and Rodríguez (2018)
0.5 mg/L aqueous ozone	Fresh cut lettuce and green bell pepper	2 log reduction of microbial load in the first 15 min and 3.5 log reduction after 30 min of exposure	Alexopoulos et al. (2013)
10,000 ppm of gaseous ozone for 30 min under vacuum	Fresh-cut cantaloupe	4.2 and 2.8 log cfu/rind-disk (12.6 cm^2), reduction of *Salmonella* counts in mature nonripe and ripe melons, respectively	Selma, Ibáñez, Cantwell, and Suslow (2008)
Ozone flow of 150 L/h dissolved in deionized water	Shredded lettuce	Reduction of *S. sonnei* counts by 1.8 log units	Selma, Beltrán, Allende, Chacón-Vera, and Gil (2007)
3 ppm for 5 min (bubbling)	Apples, lettuce, strawberries and cantaloupe	4–5 log reduction of mesophilic bacteria	Rodgers et al. (2004)
2.5 ppm of ozone, stirred for 10 min	Iceberg lettuce	0.6–0.8 log reduction in aerobic plate count	Garcia, Mount, and Davidson (2003)
5 ppm of ozone for 10 min	Lettuce	1.5 log reduction of microbial load	Koseki and Itoh (2001)

deterioration and corrosion on metal and other types of surfaces. Table 6 elucidates the usefulness of ozone on fresh-cut fruits and vegetables. Ozone may be applied in both aqueous phase and gaseous phase to be used as a sanitizer.

Ozone is most effective in a pH range of 6–8. An exposure of 1 ppm for 8 h is supposed to be toxic (OSHA, 1988). Recently, generation of micro bubbles of O_3 in the wash water has helped in increasing the effectiveness of O_3 with greater log reductions in microbial populations (Lee, Song, Manna, & Ha, 2008).

3.2.2 Irradiation

Ionizing irradiation, such as gamma-rays, produce electrically charged ions by splitting water molecules (Ramos, Pilawa, & Stroka, 2013). Irradiation using gamma rays is a well-established

Table 7 Application of gamma radiation as sanitizing agent on minimally processed foods

Dose (kGy)	Product	Microbial reduction	Reference
0.2–1	Sliced fresh cucumber	Elimination of *Salmonella poona* in sliced cucumber	Joshi, Moreira, Omac, and Castell-Perez (2018)
2	Fresh-cut watercress	Better preservation of overall quality of fresh-cut watercress with 2 kGy dose	Pinela et al. (2016)
1	Fresh-cut lettuce	Total coliform group were lowered to less than 30 MPN/100 g	Zhang, Lu, Lu, and Bie (2006)
1	Broccoli, mung beans, cabbage, and tomato	4.14–5.25 log reduction of *L. monocytogenes*	Bari et al. (2005)
0.35	Cut romaine lettuce	1.5 log reduction of aerobic plate count	Prakash, Guner, Caporaso, and Foley (2000)

process, its activity is not limited to the surface, it can penetrate into the product and eliminate microorganisms that are present in crevices and creases of fruits and vegetables (Husman et al., 2004). Collisions between the ionizing radiation and food particles at the atomic and molecular level results in the production of ion pairs and free radicals, which inactivate the microorganisms (Ramos et al., 2013). A maximum dose of 1 kGy has been approved by FDA to decontaminate fruits and vegetables. The main advantages of using ionizing radiations are the good disinfecting ability of the produce under very low energy requirement and temperature (Ramos et al., 2013). At higher concentration though, irradiation may affect the flavor of the produce (Goodburn & Wallace, 2013). Table 7 summarizes a few reports of use of radiations as a sanitizing treatment.

3.2.3 Ultraviolet light

Ultraviolet (UV) light is an electromagnetic radiation with wavelengths ranging between 100 and 400 nm. Within this range, the wavelength of 190–280 nm (UV-C light) is used for antimicrobial application as it damages DNA that causes cell death (Birmpa, Sfika, & Vantarakis, 2013). UV-C radiation is produced through UV lamps that consist of a tube with xenon or krypton gas, a mercury lamp, and an electrode at each side of the tube. When an electrical current is passed through, the mercury atoms

Table 8 Use of UV on minimally processed fruits and vegetables

Disinfecting treatment	Product	Microbial reduction	Reference
UV-C (2.5 kJ/m^2)	Minimally processed mango	2.4–2.6 log cfu/g reduction of *C. sakazakii*	Santo, Graça, Nunes, and Quintas (2018)
3.2 kJ/m^2 UV-C	Fresh-cut dragon fruit	Significantly lower total aerobic bacteria, coliforms, yeast and mold during storage	Nimitkeatkai and Kulthip (2016)
1.2 kJ/m^2 UV-C	Fresh-cut endive	Significantly improved microbiological and sensory quality during the entire storage period	Hägele et al. (2016)
1.4–13.7 kJ m^2 UV-C	Fresh-cut watermelon	>1 log reduction in microbial populations at 4.1 kJ/m^2	Fonseca and Rushing (2006)

become excited. UV light is produced when the atoms return to their ground state (Gray, 2014). UV light is highly effective for surface applications and also reduces process time (Birmpa et al., 2013). UV technology has been FDA approved (21CFR179.39) for use as a disinfectant to treat food as long as the proper wavelength of energy is maintained (200–300 nm). However, its prolonged use can alter the organoleptic properties of the food (Gadgil, 1997). Table 8 summarizes the application of UV light in few fresh-cut products.

4 Synergistic effects of disinfectants

The concept of using multiple methods is analogous to hurdle technology, where two or more preservation technologies are used to prevent growth of microorganisms in or on foods (Leistner, 1994a, 1994b). To combat the ill-effects of a particular sanitizing agent at a higher level, the combination of two interventions is much more useful and effective than each used alone. It is a multi-targeted approach to kill the microbes since different approaches target different cell systems, e.g., cell membrane, enzyme systems, DNA, pH, E_h, a_w, within the microbial cell and contribute to disturbing the homeostasis of the microorganisms (Leistner, 1994a, 1994b). Since different hurdles have different spectra of antimicrobial action, the combined hurdles could attack microorganisms in different ways and may increase the effectiveness of preservation

Table 9 Sanitizers used in combination on minimally processed fruits and vegetables

Disinfecting treatment	Product	Effect on microbes	References
0.3 kJ/m^2 of UV-C + pulsed light 15 kJ/m^2	Fresh-cut broccoli	2.4 log reduction of *L. innocua*	Collazo et al. (2019)
0.3 kJ/m^2 UV-C + 50 mg/L peracetic acid	Fresh-cut broccoli	Reduced mesophile by 2 log_{10} in broccoli	Collazo et al. (2018)
7.5 ppm ozone + 150 ppm of chlorine	Shredded lettuce	1.4 log reduction in APC compared to untreated	Garcia et al. (2003)
0.5% citric acid + 0.5–20 kJ/m^2 UV-C	Fresh-cut apples	Reduction in bacterial count of fresh-cut apples by 2.6 log cfu/g	Chen, Hu, He, Jiang, and Zhang (2016)
20 kJ/m^2 UV-C radiation + 5 ppm gaseous ozone	Fresh-cut rocket	Reduced microbial load during 8 days of storage at 5°C	Gutiérrez, Chaves, and Rodríguez (2016)
Acidic electrolyzed water (AEW) + aqueous ozone	Fresh-cut cilantro	Reduced initial total aerobic plate count and maintenance of low microbial count during storage	Wang, Feng, and Luo (2004)
Chlorination + 0.55 kGy irradiation	Shredded iceberg lettuce	5.4 log reduction of *E. coli* O157:H7	Foley et al. (2002)
1.5% lactic acid + 1.5% H_2O_2	Apples, oranges, and tomatoes	>5 log reduction of *Salmonella* and *E. coli* O157:H7	Venkitanaraynana, Lin, Bailey, and Doyle (2002)

synergistically. Many studies have demonstrated drastic reduction in microbial populations in fresh-cut products when treated with two or more sanitizing agents simultaneously (Table 9).

5 Safety concerns

Any sanitizing treatment applied must not only be effective, but also must be compatible with commercial packing and processing practices and technical capabilities. The treatment must be affordable and safe to carry out, have no adverse effect on quality, and be approved by applicable regulatory agencies. All sanitizers must have Environmental Protection Agency (EPA) registration and Food and Drug Administration (FDA) clearance (FDA, 2000) or Generally Recognized as Safe (GRAS) status for use in washing produce. Only food grade, EPA registered sanitizers should be used during produce washing or equipment sanitizing. The recommended levels for usage should be taken into consideration during application, along with the conditions in which the maximum potential of the sanitizer can be extracted.

6 Conclusions

The market for minimally processed foods has grown rapidly in recent years due to the fresh-like characteristics and convenience associated with these foods. Its growth has increased the awareness regarding microbiological and physiological aspects associated with the quality. The advances in alternative, cost-efficient, and environmentally friendly sanitizers has improved the quality and safety of minimally fresh processed fruits and vegetables. Washing and disinfection are the most important stages of the production chain where a reduction in the microbial load can be obtained. Since, the activity of the disinfecting agents depends on various physico-chemical process parameters, their application should be done under conditions that are most favorable to get the best sanitizing effect. The combined application of multiple sanitizing agents has shown advantages over individual treatments resulting in better microbial reduction, enhanced shelf-life, and food quality maintenance.

References

21CFR173.325. Secondary direct. Food additives permitted in food for human consumption: Acidified sodium chlorite solutions, Code of Federal Regulations. Title 21, Part 173.325.

21CFR179.39. Ultraviolet radiation for the processing and treatment of food. Code of Federal Regulations 21, Part 179, Section 179.39.

21CFR184.1366. Hydrogen peroxide. Code of Federal Regulations 21, Part 170–199, Section 184.1366, 463.

Akbas, M. Y., & Ölmez, H. (2007). Effectiveness of organic acid, ozonated water and chlorine dipping on microbial reduction and storage quality of fresh-cut iceberg lettuce. *Journal of the Science of Food and Agriculture, 87*, 2609.

Alexandre, E. M. C., Brandao, T. R. S., & Silva, C. L. M. (2012). Assessment of the impact of hydrogen peroxide solutions on microbial loads and quality factors of red bell peppers, strawberries and watercress. *Food Control, 27*, 362–368.

Alexopoulos, A., Plessas, S., Ceciu, S., Lazar, V., Mantzourani, I., Voidarou, C., et al. (2013). Evaluation of ozone efficacy on the reduction of microbial population of fresh cut lettuce (*Lactuca sativa*) and green bell pepper (*Capsicum annuum*). *Food Control, 30*, 491–496.

Alicea, C., Annous, B. A., Mendez, D. P., Burke, A., & Orellana, L. E. (2018). Evaluation of hot water, gaseous chlorine dioxide, and chlorine treatments in combination with an edible coating for enhancing safety, quality, and shelf life of fresh-cut cantaloupes. *Journal of Food Protection, 81*, 534–541.

Allende, A., Selma, M. V., López-Gálvez, F., Villaescusa, R., & Gil, M. I. (2008). Impact of washwater quality on sensory and microbial quality, including *Escherichia coli* cross-contamination, of fresh-cut escarole. *Journal of Food Protection, 71*, 2514–2518.

Artés, F., & Allende, A. (2005). Processing lines and alternative preservation techniques to prolong the shelf-life of minimally fresh processed leafy vegetables. *European Journal of Horticultural Science, 70*, 231–245.

Back, K. H., Ha, J. W., & Kang, D. H. (2014). Effect of hydrogen peroxide vapor treatment for inactivating *Salmonella typhimurium, Escherichia coli O157:H7* and *Listeria monocytogenes* on organic fresh lettuce. *Food Control, 44*, 78–85.

Banach, J. L., Overbeek, L. S., Groot, M. N., Zouwen, P. S., & Fels-Klerx, H. J. (2018). Efficacy of chlorine dioxide on Escherichia coli inactivation during pilot-scale fresh-cut lettuce processing. *International Journal of Food Microbiology, 269*, 128–136.

Bari, M. L., Nakauma, M., Todoriki, S., Juneja, V. K., Isshiki, K., & Kawamoto, S. (2005). Effectiveness of irradiation treatments in inactivating *Listeria monocytogenes* on fresh vegetables at refrigeration temperature. *Journal of Food Protection, 68*, 318–323.

Baur, S., Klaiber, R., Hammes, W. P., & Carle, R. (2004). Sensory and microbiological quality of shredded, packaged iceberg lettuce as affected by pre-washing procedures with chlorinated and ozonated water. *Innovative Food Science and Emerging Technologies, 5*, 45–55.

Baur, S., Klaiber, R., Wei, H., Hammes, W. P., & Carle, R. (2005). Effect of temperature and chlorination of pre-washing water on shelf-life and physiological properties of ready-to-use iceberg lettuce. *Innovative Food Science and Emerging Technologies, 6*, 171–182.

Beuchat, L. R. (2000, chapter 4, Aspen Publ., Gaithersburg, MD. pp. 63–78). Use of sanitizers in raw fruit and vegetable processing. In *Minimally processed fruits and vegetables: Fundamental aspects and applications*.

Beuchat, L. R., Adler, B. A., & Lang, M. M. (2004). Efficacy of chlorine and a peroxyacetic acid sanitizer in killing *Listeria monocytogenes* on iceberg and romaine lettuce using simulated commercial processing conditions. *Journal of Food Protection, 67*, 1238–1242.

Birmpa, A., Sfika, V., & Vantarakis, A. (2013). Ultraviolet light and ultrasound as non-thermal treatments for the inactivation of microorganisms in fresh ready-to-eat foods. *International Journal of Food Microbiology, 167*, 96–102.

Burnett, A. B., Iturriaga, M. H., Escartin, E. F., Pettigrew, C. A., & Beuchat, L. R. (2004). Influence of variations in methodology on populations of *Listeria monocytogenes* recovered from lettuce treated with sanitizers. *Journal of Food Protection, 67*, 742–750.

Carpenter, C. E., & Broadbent, J. R. (2009). External concentration of organic acid anions and pH: Key independent variables for studying how organic acids inhibit growth of bacteria in mildly acidic foods. *Journal of Food Science, 74*, 12–15.

CFR. (2005). *Secondary direct food additives permitted in foods for human consumption. Code of Federal Regulations. Title 21, Part 173.* http://www.gpoaccess.gov/cfr/index.html.

Chen, C., Hu, W., He, Y., Jiang, A., & Zhang, R. (2016). Effect of citric acid combined with UV-C on the quality of fresh-cut apples. *Postharvest Biology and Technology, 111*, 126–131.

Chen, X., Xue, S. J., Shi, J., Kostrzynska, M., Tang, J., Guévremont, E., et al. (2018). Red cabbage washing with acidic electrolysed water: Effects on microbial quality and physicochemical properties. *Food Quality and Safety, 2*, 229–237.

Chen, Z., Zhu, C., Zhang, Y., Niu, D., & Du, J. (2010). Effects of aqueous chlorine dioxide treatment on enzymatic browning and shelf-life of fresh-cut asparagus lettuce (*Lactuca sativa* L.). *Postharvest Biology and Technology, 58*, 232–238.

Cherry, J. P. (1999). Improving the safety of fresh produce with antimicrobials. *Food Technology, 53*, 54–58.

Collazo, C., Charles, F., Aguiló-Aguayo, I., Marín-Sáez, J., Lafarga, T., Abadias, M., et al. (2019). Decontamination of listeria innocua from fresh-cut broccoli using UV-C applied in water or peroxyacetic acid, and dry-pulsed light. *Innovative Food Science and Emerging Technologies, 52*, 438–449.

Collazo, C., Lafarga, T., Aguiló-Aguayo, I., Marín-Sáez, J., Abadias, M., & Viñas, I. (2018). Decontamination of fresh-cut broccoli with a water-assisted UV-C technology and its combination with peroxyacetic acid. *Food Control, 93*, 92–100.

Ding, T., Ge, Z., Shi, J., Xu, Y.-T., Jones, C. L., & Liu, D.-H. (2015). Impact of slightly acidic electrolyzed water (SAEW) and ultrasound on microbial loads and quality of fresh fruits. *LWT-Food Science and Technology, 60*, 1195–1199.

Du, J., Fu, Y., & Wang, N. (2009). Effects of aqueous chlorine dioxide treatment on browning of fresh-cut lotus root. *LWT-Food Science and Technology, 42*, 654–659.

EPRI. (1997). Ozone-GRAS affirmation for use in food. *Food Industry Currents, 1*, 1–6.

Ergun, M., & Dogan, E. (2018). Use of hydrogen peroxide, citric acid and sodium hypochlorite as sanitizer for minimally processed table grapes. *Ciência e Técnica Vitivinícola, 33*, 58–65.

FDA. (2000). Food and drug administration, center for food safety and applied nutrition. Kinetics of microbial inactivation for alternative food processing technologies. http://vm.cfsan.fda.gov/~comm/ift-toc.html. Accessed 15 August 2001.

Foley, D. M., Dufour, A., Rodriguez, L., Caporaso, F., & Prakash, A. (2002). Reduction of *Escherichia coli 0157:H7* in shredded iceberg lettuce by chlorination and gamma irradiation. *Radiation Physics and Chemistry, 63*, 391–396.

Fonseca, J. M., & Rushing, J. W. (2006). Effect of ultraviolet-C light on quality and microbial population of fresh-cut watermelon. *Postharvest Biology and Technology, 40*, 256–261.

Francis, G. A., & O'Beirne, D. (2002). Effects of vegetable type and antimicrobial dipping on survival and growth of Listeria innocua and E. coli. *International Journal of Food Science and Technology, 37*, 711.

Gadgil, A. (1997). *Field-testing UV disinfection of drinking water.* United Kingdom: Water Engineering Development Center, University of Loughborough.

Gao, J., Zhu, Y., & Luo, F. (2018). Effects of ethanol combined with ascorbic acid and packaging on the inhibition of browning and microbial growth in fresh-cut Chinese yam. *Food Science & Nutrition, 6*, 998–1005.

Garcia, A., Mount, J. R., & Davidson, P. M. (2003). Ozone and chlorine treatment of minimally processed lettuce. *Journal of Food Science, 68*, 2747–2751.

Gawande, P. V., & Bhagwat, A. A. (2002). Protective effect of cold temperature and surface contact on acid tolerance of *Salmonella* spp. *Journal of Applied Microbiology, 93*, 689–696.

Gómez-López, V. M., Lannoo, A. S., Gil, M. I., & Allende, A. (2014). Minimum free chlorine residual level required for the inactivation of *Escherichia coli O157:H7* and trihalomethane generation during dynamic washing of fresh-cut spinach. *Food Control, 42*, 132–138.

Goodburn, C., & Wallace, C. A. (2013). The microbiological efficacy of decontamination methodologies for fresh produce: A review. *Food Control, 32*, 418–427.

Gray, N. F. (2014). Ultraviolet disinfection. In *Microbiological aspects and risks. Microbiology of waterborne diseases* (2nd ed., pp. 617–630) [chapter 34], Academic Press.

Gutiérrez, D. R., Chaves, A. R., & Rodríguez, S. D. (2016). Use of UV-C and gaseous ozone as sanitizing agents for keeping the quality of fresh-cut rocket (*Eruca sativa* mill). *Journal of Food Processing & Preservation, 41*, 12968.

Gutiérrez, D. R., Chaves, A. R., & Rodríguez, A. C. (2018). UV-C and ozone treatment influences on the antioxidant capacity and antioxidant system of minimally processed rocket (*Eruca sativa* mill.). *Postharvest Biology and Technology, 138*, 107–113.

Hägele, F., Nübling, S., Schweiggert, R. M., Baur, S., Weiss, A., Schmidt, H., et al. (2016). Quality improvement of fresh-cut endive (*Cichorium endivia* L.) and recycling of washing water by low-dose uv-c irradiation. *Food and Bioprocess Technology, 9*, 1979–1990.

Han, Y., Sherman, D. M., Linton, R. H., Nielson, S. S., & Nelson, P. E. (2000). The effects of washing and chlorine dioxide gas on survival and attachment of *Escherichia coli O157:H7* to green pepper surfaces. *Food Microbiology, 17*, 521–533.

Hao, J., Li, H., Wan, Y., & Liu, H. (2015). Combined effect of acidic electrolyzed water (AcEW) and alkaline electrolyzed water (AlEW) on the microbial reduction of fresh-cut cilantro. *Food Control, 50*, 699–704.

Husman, A. M., Bijkerk, P., Lodder, W., Van Den Berg, H., Pribil, W., Cabaj, A., et al. (2004). Calicivirus inactivation by nonionizing (253.7-nanometer-wavelength [UV]) and ionizing (gamma) radiation. *Applied and Environmental Microbiology, 70*, 5089–5093.

IFPA (2001). International fresh-cut produce association. In *Food safety guidelines for the fresh-cut produce industry* (4th ed.). Alexandria, VA: IFPA.

Inatsu, Y., Bari, M. L., Kawasaki, S., Isshiki, K., & Kawamoto, S. (2005). Efficacy of acidified sodium chlorite treatments in reducing *Escherichia coli* O157:H7 on Chinese cabbage. *Journal of Food Protection, 68*, 251–255.

Jiang, L., Chen, Z., Liu, L., Wang, M., Liu, Y., & Yu, Z. (2017). Effect of chlorine dioxide on decontamination of fresh-cut coriander and identification of bacterial species in fresh-cutting process. *Journal of Food Processing & Preservation, 16*, 1027.

Joshi, B., Moreira, R. G., Omac, B., & Castell-Perez, M. E. (2018). A process to decontaminate sliced fresh cucumber (*Cucumis sativus*) using electron beam irradiation. *LWT-Food Science and Technology, 91*, 95–101.

Khadre, M. A., Yousef, A. E., & Kim, J. G. (2001). Microbiological aspects of ozone applications in food: A review. *Journal of Food Science, 66*, 1242–1252.

Koseki, S., & Itoh, K. (2001). Prediction of microbial growth in fresh-cut vegetables treated with acidic electrolyzed water during storage under various temperature conditions. *Journal of Food Protection, 64*, 1935–1942.

Koseki, S., Yoshida, K., Isobe, S., & Itoh, K. (2001). Decontamination of lettuce using electrolyzed water. *Journal of Food Protection, 64*, 652–658.

Koseki, S. K., Yoshida, Y., Isobe, S., & Itoh, K. (2004). Effect of mild heat pre-treatment with alkaline electrolyzed water on the efficacy of acidic electrolyzed water against *Escherichia coli O157:H7* and *Salmonella* on lettuce. *Food Microbiology, 21*, 559–566.

Lang, M. M., Harris, L. J., & Beuchat, L. R. (2004). Survival and recovery of *Escherichia coli* O157:H7, *Salmonella*, and *Listeria monocytogenes* on lettuce and parsley as affected by method of inoculation, time between inoculation and analysis, and treatment with chlorinated water. *Journal of Food Protection, 67*, 1092–1103.

Lee, B. H., Song, W. C., Manna, B., & Ha, J. K. (2008). Dissolved ozone flotation (DOF)—A promising technology in municipal wastewater treatment. *Desalination, 225*, 260–273.

Leistner, L. (1994a). *Food design by hurdle technology and HACCP.* Kulmbach: Adalbert Raps Foundation.

Leistner, L. (1994b). Further developments in the utilization of hurdle technology for food preservation. *Journal of Food Engineering, 22*, 421–432.

Lianou, A., Koutsoumanis, K. P., & Sofos, J. N. (2012). Organic acids and other chemical treatments for microbial decontamination of food. In *Microbial decontamination in the food industry novel methods and applications* (pp. 592–664) [chapter 20].

Luo, Y., Zhou, B., Haute, S. V., Nou, X., Zhang, B., Teng, Z., et al. (2018). Association between bacterial survival and free chlorine concentration during commercial fresh-cut produce wash operation. *Food Microbiology, 70*, 120–128.

Nascimento, M. S., Catanozi, M., & Silva, K. C. (2003). Effects of different disinfection treatments on the natural microbiota of lettuce. *Journal of Food Protection, 66*, 1697–1700.

Nimitkeatkai, H., & Kulthip, J. (2016). Effect of sequential UV-C irradiation on microbial reduction and quality of fresh-cut dragon fruit. *International Food Research Journal, 23*, 1818–1822.

Olmez, H., & Kretzschmar, H. (2009). Potential alternative disinfection methods for organic fresh-cut industry for minimizing water consumption and environmental impact. *LWT-Food Science and Technology, 42*, 686–693.

OSHA. (1988). *Occupational safety and health administration. 1988 OSHA PEL Project Documentation.* https://www.cdc.gov/niosh/pel88/10028-15.html.

Parish, M. E., Beuchat, L. R., Suslow, T. V., Harris, L. J., Garret, E. H., & Farber, J. M. (2003). Methods to reduce/eliminate pathogens from produce and fresh-cut produce. *Comprehensive Reviews in Food Science and Food Safety, 2*, 161–173.

Pinela, J., Barreira, J. C. M., Barros, L., Verde, S. C., Antonio, A. L., Carvalho, A. M., et al. (2016). Suitability of gamma irradiation for preserving fresh-cut watercress quality during cold storage. *Food Chemistry, 206*, 50–58.

Pinto, L., Ippolito, A., & Baruzzi, F. (2015). Control of spoiler *Pseudomonas* spp. on fresh cut vegetables by neutral electrolyzed water. *Food Microbiology, 50*, 102–108.

Prakash, A., Guner, A. R., Caporaso, F., & Foley, D. M. (2000). Effects of low-dose gamma irradiation on shelf life and quality characteristics of cut romaine lettuce packaged under modified atmosphere. *Journal of Food Science, 65*, 549–553.

Ramos, P., Pilawa, B., & Stroka, E. (2013). EPR studies of free radicals in thermally sterilized famotidine. *Nukleonika, 58*, 413–418.

Rico, D., Martín-Diana, A. B., Barry-Ryan, C., Frías, J. M., Henehan, G. T. M., & Barat, J. M. (2008). Use of neutral electrolysed water (EW) for quality maintenance and shelf-life extension of minimally processed lettuce. *Innovative Food Science and Emerging Technologies, 9*, 37–48.

Rico, D., Martín-Diana, A. B., Frías, J. M., Henehan, G. T. M., & Barry-Ryan, C. (2006). Effect of ozone and calcium lactate treatments on browning and 1560 texture properties of fresh-cut lettuce. *Journal of the Science of Food and Agriculture, 86*, 2179–2188.

Rodgers, S. T., Cash, J. N., Siddiq, M., & Ryser, E. T. (2004). A comparison of different chemical sanitizers for inactivating *Escherichia coli O157:H7* and *Listeria monocytogenes* in solution and on apples, lettuce, strawberries, and cantaloupe. *Journal of Food Protection, 67*, 721–731.

Santo, D., Graça, A., Nunes, C., & Quintas, C. (2018). *Escherichia coli* and *Cronobacter sakazakii* in 'Tommy Atkins' minimally processed mangos: Survival, growth and effect of UV-C and electrolyzed water. *Food Microbiology, 70*, 49–54.

Sapers, G. M. (2003). Washing and sanitizing raw materials for minimally processed fruit and vegetable products. In *Microbial safety of minimally processed foods* (pp. 221–253). Boca Raton, FL: CRC Press.

Selma, M. V., Beltrán, D., Allende, A., Chacón-Vera, E., & Gil, M. I. (2007). Elimination by ozone of *Shigella sonnei* in shredded lettuce and water. *Food Microbiology, 24*, 492–499.

Selma, M. V., Ibáñez, A. M., Cantwell, M., & Suslow, T. (2008). Reduction by gaseous ozone of *Salmonella* and microbial flora associated with fresh-cut cantaloupe. *Food Microbiology, 25*, 558–565.

Seo, K. H., & Frank, J. F. (1999). Attachment of *Escherichia coli O157:H7* to lettuce leaf surface and bacterial viability in response to chlorine treatment as demonstrated by using confocal scanning microscopy. *Journal of Food Protection, 62*, 3–9.

Simons, L. K., & Sanguansri, P. (1997). Advances in the washing of minimally processed vegetables. *Food Australia, 49*, 75–80.

Sun, X., Baldwin, E., Plotto, A., Narciso, J., Ference, C., Ritenour, M., et al. (2017). Controlled-release of chlorine dioxide in a perforated packaging system to extend the storage life and improve the safety of grape tomatoes. *Journal of Visualized Experiments, 122*, 55400.

Suslow, T. V. (1997). *Postharvest chlorination: Basic properties and key points for effective disinfection.* University of California, Division of Agriculture and Natural Resources. Publication 8003.

Suslow, T. V. (2004). *Oxidation-reduction potential (ORP) for water disinfection monitoring, control, and documentation.* University of California, Division of Agriculture and Natural Resources. Publication 8149.

Sy, K., Murray, M. B., Harrison, M. D., & Beuchat, L. R. (2005). Evaluation of gaseous chlorine dioxide as a sanitizer for killing *Salmonella, Escherichia coli O157:H7, Listeria monocytogenes*, and yeasts and molds on fresh and fresh-cut produce. *Journal of Food Protection, 68*, 1176–1187.

Ukuku, D. O. (2004). Effect of hydrogen peroxide treatment on microbial quality and appearance of whole and fresh-cut melons contaminated with *Salmonella* spp. *International Journal of Food Microbiology, 95*, 137–146.

Ukuku, D. O., Mukhopadhyay, S., Geveke, D., Olanya, M., & Niemira, B. (2016). Effect of hydrogen peroxide in combination with minimal thermal treatment for reducing bacterial populations on cantaloupe rind surfaces and transfer to fresh-cut pieces. *Journal of Food Protection, 79*, 1316–1324.

Venkitanarayanan, K. S., Lin, C. H., Bailey, H., & Doyle, M. P. (2002). Inactivation of *Escherichia coli O157:H7* and *Listeria monocytogenes*, on apples, oranges, and tomatoes by lactic acid and hydrogen peroxide. *Journal of Food Protection, 56*, 100–105.

Waghmare, R. B., & Annapure, U. S. (2017). Effects of hydrogen peroxide, modified atmosphere and their combination on quality of minimally processed cluster beans. *Journal of Food Science and Technology, 54*, 3658–3665.

Wang, H., Feng, H., & Luo, Y. (2004). Microbial reduction and storage quality of fresh-cut cilantro washed with acidic electrolyzed water and aqueous ozone. *Food Research International, 37*, 949–956.

Wengert, S. L., Aw, T. G., Ryser, E. T., & Rose, J. B. (2017). Postharvest reduction of coliphage MS2 from romaine lettuce during simulated commercial processing with and without a chlorine-based sanitizer. *Journal of Food Protection, 80*, 220–224.

White, G. C. (1972). *Handbook of chlorination.* New York: Van Nostrand Reinhold Co.
WHO. (1998). *World Health Organization, food safety unit. Food safety issues: Surface decontamination of fruits and vegetables eaten raw: A review.* Geneva: World Health Organization. WHO/FSF/FOS/98.2.
Yu, K., Newman, M., Archbold, D., & Hamilton-Kemp, T. (2001). Survival of *Escherichia coli* O157:H7 on strawberry fruit and the reduction of pathogen population by chemical agents. *Journal of Food Protection, 64,* 1334–1340.
Zhang, L., Lu, Z., Lu, F., & Bie, X. (2006). Effect of γ irradiation on quality-maintaining of fresh-cut lettuce. *Food Control, 17,* 225–228.
Zhuang, R. Y., Beuchat, L. R., & Anjulo, F. J. (1995). Fate of *Salmonella montevideo* on and in raw tomatoes as affected by temperature and treatment with chlorine. *Applied and Environmental Microbiology, 62,* 2127–2131.

Further reading

Bari, M. L., Sabina, Y., Isobe, S., Uemura, T., & Isshiki, K. (2003). Effectiveness of electrolyzed acidic water in killing *Escherichia coli* O157:H7, *Salmonella enteridis,* and *Listeria monocytogenes* on the surfaces of tomatoes. *Journal of Food Protection, 66,* 542–548.

6

Texturizers for fresh-cut fruit and vegetable products

Sindhu Chinnaswamy, Shalini Gaur Rudra, Ram Roshan Sharma

Division of Food Science and Postharvest Technology, ICAR-Indian Agricultural Research Institute, New Delhi, India

1 Introduction

According to the International Fresh cut Produce Association (IFPA), fresh-cut products of fruit and vegetables have been available to consumers since the 1930s in retail supermarkets (Martin-Diana et al., 2007). Fresh-cut produce offers significant convenience and nutritious salad on the go for all segments of the working population, be it single people, dual earning partner families, institutional purposes, canteens, etc. Nowadays, supermarkets play a major role in fresh-cut fruit and vegetable marketing. Recent years have witnessed a tremendous increase in demand for ready-to-use or ready-to-consume fresh-cut or minimally processed fruit and vegetables, which are fresh, convenient, and safe for direct consumption. They are easy to handle and time-saving, besides being nutritious (Ma, Zhang, Bhandari, & Gao, 2017).

These fruits and vegetables are minimally processed and include physical modifications to their structure through operations like peeling, trimming, sizing, cutting, coring, etc. These operations result in alteration of tissue structures, exposure of much higher cut surface to atmosphere, and consequently, a much shorter shelf-life. Such pre-cut fresh produce is highly vulnerable in terms of physiological and microbiological agents leading to a shorter shelf-life and poor quality. This leads to reduction of quality attributes such as color, firmness, juiciness, and flavor, owing to excessive moisture loss because of alteration in natural covering of produce. Fruits and vegetables are living

Fresh-cut Fruits and Vegetables. https://doi.org/10.1016/B978-0-12-816184-5.00006-9

commodities, which have high rates of respiration, ethylene production, and biochemical reactions (pectin degradation, organic acids, soluble solids, enzyme activity).

Firmness of fruits and vegetables is a key factor for consumer acceptability. Shriveling of fresh produce, decreased turgidity, and wrinkling of outer skin/outermost layer of fresh-cut produce are the main factors leading to rejection of packaged fresh-cut produce by consumers. Fresh-cut fruit and vegetables (FCFV) are highly perishable and more prone to deterioration and contamination from microbiological and other sources than whole fruits. Considering the interests of all stakeholders, there is a need to set up protocols or procedures for processing operations to provide good quality and safe end products to consumers. Many preharvest practices have a bearing on the quality of FCFV, which include Good Agricultural Practices, organic farming, spraying with micro-nutrients, adaptation of precision farming. Similarly, postharvest considerations like harvesting of produce at proper maturity in case of fruits and tenderness for vegetables, proper handling, removal of field heat, and cold chain transportation by cooling before subjecting to processing can reduce postharvest losses and enhance the shelf-life of FCFV.

2 Textural aspects in fresh-cut fruits and vegetables

2.1 What is texture?

Texture is a set of physical attributes of a product that can be measured both qualitatively and quantitatively using destructive or nondestructive methods. It can measure the sensory manifestation of structure or inner structure of a fruit/vegetable, in terms of types of cells, their reaction to stress, and amount and type of moisture/oil (Meilgaard, Civille, & Carr, 2006). Texture of fresh-cut produce is dependent on complex properties, which include cell wall composition, tissue type, water content, and water stress, etc.

Textural components of a produce can serve to indicate both sensory and physical attributes.

- Sensory attributes: firmness or touch, crunch or acoustics on bite
- Physical attributes: mechanical properties, crunchiness, succulence, chewiness, smoothness, grittiness, internal flesh firmness, etc.

Firmness is the most commonly explored textural parameter employed to measure the textural attributes of fruits. It is associated with turgidity, physical structure of cells in the tissues, their size, shape, arrangement, and thickness. Many of these factors are interrelated, for example, tissues with small cells tend to have a greater content of cell walls and lower relative amount of cytoplasm and vacuole (cell sap), greater area for cell-to-cell contact, and lower amounts of intercellular air spaces, making the tissue firmer and apparently less juicy (Siddiqui, Chakraborty, Zavala, & Dhua, 2011).

2.2 Measurement of texture of fresh-cut fruits and vegetables

Texture of any fresh-cut or whole fruits and vegetables can be evaluated by using an instrument or by sensory methods. Sensory evaluation can be done qualitatively and involves sensing (touch, sight) of produce by a trained or semitrained sensory panel, whereas instrumental measurement involves destructive and non-destructive methods.

i. **Destructive method** includes deformation or fragmentation of a fruit/vegetable when a force is applied and quantitatively measured. Tests employed for these include the puncture test, Magness Taylor test, Kramer multiblade shear test, compression, cutting, and texture profile analysis. These tests may be conducted using a texture-meter, such as the Texture Analyzer.
ii. **Tensile tests** are used to determine the firmness or crispness of fresh-cut fruits and vegetables. The undesirable texture of apple and watermelon can be examined and is determined to be less crisp (lesser number of peaks upon puncture) due to the tissue rupture leading to splitting of cells and openings across the primary cell walls. For example, cucumbers and carrots show some elasticity and bending followed by failure, muskmelon shows a mixture of cell rupture and cell separation along the middle lamellae at the fracture surface, while banana shows almost exclusively cell separation without cell rupture (Toivonen & Brummell, 2008). Since the firmness of fresh-cut fruits and vegetables changes rapidly with maturity, processing operations followed by packaging and storage necessitate augmentation and proper treatment with texturizers.

iii. **Nondestructive methods** involve measurement of sensory attributes like firmness or texture of whole or fresh-cut fruits and vegetables without deforming the fruit surface. Such methods as evaluation of the variables obtained from force-deformation curves, examination of collision forces, Ricochet's method, nuclear magnetic resonance (NMR), optical sensors like visual (VIS), and near-infra-red (NIR) can be used to evaluate the firmness of fruit and vegetables. A durometer is another such instrument to measure the force deformation of the produce (García-Ramos, Valero, Homer, Ortiz-Cañavate, & Ruiz-Altisent, 2005).

3 Critical factors affecting the quality of fresh-cut fruits and vegetables

The quality of fresh-cut fruits and vegetables is greatly affected by physiological changes like enzymatic browning due to tissue damage and high respiration rate. Physical and pathological factors for loss of quality include various mechanical injuries, removal of outer naturally protective covering, culminating in faster weight loss, shriveling, loss of color, appearance, and short shelf-life (Siddiqui et al., 2011). For fresh-cut fruits and vegetables like apple, mango, papaya, potato and tomato slices, onion dices, trimmed leafy vegetables, carrot and celery sticks, shredded cabbage, pineapple and melon cubes, etc., proper care needs to be taken to maintain its quality attributes like color, texture, appearance, and shelf-life extension. Thus by regulating temperature, relative humidity, storage in refrigerated conditions (below 4°C), pretreatments with texturizers, antibrowning agents, natural acidulants, chelating agents, and packaging with suitable films and trays may help to prolong the shelf-life of the produce.

Rapid textural changes can occur in fresh-cut fruits and vegetables immediately after processing. The increased rate of respiration, ethylene production, and higher transpiration losses will lead to the structural degradation of the tissues, loss of turbidity, and adverse effects of enzymes due to biochemical changes within the cell wall of the produce. This deterioration can be reduced or delayed by applying texturizers/firming agents externally as pretreatment. They will act as a barrier against physical, biochemical, and microbiological damage.

Factors affecting the quality of fresh whole as well as cut produce during postharvest processing

Critical factors	Whole produce	Minimally processed produce
1. Physical environment	Presence of protective/outer layer leads to less contamination	Exposure of cut surface area to the environment increases the micro flora
2. Processing chain	Less operations involved	Series of operations like cutting, trimming, peeling, shredding, etc.
3. Packaging	Usually carton boxes are used	Special methods like modified atmospheric package with varied gas composition
4. Transportation	Reefer vans	It requires cold chain management from processing to the consumer's table

Hence, proper handling is required by the pre-cut produce. Treatments such as antibrowning agents, texturizers, or firming agents like calcium derivatives, acidulates like citric acid, malic acid, etc., and edible coating with preservatives serve to enhance the structural and physical sensory properties (Qadri, Yousuf, & Srivastava, 2015).

4 Treatments to enhance the textural properties of fresh-cut produce

Textural attributes are the physical tests that help quantitatively determine the indicators for either purchasing behavior through touch or pressing by fingers to judge the firmness of produce, or snap tests by consumers for some particular produce, or perception upon bite or mastication and thereby useful information on physical state of fresh-cut fruit and vegetables. Owing to alteration in textural attributes, use of texturizers, antibrowning agents, and antimicrobial agents has been in practice in processing industries to maintain freshness and texture in produce for longer durations of storage and retail shelf. In this chapter, various approaches, methodologies, and the efficacy of various approaches to preserve or maintain the textural attributes of fresh-cut fruits and vegetables will be discussed.

4.1 Texturizers

Texturizers are hydrocolloid substances to maintain or to improve the texture or firmness of commodities in the minimal food processing industry. In addition, reinforcement of cell walls in fresh-cut produce also enhances the shelf-life of the cut produce. Over the years, use of polysaccharides, proteins, and lipid based edible coatings and calcium salts as firmness improving agents have served to enhance the texture and retain the quality of produce. Incorporation of firming agents helps in reduction of postharvest losses in soft fruits like strawberries, pears, apples, etc., since they are highly vulnerable to microbial and pathological damage. Most texturizers are rich sources of calcium, which is well known for its healing property by forming a thin layer on the applied area in fresh-cut and whole fruit and vegetables. Varied derivatives of calcium can be used as firmness enhancers in fresh-cut processing.

4.1.1 Postharvest dip with calcium derivatives

Calcium is a fruit cell wall stabilizer that helps in holding firmness of cell walls in produce and diminishing the internal breakdown, bitter pits, and water core in whole fruits. The decrease in cell-to-cell adhesion during storage is the major reason for decrease in firmness, the middle lamella responsible for the cell-to-cell addition and primary cell wall being composed of rigid cellulose micro fibrils held together by a network of matrix glycones (hemicelluloses and pectins). During ripening the de-esterification of pectin leads to loss of integrity of cell walls, decrease in adhesion, increase in intercellular spaces, and changes in tissue structure. Calcium ions can passively diffuse within the cell wall structure because the porosity of plant cell walls is approximately 3.5–9.2 nm, while calcium salts are about 0.1 nm (Ngamchuachit, Sivertsen, Mitcham, & Barrett, 2014).

When fruit parenchyma cells come into contact with calcium salt solutions, calcium ions are transported primarily through the apoplast, or intercellular spaces. Then these de-esterified negatively charged pectin molecules can cross link with divalent cations such as calcium, which provides rigidity to cell wall and reduces its porosity (Cybulska, Pieczywek, & Zdunek, 2012; Ortiz, Graell, & Lara, 2010). Negatively charged chloride or lactate ions remain unbound in solutions.

Another mechanism for uptake of calcium ions relates to cross linking of calcium ions to homogalacturonan in the middle lamella and cell walls, which leads to the initial firming effect of calcium treatments. The subsequent firming during storage may be due to interaction of calcium ions with negatively charged head groups of phospholipids and proteins of plasma membrane. Thus, calcium ions protect the membrane from lipid degradation by stabilization and degradation by lipolytic enzymes. In the cell walls, formation of cell bridges also leads to reduced accessibility to decay causing fungal or bacterial hydrolases (Mignani et al., 1995; Picchioni, Watada, Whitaker, & Reyes, 1996).

This mineral (Ca) plays a vital role in plant or plant produce in both pre- and postharvest applications. External application of calcium is very helpful to enhance fruit quality, and to diminish softening, respiration rate, and textural attributes of the produce. Application of calcium salts like calcium chloride, calcium lactate, calcium propionate, calcium pectate, and calcium carbonate have been known to reduce the undesirable changes in minimally processed produce. Calcium is one of the major elements responsible for maintaining structural integrity of cells and cell wall membrane. Postharvest treatments with calcium salts can effectively increase the calcium content in tissue and delay the senescence and, ethylene production by reducing respiration rate, resulting in firmer, higher quality fruits (Sams, Conway, Abbott, Lewis, & Ben-Shalom, 1993). It can alter the function of membranes by interacting with pectic acid present in the cell wall by forming calcium pectate. Thus, it acts as firming agent or firm enhancer or texturizer. Calcium salts help to reduce the tissue damage, avoid browning by reducing the oxidation process and also off-flavors (Soliva-Fortuny & Martin-Belloso, 2003). Various calcium salts like calcium chloride, calcium lactate, calcium propionate, and calcium gluconate have been reported to texturize the fresh-cut produce and effect improvement in their texture, acceptance, and storability (Alzamora et al., 2005). Further, combining the calcium salt treatment with heat application helps to trigger the action of heat-activated pectin methylesterase and also improves the diffusion of calcium into tissues of treated produce.

Another benefit derived from calcium treatments is the incorporation of significant quantities of this nutrient into the fruit or vegetable matrix. Calcium content in diet is critical to most biological processes. Increasingly, food products are being fortified with calcium due to increased incidence of lifestyle disorder related diseases like hypertension, calcium deficiency, and osteoporosis. There are evidences to support that increased calcium intake might reduce the incidence of such diseases. Use of phosphorous free sources of calcium such as gluconate, citrate, lactate, acetate, and carbonate salts of calcium can thus help to fortify minimally processed fruits and vegetables. Various authors (Anino, Salvatori, & Alzamora, 2006) have explored the possibility of obtaining products fortified with calcium through use of salts like calcium lactate and calcium gluconate combined with impregnation techniques.

4.1.2 Calcium chloride

Enhancement of the shelf-life of fresh-cut fruits and vegetables can be achieved through dipping in calcium derivatives or incorporating them in edible coatings. The most common calcium salt use is calcium chloride because it acts as sanitizer as well as texturizer, besides being easily available and less costly compared to other salts.

Calcium chloride has been successfully used as a firming agent for strawberries, pears, apples, and peaches (Giacalone & Chiabrando, 2013). Calcium chloride reduces the microbial contamination on the surface as it acts as a surface sanitizer and delays softening of tissue by reducing respiration rate. Thus, it has been found to be an effective treatment to minimize postharvest decay. In a study by Albertini, Reyes, Trigo, Sarriés, and Spoto (2016), dipping of papaya slices in 0.75% calcium chloride solution containing 0.1% cinnamaldehyde gave up to 12 days extension of shelf-life at 5°C storage with lower microbial count and 100% acceptance upon sensory evaluation. The calcium chloride treatment yielded increased firmness until 12 days storage. While control fruit decreased in firmness from 1.75 to 1.25 N after 12 days, calcium chloride and cinnamaldehyde treated fruits showed 1.5 N firmness. They attributed this to creation of calcium pectate through covalent linkages between pectin molecules in the cell wall and the middle lamella. This limits the action of enzymes responsible for the loss of texture and firmness (Salunke, Bolin, & Reddy, 1991).

Giacalone and Chiabrando (2013) compared the effect of calcium propionate (1%) and citric acid (1%) treatment with calcium chloride and citric acid (at same concentrations) for fresh-cut apple dices stored at 1°C for 5 days. Their results indicated high value of firmness for calcium chloride and citric acid treated fruits. The firmness measured in terms of durofel index of apple dices (manual penetrometer, Fruit Pressure Tester Model 327 FT^2; Facchini Srl; Alfonsine, Italy) was found to decrease, for calcium chloride and citric acid treated fruits, from 80 on day 0 to 76 on day 5. The calcium propionate treated fruits had initial firmness of 69, which decreased to 66 on day 5. The reduced pectin methylesterase activity due to inhibiting effect of the propionate ions in fresh-cut apples treated with calcium propionate has also been reported by Quiles, Hernando, Pérez-Munuera, and Lluch (2007). The citric acid and calcium chloride treatment was found better over calcium propionate treatment as higher browning and off-flavors were associated with the propionate salts. On the other hand, calcium chloride and citric acid reduced browning owing to polyphenol oxidase inhibition by the chloride ions.

Ngamchuachit et al. (2014) subjected mango cubes of Kent and Tommy Atkins varieties to calcium treatment from two salts (calcium chloride and calcium lactate) at concentrations up to 0.204 M and 4 dipping times (0–5 min). The cubed mangoes were stored at 5°C in lidded polystyrene containers. While increase in Calcium concentrations lead to higher firmness of fresh fruits, dip time had no significant effect for Kent variety in treatment beyond 1 min; however, in the case of Tommy Atkins, the highest firming was found at 2.5 min of dip time. A similar firming effect was initially found with both calcium chloride and calcium lactate treatments; however, during storage, a higher rate of softening was found in calcium lactate treated mangoes compared to calcium chloride treated samples. In the case of mangoes given the calcium chloride treatment at 0.204 M concentration, firmness after 9 days of cutting decreased by 18.27% in Kent variety, 20.05% in Tommy Atkins variety.

Mild hot water treatment followed by edible coating generally hastens the damaged tissue repair by activating the pectin methylesterase (PME) activity in fresh-cut produce. Moderate heat treatment in combination with post-cut calcium treatment has been found effective for fresh-cut mango cubes (Trindade et al., 2003). Martin-Diana et al. (2006) reported 1.5% calcium lactate at 50°C wash treatment for 1 min as the best treatment to inhibit the browning and preserve firmness of minimally processed lettuce due to higher turgor pressure and enhanced PME activity. Other authors, however, prefer lower temperatures (4°C for 3 min) for calcium treatment to be beneficial compared to room temperature.

Costa, Cardoso, Martins, Empis, and Martins (2008) studied the effect of heat treatment at 45°C in combination with calcium chloride solution on the quality of minimally processed kiwifruit. They studied the effect on firmness, calcium content, PME activity, and heat shock proteins accumulation over a period of 8 days storage at 4°C. They observed beneficial effects of heat treatment on firming of fruit tissues (24%) while calcium dips had marginal effect. Further applied heat treatments did not allow the diffusion of calcium into the fruit and instantly led to loss of calcium ions. This phenomenon was however lowest at 3% calcium chloride treatment where osmotic pressure of solution could have balanced the effect of temperature gradient.

Similar findings have been reported by Lurie, Fallik, and Klein (1996), who reported that heat treatment fills in cracks present in apples' epicuticular wax and decreases the transportation ability of applied calcium in the fruit tissues. Garcia, Herrera, and Morilla (1996), on the contrary, reported the heating of strawberries at 45°C during calcium chloride dip treatment enhanced the penetration of ions into the fruit. Luna-Guzmán and Barrett (2000) reported improvement in firmness of fresh-cut cantaloupe melons upon dipping in 2.5% calcium chloride solution for 1 min at 60°C.

Fresh-cut cabbage shreds were treated with calcium chloride in combination with ascorbic acid and citric acid (chelating agents) and stored for 14 days at 0°C (Manolopoulou & Varzakas, 2011). They found improved texture and overall quality of the produce due to reduced activity of polyphenol oxidase and the rupture force was found to increase significantly from 257 to 290 N upon storage at 0°C for 14 days. They recommended calcium chloride and citric acid treatment combined with low temperature storage as an effective treatment for maintaining acceptable marketing conditions of minimally processed cabbage for 22 days. Kim and Klieber (1997) also reported use of calcium chloride to reduce browning in minimally processed Chinese cabbage.

Shih-Ying Pan, Chen, and Lai (2013) determined the effect of ascorbic acid, calcium chloride, and cinnamon oil in combination with tapioca starch and decolorized Hsian-Tsao leaf gum (dHG) based edible coatings on fresh-cut Fuji apple slices. They found that inclusion of 1% calcium chloride in starch dHG in the coatings did not show extra beneficial effect in terms of maintaining the texture of apple pieces. This was attributed to the possibility of uronic acid in Hsian-Tsao leaf gum to be cross-linked with calcium ions to enhance the network structure of the coatings. As a result, availability of calcium ions for texture enhancement of apple pieces was limited.

In a similar study by Sanchísa et al. (2016), apple pectin coatings were combined with antioxidant aqueous solution (calcium chloride 1% and citric acid 1%) to enhance the quality of fresh-cut 'Rojo Brilliante' persimmon during storage at 5°C. These coatings could effectively control the enzymatic browning and reduce *Escherichia coli*, *Salmonella enteritidis*, and total aerobic mesophilic bacteria and helped to achieve accomplish a commercial shelf-life up to 7 days. They found that combination of calcium chloride with ascorbic acid or citric acid prevented excessive softening of persimmon slices and helped maintain their firmness. The antioxidant solution treatment led to an initial increase in firmness of persimmon from 3.5 to 3.67 N. Subsequently, on storage for 9 days, the firmness decreased to 3.36 N for treated fruits compared to 3.21 N for control fruits. Combined pectin coating of calcium salts with citric acid enhanced the antimicrobial activity by reducing the internal pH of microbial cells by ionization of the undissociated acid molecules, and disruption of substrate transport by altering cell membrane permeability (Valencia-Chamorro, Palou, Del Río, & Pérez-Gago, 2011).

Martin-Diana et al. (2007) reviewed the use of various calcium salts to preserve fresh and minimally processed fruits and vegetables. They reported improved firmness in various fresh-cut fruits and vegetables like carrot, diced tomatoes, iceberg lettuce, grapefruit, strawberry, honeydew, and peaches. In addition, they emphasized a need for further research on bioavailability of calcium incorporated into the product. Since bioavailability depends on many factors such as commodity pH and fiber content, each particular case should be studied when obtaining a fortified product.

4.1.3 Calcium lactate

Calcium lactate is a biopreservative, and a well-known firming agent, used as an alternative for calcium chloride as it is devoid of the off-flavor and bitterness associated with calcium chloride. Recently, use of calcium salts other than calcium chloride have been linked to reduction in formation of carcinogenic chlorinated compounds like chloramines and trihalomethanes (Martin-Diana et al., 2005) which calcium chloride may release in water during treatment. Rico, Martin-Diana, Frias, Henehan, and Barry-Ryan (2007) explored the use of calcium treatment with heat shock to improve the quality of fresh-cut lettuce. The calcium content in lettuce treated with 1.5% calcium lactate solution was improved from 0.162 mg/100 g to 0.191 mg/100 g as the temperature of treatment was increased from 25°C to 50°C. The effect of calcium lactate treatment on fresh-cut 'Flor de Invierno' pears structural integrity, texture, PME, and polygalacturonase activity have been reported by Alandes, Munuera, Llorca, Quiles, and Hernando (2009). They submerged cubes of pear in 0.5% calcium lactate solution with sodium benzoate and potassium sorbate (300 mg/L each) at 10°C for 1 min. While control fruits firmness decreased by 18.2%, the calcium treated fruits firmness decreased by only 8.78% during 2 weeks of storage. Similar findings have been reported by Dong et al. (2000) for fresh-cut Bartlett, Anjou and Bosc pears treated with calcium lactate. This was attributed to strengthening of structure of the pears by maintaining fibrilar packing in the cell walls and reinforcing cell-to-cell contacts. Formation of calcium pectate and counteractive effects for PME activity increase were also considered major factors for the quality improvement.

An interesting study was conducted by Silveira, Aguayo, Chisari, and Artés (2011) to discern the effect of various calcium salts—chloride, citrate, lactate, ascorbate, tartarate, silicate, propionate, or acetate—using a calcium concentration equivalent of 0.4% pure calcium chloride combined with 50 mg/L hydrogen peroxide. Trapezoidal shaped sections of Galia melon were dipped in calcium solutions for 1 min at 60°C and stored at 5°C for 10 days under passive modified atmosphere reaching 4.5 kPa oxygen and 14.7 kPa carbon di oxide. At the end of the shelf-life ascorbate chloride and lactate salts of calcium provided melon pieces with good firmness and increased tissue total calcium content. Synergistic effects of combining heat treatment with calcium salts were also reported. While control calcium citrate and silicate showed firmness loss of 31%, 23%, and 24% respectively, remaining treatments led to firmness decrease between 17% and 20%.

Ozone is a strong antimicrobial agent with high reactivity and penetrability and spontaneous decomposition to a non-toxic product. Combination of both calcium lactate and ozone resulted in reduction of polyphenol oxidase and peroxidase enzyme activity in fresh-cut lettuce. It also improved the texture, as calcium could maintain cell wall structure by interacting with the pectic acid present in cell walls to form calcium pectate, thereby maintaining water activity and thus reducing the stress usually associated with higher respiration rates.

Rico, Martin-Diana, Frias, et al. (2007) studied the effect of ozone treatment, calcium lactate (1.5%, 50°C), and their combination on the quality of fresh-cut lettuce over 10 days of refrigerated storage. They analyzed the leaf texture using Kramer shear cell with an 8-blade probe. Since lettuce has both vascular and photosynthetic types of tissues, which are irregularly distributed, and relative position of vascular packages (parallel, oblique, or perpendicular orientation) with respect to shear blades directly affects the measurements, perpendicular orientation of photosynthetic tissue was selected for experiments. Textural values were expressed in terms of crispiness coefficient, which was determined by dividing the difference between minimum and maximum load with the maximum load. During storage, PME activity increased significantly during storage for all treatments. Fresh-cut lettuce treated with calcium lactate had significantly higher PME activity than samples treated with ozone or combined treatment. A significant reduction in crispiness coefficient (CC) was observed during storage, which was associated with loss of turgid and fresh like textural properties. Samples treated with ozone showed lowest value of CC, while samples treated with calcium lactate had the highest values. The use of heat shock and addition of calcium lactate could make the tissues firmer by binding to pectin carboxyl groups generated through the action of PME. Samples subjected to combined treatment and showed similar values to samples treated with ozone alone. Hence, calcium lactate treatment was considered a more suitable treatment than ozone to extend the quality of fresh-cut lettuce.

Not only fresh produce but also processed products derived from calcium treated produce have shown improved firmness and color; e.g., apple sauce produced from calcium-treated apples shows higher viscosity and lighter color (Manganaris, Vasilakakis, Diamantidis, & Mignani, 2007).

Lovera, Ramallo, and Salvadori (2014) determined the effect of process temperature calcium source (calcium gluconate and calcium lactate), calcium concentration (0.5% and 1.5%), and pH on the quality of papaya fruit in syrup. In fresh fruit, average calcium content was 18.6 mg/100 g, which increased up to 240 mg/100 g after treatment for 8 h at 45°C in 1.5% calcium solution. The effect of pH on calcium uptake in papaya was not clear but highest calcium concentrations were achieved by impregnation treatment with calcium lactate at pH 4.2. Fruit firmness was significantly affected by a source of calcium, with calcium lactate impregnated samples showing higher firmness. Relative firmness F_{max}/F^0_{max} increased from 1 to 5 when calcium gain increase from 0 to 70 mg/100 g; however, further increases up to 200 mg/100 g calcium retained this ratio to approximately 5.

4.2 Methods for calcium application on fresh-like minimally processed fruits and vegetables

As per literature reports, there are two main methods of calcium application on fresh-cut fruits and vegetables:

1. washing and dipping; and
2. impregnation process.

4.2.1 *Washing and dipping treatment*

This is the most common and widely used technique for imbibing calcium salts in fresh-cut fruit and vegetables. Washing with calcium salts is a primary process, whereas dipping or coating treatments are usually followed by fresh-cut processing (peeling, cutting, shredding, slicing, coring, dicing, etc.), depending on produce type.

Dipping treatments cover the maximum surface area of fresh-cut produce without any damage and stress. Dipping time may vary from 1 to 5 min, depending on the commodities. Fresh-cut cantaloupe, carrot, and lettuce are usually dipped in calcium solution for 1 to 5 min for better dispersion (Martin-Diana et al., 2005). Washing with calcium chloride solution helps to reduce microbial load as it acts as a surface sterilizer. While most studies employed cool or room temperatures for calcium treatments, use of warm temperatures (40–60°C) increased the beneficial effect of treatments due to higher penetration of washing solutions inside the product and filling of cracks present in epicuticular wax (Lurie et al., 1996). However, contradictory reports also exist, which highlight the variation of effect of treatment for each specific fruit and/or cultivar.

4.2.2 *Impregnation*

Impregnation is the process of modification of the composition of food material through partial water removal and impregnation with solutes, without affecting the material's integrity. The driving forces for the processes can be osmotic gradient between the sample and solution, application of vacuum followed by restoration of atmospheric condition, or both osmotic gradient and vacuum conditions. Depending on the time (minutes to days) of the process and the magnitude of the vacuum (5–200 mbar) or solution concentration (20–75°Brix) used, the product obtained might lack the characteristics of a minimally processed product and rather be considered a dehydrated product.

Impregnation provides broad applications in fruits and vegetables processing (Radziejewska-Kubzdela, Biegańska-Marecik, & Kidoń, 2014). Some of the applications include development of a wide range of re-formulated products by impregnation with different sources of calcium, and improving the final product quality (Fito et al., 2001).

Gras, Vidal, Betoret, Chiralt, and Fito (2003) applied vacuum impregnation to enhance the rigidity and brittleness of tissues in eggplant, carrot, and oyster mushrooms, through the use of sucrose and calcium lactate solution seeking textural improvements. Anino et al. (2006) have done extensive work on the application of vacuum impregnation to fresh-cut apples. Strengthening of raw material structure of zucchini with calcium chloride solution by vacuum impregnation has been reported by Occhino, Hernando, Liorca, Neri, and Pittia (2011), with a mixture of calcium chloride, maltodextrin, and sodium chloride.

Biegańska-Marecik and Czapski (2007) applied calcium chloride in combination with ascorbic acid, 4 hexylresorcinol and sucrose to improve the texture of minimally processed apples. Anino et al. (2006) used an isotonic solution of glucose with an addition of calcium lactate and calcium gluconate in vacuum impregnation of apples. Greater rigidity of tissues in apples was observed after vacuum impregnation than after the process conducted at atmospheric pressure. Another interesting approach of counteracting changes in minimally processed produce tissues is through combination of PME with calcium ions in the impregnating solutions. PME catalyzes hydrolysis of methyl groups esterifying the pectin. Its action releases free carboxylic acids, which can chelate with calcium to form a gel that strengthens the cell wall structure and increases the firmness of the tissues. Guillemin, Degraeve, Noel, and Saurel (2008) demonstrated the use of vacuum impregnation to add exogenous PME to fruit pieces more rapidly and homogenously to improve the firmness of thermally treated fruit. Similar findings have been reported for pasteurized apples, strawberries, and raspberries (Degraeve, Saurel, & Coutel, 2003a, 2003b; Suutarinen, Honkapaa, Heinio, Autio, & Mokkila, 2000; Van Buggenhout et al., 2006).

Javeri, Toledo, and Wicker (1991), reported an improvement in texture during the impregnation of vacuum blanched peach halves, which were pasteurized afterwards. Ponappa, Scheerens, and Miller (1993) reported use of a vacuum impregnation technique to introduce spermine and spermidine to improve the hardness of strawberries stored at 1°C; however, no such effect was recorded in the case of putrescine. Vargas, Chiralt, Albors, and Gonzalez-Martinez (2009) applied vacuum impregnation to obtain better adherence of chitosan based edible coating for carrots. The film produced was thicker and more uniformly distributed over sample surface leading to improved mechanical properties and color of carrot slices during storage. Another probable technique for application of calcium solutions could be spraying over the fresh-cut produce surface. Spraying or fluxing, however, has been little researched to date.

4.3 Use of edible coatings

A group of food grade additives are derived from either natural or synthetic materials composed of edible components, a thin layer of which, upon application on surface of the produce, helps to prolong the shelf-life by maintaining its textural and sensory parameters (Pascall & Lin, 2013). Edible coatings are generally in liquid emulsion state; thus, its application on produce is easy. Edible coatings are applied to food by spraying, dipping (at atmospheric pressure or in vacuum), and most recently, through electro-spraying, which produces thin and uniform coating (Hassan et al., 2018).

4.3.1 Polysaccharides

Polysaccharides are naturally occurring polymers, widely used to prepare edible films or coatings including starch, cellulose, pectin, and derivatives of all these, pullulan, alginates, and chitosan. Coating of pectin on carrot slices showed significantly lower production of lignin precursors, and off-taste causing flavonoids, such as myricetin and naringenin (Ranjithaa et al., 2017).

Brasil, Gomes, Puerta-Gomez, Castell-Perez, and Moreira (2012) incorporated microencapsulated beta cyclodextrin and trans cinnamaldehyde complex (2%) into multilayered edible coating of chitosan and pectin for fresh-cut papaya. They evaluated the quality of fruits stored at 4°C in Ziploc trays with smart snap seal, saran wraps, and cheese cloth for 15 days. They found that while firmness of control fruits decreased almost twofold on storage for 15 days, the coated samples were significantly firmer. The coating prevented tissue breakdown independent of how the fruit was packed. Uncoated fruits without any packaging (cheese cloth cover) were worst in terms of firmness due to surface dehydration and increased respiration rates. Samples packed in trays with smart snap seal maintained their quality more effectively during storage followed by saran wrapped fruits. The coated fruits packed in Ziploc recorded increase in firmness from 48 g on 0 day to 105 g on 12 days of storage, while coated fruits in saran wrap had recorded firmness of 82 g.

Employing edible coating enriched with calcium salts can also be used to enrich minimally processed fruits and vegetables with calcium. Chaiprasart, Handsawasdi, and Pipattanawong (2006), working with strawberries and red raspberries, used chitosan-based coating formulated with calcium lactate and calcium gluconate. Adding calcium did not alter the positive effects of coating, and proved to extend the shelf-life by decreasing the incidence of decay and weight loss, changes in color, titratable acidity, and pH.

Song, Jo, Song, Min, and Song (2013) applied aloe vera gel with and without 0.5% cystein to fresh-cut apples stored at 4°C for 16 days. *Aloe vera* gel coating significantly reduces the degree of softening during storage compared to control as it provided barrier to water diffusion and delayed dehydration. While firmness of apple slices decreased by 6.89% in control upon storage at 4°C for 16 days, aloe vera coated fruits firmness decreased by 1.35%, and incorporation of cystein had the highest antibrowning effect. *Aloe vera* gel in combination with ascorbic and citric acids was also found effective in retaining texture of pomegranate arils during storage for 12 days at 3 °C (Martínez-Romero et al., 2013).

Mohebbi, Ansarifar, Hasanpour, and Amiryousefi (2011) also reported effectiveness of *Aloe vera* gel coating for reducing loss of firmness of table grapes and mushrooms during storage. Xiao, Luo, and Wang (2011) evaluated the effect of sodium chlorite and its sequential treatment with chitosan coating on quality of fresh-cut d'Anjou pears. They reported no significant loss in firmness during storage amongst all treatments. In some treatments, increased firmness was also reported, which was attributed to dehydration on surface of pear cubes since they were stored in unsealed bags.

Martiñon, Moreira, Elena Castell-Perez, and Gomes (2014) developed multilayered antimicrobial edible coating with chitosan, pectin, and encapsulated trans-cinnamaldehyde to extend the shelf-life of fresh-cut cantaloupe stored at 4°C. The firmness is defined as maximum force required to compress the sample to 50% of original size, and was determined using 5.2 cm diameter cylindrical probe through uniaxial compression. While uncoated samples were soft and mushy (9 N force) by day 15, 2 g/100 g chitosan coating yielded fruits with highest values of force (45–54 N) after storage.

Saba and Sogvar (2016) determined the effect of carboxymethyl cellulose in combination with calcium chloride and ascorbic acid on browning and quality of fresh-cut apples. During storage for 12 days, the firmness of uncoated apple slices decreased from 23.4 to 17 N, but coated fruits could retain 99.88% of their firmness.

Poverenov et al. (2014) used layer-by-layer electrostatic deposition of oppositely charged natural polysaccharide, alginate, and chitosan to extend the shelf-life of fresh-cut melon. After 11 days of storage, alginate coating had no more beneficial effect on fruit texture, while chitosan and especially alginate-chitosan reduced the texture degradation probably due to their antimicrobial effect, which could inhibit production of hydrolytic enzymes by microbes. The combined layer-by-layer coating had advantages of both coating materials when internal alginate layer provided perfect addition and external chitosan layer mitigated structural degradation due to microbial enzymes. In addition, due to alginate component, LbL coating included calcium ions, which are known to be effective texture enhancers. In another interesting study by Yousuf and Srivastava (2015), pretreatment of fresh-cut papaya fruit with 1% calcium chloride followed by 1% psyllium gum coating was found effective to maintain firmness during 16 days of storage at 4°C in polypropylene trays wrapped with transparent poly ethylene film. Yousuf et al. (2018) have recently reviewed the application of different edible coatings on shelf-life extension of fresh-cut fruits and vegetables.

4.3.2 Protein and lipids

The whey protein and pectin transglutaminase edible coating has shown effective response in reduction of microbial count and weight loss, and prevents degradation of antioxidants in fresh-cut apple, potato, and carrot during 10 days of storage (Marquez et al., 2016).

Yousuf and Srivastava (2017) employed treatment with honey (0–150 mL/L) followed by coating with soy protein isolate (5%) on fresh-cut kajari melons. An integrated approach of using 15% honey 5% soy protein isolate after calcium chloride pretreatment was able to retain higher acceptability and better texture after storage for 12 days in polypropylene trays and wrapped with PVC stretch wrapping films.

Ghidellia, Mateos, Rojas-Argudoa, and Pérez-Gago (2014) determined the effect of soy protein cysteine-based edible coating and modified atmosphere packaging of two gas mixtures (MA—15 kPa CO_2 + 5 kPa O_2 and MA—80 kPa O_2) at 5°C for 8 days on quality of fresh-cut Telma eggplants. The firmness of fresh-cut eggplant diminished from 12 to 8 N after 9 days' storage. Edible coating and modified atmosphere packing were not very effective to retain firmness. Highest firmness was recorded for samples packed under super atmospheric oxygen conditions. Firmness retention at high oxygen levels could be related to lower activities of the cell wall hydrolytic enzymes (Deng, Wu, & Li, 2005). Similar effects of super atmosphere oxygen conditions to retain firmness have been reported for sliced carrots, fresh-cut spinach, iceberg lettuce, and melons (Allende, Luo, McEvoy, Artés, & Wang, 2004; Amanatidou, Slump, Gorris, & Smid, 2000; Oms-Oliu, Raybaudi-Massilia, Soliva-Fortuny, & Martín-Belloso, 2008).

Coating with lipids is an excellent barrier against moisture migration. Lipids, mixed with proteins and polysaccharides, produce coatings with higher mechanical and barrier residences. The simplest lipid compounds are paraffin and beeswax. Lipid based films and coating are considered to be highly effective to block the delivery of moisture due to their low polarity ex., paraffin, shellac, acetoglycerides (Hassan et al., 2018). Chauhan, Raju, Singh, and Bawa (2010) applied edible surface coating made of shellac and aloe gel to apple slices after ozonation and soaking in ascorbic acid, citric acid, and sodium benzoate. The firmness was found highest for shellac-coated slices followed by aloe gel after 30 days of storage (12.5% decrease).

Cortez-Vega, Pizato, de Souza, and Prentice (2014) applied edible coatings from protein isolate of whitemouth croaker with oregano clay montmorillonite on minimally processed papaya slices. While the firmness of control papaya decreased from 31 to 11 N after storage for 12 days, the coated papaya firmness decreased marginally from 36 to 34 N.

4.4 Novel technologies combined with texturizers

4.4.1 Modified atmosphere packaging

Amanatidou et al. (2000) studied the combined effect of high oxygen and high CO_2 concentration on sliced carrots, which were subjected to citric acid dip followed by alginate coating. Amongst these, gas composition of 50% O_2 + 30% CO_2 with citric acid (0.1%) treatment yielded improved slice firmness of 880 N compared to control sample (690 N) on 8th day of storage. Sodium alginate reacted with polyvalent cations to form a gel that acted as a barrier against water vapor loss and maintained cell structure integrity to enhance firmness.

Waghmare and Annapure (2013) evaluated the firmness of fresh-cut papaya by giving chemical dip treatment (calcium chloride 1%, citric acid 2%) followed by modified atmospheric packing of 5% O_2, 10% CO_2, 85% N_2 and storage up to 25 days at 5°C. Treated fresh-cut papaya required 1.86 N of force to penetrate whereas non-treated papaya required 1.08 N at 10 days of storage. The combined effect of pretreatment and packaging atmosphere showed an improved firmness due to interaction of calcium ions with pectic polymers and formation of cross-linked polymer network, which increased the mechanical strength.

4.4.2 Pulsed light

Avalos Llano, Marsellés-Fontanet, Martín-Belloso, and Soliva-Fortuny (2016) determined the effect of pulsed light treatments on apple wedges, which were dipped into 1% *N*-acetylcysteine and 0.5% calcium chloride solution. This was followed by 10–40 flashes of light with intensities of 0.4 J/cm^2 per pulse and storage up to 15 days at 5°C. Apple wedges exposed to 4 J/cm^2 along with dip showed an improved firmness of 11 N compared to untreated (8.5 N) at the end of storage. This could be due to modulation of pectinolytic enzyme activities, i.e., pectin methyl esterase and polygalacturonase, which exhibited increased activity upon subjecting to the sequence of pulsed light treatment (Denès, Baron, & Drilleau, 2000; Jurick et al., 2009). Similar findings have been reported in fresh-cut strawberries treated with ascorbic acid (10 g/L), calcium lactate (35.3 g/L), and flash light intensities of 4 and 8 J/cm^2, which resulted in firmness increase of 1.8 N compared to control (0.9 N) during 14 days of storage at 5°C (Avalos-Llano, Martín-Belloso, & Soliva-Fortuny, 2018).

Koh, Noranizan, Nur Hanani, Karim, and Rosli (2017) examined repetitive pulsed light treatment (0.9 J/cm^2) for every 48 h up to

28 days in combination with edible coating on fresh-cut cantaloupe packed in polypropylene bags at 4 ± 1°C storage. The firmness of treated cantaloupe and untreated samples at the end of storage period was 3.27 N and 1.71 N, respectively. Due to photo-thermal effect, formation of a thin dried film on the surface of treated fruit slices resulted in improved firmness (Charles, Vidal, Olive, Filgueiras, & Sallanon, 2013). Valdivia-Nájar, Martín-Belloso, and Soliva-Fortuny (2018) evaluated the effect of pulsed light fluences (4, 6, and 8 J/cm^2) on fresh-cut tomatoes during 20 days of storage at 5°C. Upon treatment with 8 J/cm^2 fluence pulsed light, a significant reduction in textural degradation (up to 55.7%) in treated tomato slices was observed in comparison to untreated samples, which showed 63% loss of firmness from their initial value. Correlation between PME activity and firmness was found to be high, which indicates that structural changes and cell wall alterations could be due to photochemical, photophysical, and photothermal effect of pulsed light treatment (Gómez-López, Ragaert, Debevere, & Devlieghere, 2007; Manzocco, Da Pieve, & Maifreni, 2011).

4.4.3 Nanoparticles

Meng, Zhang, and Adhikari (2014) subjected fresh-cut kiwifruits to combined ultrasound treatment (40 KHz, 350 W, 10 min) and nanosized zinc oxide (ZnO) coating (1.2 g/L) stored at 4°C for 10 days. They observed a significant decrease in loss of firmness of treated kiwifruit slices (8.1 N) over the untreated samples, which showed firmness reduction of 3.5 N during storage period. Similarly, Zhao, Liu, and Ma (2009) found an almost two fold increase in shelf-life of chitosan and nano-ZnO treated apricots due to decrease in water loss. Similar findings have been reported by Hu, Fang, Yang, Ma, and Zhao (2011) for effectiveness of nano zinc oxide coatings on plum and kiwi fruits to maintain the firmness during storage by controlling the decrease in water loss.

Zambrano-Zaragoza et al. (2014) examined the effect of nanoemulsion of Nopal (*Opuntia* spp.) mucilage extract on fresh-cut Red Delicious apples packed in polypropylene containers stored at 4 ± 1°C, 85% RH for 21 days. Fresh-cut apple wedges treated with nano-emulsions retained their firmness (6.2 N) over control samples (2.7 N). They concluded that the pectin present in mucilage helps to lower the water loss by interacting with calcium ions and cross-linking the polymer chains between the cell wall and surface of the fruit surface by creating modified atmosphere.

Ultra violet radiation of 100–400 nm range UV-A (315–400 nm), UV-B (280–315 nm), and UV-C (100–280 nm) gives higher microbial reduction rate. UV is mainly attributed to the direct DNA damage in living organisms. UV induces the formation of DNA photoproducts, such as cyclobutane pyrimidine dimers and pyrimidine 6–4 pyrimidone, which inhibit transcription and replication and eventually lead to mutagenesis and cell death (Gayan, Condon, & Alvarez, 2014).

4.4.4 Cold plasma technology

Cold plasma (CP) is a novel nonthermal and green food process technology with great potential applications for food preservation or decontamination. Plasma is described as the fourth state of matter following solid, liquid, and gas. It refers to a quasineutral ionized gas, which consists of particles including photons, free electrons, positive or negative ions, excited or nonexcited atoms, and molecules (Fernandez, Shearer, Wilson, & Thompson, 2012). Currently, air, oxygen, nitrogen, or a mixture of noble gases (helium, argon, and neon) are the most frequently used gases in CP. Plasma can be produced by the application of energy in several forms including electricity, lasers, microwaves, radiofrequency, magnetic field, alternating current, and direct current (Thirumdas, Sarangapani, & Annapure, 2015). When the active particles recombine with each other, the energy is released as visible and UV light in the process of recombination. Ramazzina et al. (2015) claimed that CP treatments improved color retention and reduced the darkened area formation of fresh-cut kiwifruit during storage without inducing any textural change compared with the control samples.

5 Future thrusts

Since the texture or firmness of fresh-cut produce has great importance from a consumer acceptability point of view, more attention needs to be given towards use of effective solutions while being acceptable to consumers. Combined effect of texturizers and novel techniques like nonthermal gas plasma, pulsed light, UV-C radiation, and modifications in packaging system to reduce the textural degradation of produce are interesting areas of research. Further use of newer natural extracts, additives, acidulants like citric acid from lime and lemon, coatings like *Aloe vera* gel extract, nopal mucilage extract, etc. can be explored to help reduce the consumption of synthetic chemicals. Fortification of fresh-cut fruits and vegetables can be an interesting avenue to improve nutritional profile while maintaining functionality of texture retention.

6 Conclusion

Use of texturizing agents and treatments alone or in combination with different technologies plays a vital role in maintaining, as well as enhancing, physical attributes of fresh-cut fruits and vegetables. These treatments may act by physically strengthening the structure of cell wall materials or by activating pectin methyl esterase to maintain the structural integrity within the cell wall of the produce. Many treatments to counter microbial growth also serve to keep cell integrity and structure intact, thus maintaining textural integrity of produce. Higher consideration should be given to improve such techniques by adopting safe and protective measures. Care should be taken while selecting texturizers for nutritional enhancement considering bioavailability of the minerals, and interactions of constituents in coatings and texturizers.

References

Alandes, L., Munuera, I. P., Llorca, E., Quiles, A., & Hernando, I. (2009). Use of calcium lactate to improve structure of "Flor de Invierno"fresh-cut pears. *Postharvest Biology and Technology, 53*, 145–151.

Albertini, S., Reyes, A. E. L., Trigo, J. M., Sarriés, G. A., & Spoto, M. A. F. (2016). Effects of chemical treatments on fresh-cut papaya. *Journal of Food Chemistry, 190*, 1182–1189.

Allende, A., Luo, Y., McEvoy, J. L., Artés, F., & Wang, C. Y. (2004). Microbial and quality changes in minimally processed baby spinach leaves stored under super atmospheric oxygen and modified atmosphere conditions. *Postharvest Biology and Technology, 33*, 51–59.

Alzamora, S. M., Salvatori, D., Tapia, M. S., Lopez-Malo, A., Welti- Chanes, J., & Fito, P. (2005). Novel functional foods from vegetable matrices impregnated with biologically active compounds. *Journal of Food Engineering, 67*, 205–214.

Amanatidou, A., Slump, R. A., Gorris, L. G. M., & Smid, E. J. (2000). High oxygen and high carbon dioxide modified atmospheres for shelf-life extension of minimally processed carrots. *Journal of Food Science, 1*, 61–66.

Anino, S. V., Salvatori, D. M., & Alzamora, S. M. (2006). Changes in calcium level and mechanical properties of apple tissue due to impregnation with calcium salts. *Food Research International, 39*, 154–164.

Avalos Llano, K. R., Marsellés-Fontanet, A. R., Martín-Belloso, O., & Soliva-Fortuny, R. (2016). Impact of pulsed light treatments on antioxidant characteristics and quality attributes of fresh-cut apples. *Innovative Food Science and Emerging Technologies, 33*, 206–215.

Avalos-Llano, K. R., Martín-Belloso, O., & Soliva-Fortuny, R. (2018). Effect of pulsed light treatments on quality and antioxidant properties of fresh-cut strawberries. *Food Chemistry, 264*, 393–400.

Biegańska-Marecik, R., & Czapski, J. (2007). The effect of selected compounds as inhibitors of enzymatic browning and softening of minimally processed apples. *Acta Scientiarum Polonorum. Technologia Alimentaria, 6*, 37–49.

Brasil, I. M., Gomes, C., Puerta-Gomez, A., Castell-Perez, M. E., & Moreira, R. G. (2012). Polysaccharide-based multilayered antimicrobial edible coating enhances quality of fresh-cut papaya. *LWT-Food Science and Technology, 47*, 39–45.

Chaiprasart, P., Handsawasdi, C., & Pipattanawong, N. (2006). The effect of chitosan coating and calcium chloride treatment on postharvest qualities of strawberry fruit (fragaria × ananassa). *Acta Horticulturae*. 708, ISHS.

Charles, F., Vidal, V., Olive, F., Filgueiras, H., & Sallanon, H. (2013). Pulsed light treatment as new method to maintain physical and nutritional quality of fresh-cut mangoes. *Innovative Food Science and Emerging Technologies, 18*, 190–195.

Chauhan, O. P., Raju, P. S., Singh, A., & Bawa, A. S. (2010). Shellac and aloe-gel-based surface coatings for maintaining keeping quality of apple slices. *Food Chemistry, 126*, 961–966.

Cortez-Vega, W. R., Pizato, S., de Souza, J. T.-A., & Prentice, C. (2014). Using edible coatings fromWhitemouth croaker (*Micropogonias furnieri)* protein isolate and organo-clay nanocomposite for improve the conservation properties of fresh-cut 'Formosa' papaya. *Innovative Food Science & Emerging Technologies, 22*, 197–202.

Costa, S. B., Cardoso, A., Martins, L. L., Empis, J., & Martins, M. M. (2008). The effect of calcium dips combined with mild heating of whole kiwifruit for fruit slices quality maintenance. *Food Chemistry, 108*, 191–197.

Cybulska, J., Pieczywek, P. M., & Zdunek, A. (2012). The effect of Ca2+ and cellular structure on apple firmness and acoustic emission. *European Food Research and Technology, 235*, 119–128.

Degraeve, P., Saurel, R., & Coutel, Y. (2003a). Vacuum-impregnation pre-treatment to improve firmness of pasteurized fruits. *Journal of Food Science, 68*, 716–721.

Degraeve, P., Saurel, R., & Coutel, Y. (2003b). Vacuum impregnation pretreatment with pectin methylesterase to improve firmness of pasteurized fruits. *Journal of Food Science, 68*, 716–721.

Denès, J. M., Baron, A., & Drilleau, J. F. (2000). Purification, properties and heat inactivation of pectin methylesterase from apple (cv. Golden delicious). *Journal of the Science of Food and Agriculture, 80*(10), 1503–1509.

Deng, Y., Wu, Y., & Li, Y. F. (2005). Effects of high O_2 levels on post-harvest quality and shelf life of table grapes during long-term storage. *European Food Research and Technology, 221*, 392–397.

Dong, X., Wrolstad, R. E., & Sugar, D. (2000). Extending shelf life of fresh-cut pears. *Journal of Food Science, 65*(1), 181–186.

Fernandez, A., Shearer, N., Wilson, D. R., & Thompson, A. (2012). Effect of microbial loading on the efficiency of cold atmospheric gas plasma inactivation of Salmonella enterica serovar Typhimurium. *International Journal of Food Microbiology, 152*, 175–180.

Fito, P., Chiralt, A., Betoret, N., Gras, M., Chafer, M., Martinez-Monzo, J., et al. (2001). Vacuum impregnation and osmotic dehydration in matrix engineering: Application in functional freshfood development. *Journal of Food Engineering, 49*, 175–183.

Garcia, J. M., Herrera, S., & Morilla, A. (1996). Effects of postharvest dips in calcium chloride on strawberry. *Journal of Agriculture and Food Chemistry, 44*(1), 30–33.

García-Ramos, F. J., Valero, C., Homer, I., Ortiz-Cañavate, J., & Ruiz-Altisent, M. (2005). Nondestructive fruit firmness sensors: A review. *Spanish Journal of Agricultural Research, 3*(1), 61–73.

Gayan, E., Condon, S., & Alvarez, I. (2014). Biological aspects in food preservation by ultraviolet light: A review. *Food and Bioprocess Technology, 7*, 1–20.

Ghidellia, C., Mateos, M., Rojas-Argudoa, C., & Pérez-Gago, M. B. (2014). Extending the shelf life of fresh-cut eggplant with a soy protein–cysteine based edible coating and modified atmosphere packaging. *Postharvest Biology and Technology, 95*, 81–87.

Giacalone, G., & Chiabrando, V. (2013). Effect of different treatments with calcium salts on sensory quality of fresh-cut apple. *Journal of Food and Nutrition Research, 52*(2), 79–86.

Gómez-López, V. M., Ragaert, P., Debevere, J., & Devlieghere, F. (2007). Pulsed light for food decontamination: A review. *Trends in Food Science and Technology, 18* (9), 464–473.

Gras, M. L., Vidal, D., Betoret, N., Chiralt, A., & Fito, P. (2003). Calcium fortification of vegetables by vacuum impregnation. Intercations with cellular matrix. *Journal of Food Engineering, 56*, 279–284.

Guillemin, A., Degraeve, P., Noel, C., & Saurel, R. (2008). Influence on impregnation solution viscosity and osmolarity on solute uptake during vacuum impregnation of apple cubes (var. Granny Smith). *Journal of Food Engineering, 86*, 475–483.

Hassan, B., Ali, S., Chatha, S., Hussain, A. I., Zia, K. M., & Akhtar, N. (2018). Recent advances on polysaccharides, lipids and protein based edible films and coatings: a review. *International Journal of Biological Macromolecules, 109*, 1095–1107.

Hu, Q., Fang, Y., Yang, Y., Ma, N., & Zhao, L. (2011). Effect of nanocomposite-based packaging on postharvest quality of ethylene-treated kiwifruit (Actinidia deliciosa) during cold storage. *Food Research International, 44*(6), 1589–1596.

Javeri, R. H., Toledo, R., & Wicker, L. (1991). Vacuum infusion of citrus pectin methylesterase and calcium effects on firmness of peaches. *Journal of Food Science, 56*, 739–742.

Jurick, W. M., Vico, I., McEvoy, J. L., Whitaker, B. D., Janisiewicz, W., & Conway, W. S. (2009). Isolation, purification, and characterization of a polygalacturonase produced in Penicillium solitum-decayed 'Golden Delicious' apple fruit. *Phytopathology, 99*(6), 636–641.

Kim, S., & Klieber, A. (1997). Quality maintenance of minimally processed Chinese cabbage with low temperature and citric acid dip. *Journal of the Science of Food and Agriculture, 75*(1), 31–36.

Koh, P. C., Noranizan, M. A., Nur Hanani, Z. A., Karim, R., & Rosli, S. Z. (2017). Application of edible coatings and repetitive pulsed light for shelf life extension of fresh-cut cantaloupe (Cucumis melo L. reticulatus cv. Glamour). *Postharvest Biology and Technology, 129*, 64–78.

Lovera, N., Ramallo, L., & Salvadori, V. (2014). Effect of processing conditions on calcium content, firmness, and color of papaya in syrup. *Journal of Food Processing, 2014.* 8 pp.

Luna-Guzmán, I., & Barrett, D. M. (2000). Comparison of calcium chloride and calcium lactate effectiveness in maintaining shelf stability and quality of fresh-cut cantaloupes. *Postharvest Biology and Technology, 19*, 61–72.

Lurie, S., Fallik, E., & Klein, J. D. (1996). The effect of heat treatment on apple epicuticular wax and calcium uptake. *Postharvest Biology and Technology, 8*(4), 271–277.

Ma, L., Zhang, M., Bhandari, B., & Gao, Z. (2017). Recent developments in novel shelf life extension technologies of fresh-cut fruits and vegetables. *Trends in Food Science and Technology, 64*, 23–38.

Manganaris, G. A., Vasilakakis, M., Diamantidis, G., & Mignani, I. (2007). The effect of postharvest calcium application on tissue calcium concentration, quality attributes, incidence of flesh browning and cell wall physicochemical aspects of peach fruits. *Journal of Food Chemistry, 4*, 1385–1392.

Manolopoulou, E., & Varzakas, T. (2011). Effect of storage conditions on the sensory quality, colour and texture of fresh-cut minimally processed cabbage with the addition of ascorbic acid, citric acid and calcium chloride. *Journal of Food and Nutrition Sciences, 2*, 956–963.

Manzocco, L., Da Pieve, S., & Maifreni, M. (2011). Impact of UV-C light on safety and quality of fresh-cut melon. *Innovative Food Science and Emerging Technologies, 12*, 13–17.

Marquez, G. R., Di Pierro, P., Mariniello, L., Esposito, M., Giosafatto, C. V. L., & Porta, R. (2016). Effects of GA3 and divalent cations on aspects of pectin metabolism and tissue softening in fresh-cut fruit and vegetable coatings by transglutaminasecrosslinked whey protein/pectin edible films. *LWT- Food Science and Technology, 75*(2017), 124–130.

Martin-Diana, A. B., Rico, D., Barry-Ryan, C., Frias, J. M., Mulcahy, J., & Henehan, G. T. M. (2005). Comparison of calcium lactate with chlorine as a washing treatment for fresh-cut lettuce and carrots: Quality and nutritional parameters. *Journal of the Science of Food and Agriculture, 85*, 2260–2268.

Martin-Diana, A. B., Rico, D., Frı'as, J. M., Henehan, G. T. M., Mulcahy, J., Barat, J. M., et al. (2006). Effect of calcium lactate and heat-shock on texture in fresh-cut lettuce during storage. *Journal of Food Engineering, 77*(4), 1069–1077.

Martin-Diana, A., et al. (2007). Calcium for extending the shelf life of fresh whole and minimally processed fruits and vegetables: A review. *Trends in Food Science and Technology, 18*(4), 210–218.

Martínez-Romero, D., Castillo, S., Guilléna, F., Díaz-Mula, H. M., Zapata, P. J., Valero, D., et al. (2013). *Aloe vera* gel coating maintains quality and safety of ready-to-eat pomegranate arils. *Postharvest Biology and Technology, 86*, 107–112.

Martiñon, M. E., Moreira, R. G., Elena Castell-Perez, M., & Gomes, C. (2014). Development of a multilayered antimicrobial edible coating for shelflife extension of fresh-cut cantaloupe (*Cucumis melo* L.) stored at 4°C. *LWT-Food Science and Technology, 56*, 341–350.

Meilgaard, M. C., Civille, G. V., & Carr, B. T. (2006). *Sensory evaluation techniques* (4th ed.). Boca Raton, FL: CRC Press.

Meng, X., Zhang, M., & Adhikari, B. (2014). The effects of ultrasound treatment and nano-zinc oxide coating on the physiological activities of fresh-cut kiwifruit. *Food and Bioprocess Technology, 7*, 126–132.

Mignani, I., Greve, L. C., Ben-Arie, R., Stotz, H. U., Li, C., Shackel, K. A., et al. (1995). *The quality of fruits and vegetables: Fresh fruits and vegetables* (2nd ed.). Boca Raton, FL: CRC Press.

Mohebbi, M., Ansarifar, E., Hasanpour, N., & Amiryousefi, M. R. (2011). Suitability of aloe vera and gum tragacanth as edible coatings for extending the shelf life of button mushroom. *Food and Bioprocess Technology, 5*(8), 3193–3202.

Ngamchuachit, P., Sivertsen, H. K., Mitcham, E. J., & Barrett, D. M. (2014). Effectiveness of calcium chloride and calcium lactate on maintenance of textural and sensory qualities of fresh-cut mangoes. *Journal of Food Science, 79*, C786–C794.

Occhino, E., Hernando, I., Liorca, E., Neri, L., & Pittia, P. (2011). Effect of vacuum impregnation treatments to improve quality and texture of zucchini (*Cucurbita pepo*, L). *Procedia Food Science, 1*, 829–835.

Oms-Oliu, G., Raybaudi-Massilia, R. M., Soliva-Fortuny, R., & Martín-Belloso, O. (2008). Effect of superatmospheric and low oxygen modified atmospheres on shelf-life extension of fresh-cut melon. *Food Control, 19*, 191–199.

Ortiz, A., Graell, J., & Lara, I. (2010). Cell wall-modifying enzymes and firmness loss in ripening 'Golden Reinders' apples: A comparison between calcium dips and ULO storage. *Food Chemistry, 128,* 1072–1079.

Pascall, A., & Lin, S. J. (2013). The application of edible polymeric films and coatings in the food industry. *Journal of Food Processing & Technology, 4*(2).

Picchioni, G. A., Watada, A. E., Whitaker, B. D., & Reyes, A. (1996). Calcium delays senescence-related membrane lipid changes and increases net synthesis of membrane lipid components in shredded carrots. *Postharvest Biology and Technology, 9,* 235–245.

Ponappa, T., Scheerens, J. C., & Miller, A. R. (1993). Vacuum infiltration of polyamines increases firmness of strawberry slices under various storage conditions. *Journal of Food Science, 58,* 361–364.

Poverenov, E., Danino, S., Horev, B., Granit, R., Vinokur, Y., & Rodov, V. (2014). Layer-by-layer electrostatic deposition of edible coating on fresh cut melon model: Anticipated and unexpected effects of alginate–chitosan combination. *Food and Bioprocess Technology, 7,* 1424–1432.

Qadri, O. S., Yousuf, B., & Srivastava, A. K. (2015). Fresh-cut fruits and vegetables: Critical factors influencing microbiology and novel approaches to prevent microbial risks—A review. *Cogent Food & Agriculture, 1,* 112–160.

Quiles, A., Hernando, I., Pérez-Munuera, I., & Lluch, M. A. (2007). Effect of calcium propionate on the microstructure and pectin methylesterase activity in the parenchyma of fresh-cut Fuji apples. *Journal of the Science of Food and Agriculture, 87,* 511–519.

Radziejewska-Kubzdela, E., Biegańska-Marecik, R., & Kidoń, M. (2014). Applicability of vacuum impregnation to modify Physico-chemical, sensory and nutritive characteristics of plant origin products—A review. *International Journal of Molecular Sciences, 15*(9), 16577–16610. Sep.

Ramazzina, I., Berardinelli, A., Rizzi, F., Tappi, S., Ragni, L., & Sacchetti, G. (2015). Effect of cold plasma treatment on physico-chemical parameters and antioxidantactivity of minimally processed kiwifruit. *Postharvest Biology and Technology, 107,* 55–65.

Ranjithaa, K., Raoa, S. D. V., Shivashankara, K. S., Oberoia, H. S., Roy, T. K., & Bharathamma, H. (2017). Shelf-life extension and quality retention in fresh-cut carrots coated with pectin. *Innovative Food Science & Emerging Technologies, 42,* 91–100.

Rico, D., Martin-Diana, A. B., Frias, J. M., Henehan, G. T. M., & Barry-Ryan, C. (2007). Effect of ozone and calcium lactate treatments on browning and texture properties of fresh-cut lettuce. *Journal of Science and Food Agriculture, 86,* 2179–2188.

Saba, M. K., & Sogvar, O. B. (2016). Combination of carboxymethyl cellulose-based coatings with calcium and ascorbic acid impacts in browning and quality of fresh-cut apples. *LWT-Food Science and Technology, 66,* 165–171.

Salunke, D. K., Bolin, H. R., & Reddy, N. R. (1991). *Storage processing and nutritional quality of fruits and vegetables: Fresh fruits and vegetables* (2nd ed.). Boca Raton, FL: CRC Press.

Sams, C. E., Conway, S. W., Abbott, J. A., Lewis, R. J., & Ben-Shalom, N. (1993). Firmness and decay of apples following postharvest pressure infiltration of calcium and heat treatment. *Journal of the American Society for Horticultural Science, 118,* 623–627.

Sanchísa, E., González, S., Ghidellia, C., Sheth, C. C., Mateos, M., Paloua, L., et al. (2016). Browning inhibition and microbial control in fresh-cut persimmon (Diospyros kaki Thunb. cv. Rojo Brillante) by apple pectin-based edible coatings. *Postharvest Biology and Technology, 112,* 186–193.

Shih-Ying Pan, S. Y., Chen, C.-H., & Lai, L.-S. (2013). Effect of tapioca starch/decolorized Hsian-tsao leaf gum-based active coatings on the qualities of fresh-cut apples. *Food and Bioprocess Technology, 6*, 2059–2069.

Siddiqui, M. W., Chakraborty, I., Zavala, A., & Dhua, R. S. (2011). Advances in minimal processing of fruits and vegetables. *Journal of Scientific and Industrial Research, 70*, 823–834.

Silveira, A. C., Aguayo, E., Chisari, M., & Artés, F. (2011). Calcium salts and heat treatment for quality retention of fresh-cut 'Galia' melon. *Postharvest Biology and Technology, 62*, 77–84.

Soliva-Fortuny, R. C., & Martin-Belloso, O. (2003). New advances in extending the shelf life of fresh-cut fruits: A review. *Trends in Food Science and Technology, 14*, 341–353.

Song, H. E., Jo, W.-S., Song, N.-B., Min, S. C., & Song, K. B. (2013). Quality change of apple slices coated with *Aloe vera* gel during storage. *Journal of Food Science, 78* (6), C817–C822.

Suutarinen, J., Honkapaa, K., Heinio, R. L., Autio, K., & Mokkila, M. (2000). The effect of different prefreezing treatments on the structure of strawberries before and after jam making. *LWT-Food Science and Technology, 33*, 188–201.

Thirumdas, R., Sarangapani, C., & Annapure, U. S. (2015). Cold plasma: A novel nonthermal technology for food processing. *Food Biophysics, 10*, 1–11.

Toivonen, P. M. A., & Brummell, D. A. (2008). Biochemical bases of appearance and texture changes in fresh-cut fruit and vegetables. *Postharvest Biology and Technology, 48*, 1–14.

Trindade, P., Abreu, M., Gonc_alves, E. M., Beirao-da Costa, S., Beirao-da-Costa, M. L., & Moldao-Martins, M. (2003). The effect of heat treatments and calcium chloride applications on quality of fresh-cut mango. *Acta Horticulturae (ISHS), 599*, 603–609.

Valdivia-Nájar, G., Martín-Belloso, O., & Soliva-Fortuny, R. (2018). Impact of pulsed light treatments and storage time on the texture quality of fresh-cut tomatoes C. *Innovative Food Science & Emerging Technologies, 45*, 29–35.

Valencia-Chamorro, S. A., Palou, L., Del Río, M. A., & Pérez-Gago, M. B. (2011). Performance of hydroxypropyl methylcellulose (HPMC)-lipid edible coatings with antifungal food additives during cold storage of 'Clemenules' mandarins. *LWT-Food Science and Technology, 44*, 2342–2348.

Van Buggenhout, S., Messagie, I., Maes, V., Duvetter, T., van Loey, A., & Hendrickx, M. (2006). Minimizing texture loss of frozen strawberries: Effect of infusion with pectin methylesterase and calcium combined with different freezing conditions and effect of subsequent storage/thawing conditions. *European Food Research and Technology, 223*, 395–404.

Vargas, M., Chiralt, A., Albors, A., & Gonzalez-Martinez, C. (2009). Edible coatings applied by vacuum impregnation on quality preservation of fresh-cut carrot. *Postharvest Biology and Technology, 51*(2), 263–271.

Waghmare, R. B., & Annapure, U. S. (2013). Combined effect of chemical treatment and/or modified atmosphere packaging (MAP) on quality of fresh-cut papaya. *Postharvest Biology and Technology*, 147–153.

Xiao, Z., Luo, Y., & Wang, Q. (2011). Combined effects of sodium chlorite dip treatment and chitosan coatings on the quality of fresh-cut d'Anjou pears. *Postharvest Biology and Technology, 62*, 319–326.

Yousuf, B., & Srivastava, A. K. (2015). Psyllium (*Plantago*) gum as an effective edible coating to improve quality and shelf life of fresh-cut papaya (*Carica papaya*). *International Journal of Biological, Biomolecular, Agricultural, Food and Biotechnological Engineering, 9*(7).

Yousuf, B., & Srivastava, A. K. (2017). A novel approach for quality maintenance and shelf life extension of fresh-cut Kajari melon: Effect of treatments with honey and soy protein isolate. *LWT-Food Science and Technology, 79,* 568–578.
Yousuf, B., Qadri, O. S., & Srivastava, A. K. (2018). Recent developments in shelf-life extension of fresh-cut fruits and vegetables by application of different edible coatings: A review. *Lwt, 89,* 198–209.
Zambrano-Zaragoza, M. L., Gutiérrez-Cortez, E., Real, A. D., González-Reza, R. M., Galindo-Pérez, M. J., & Quintanar-Guerrero, D. (2014). Fresh-cut red delicious apples coating using tocopherol/mucilage nanoemulsion: Effect of coating on polyphenol oxidase and pectin methylesterase activities. *Food Research International, 62,* 974–983.
Zhao, L., Liu, L., & Ma, Y. (2009). Preservation of apricot by chitosan nano-ZnO film. *Food Research and Development, 30*(2), 126–128.

Further reading

Chulaki, M. M., Pawar, C. D., & Khan, S. M. (2017). Influence of ascorbic acid and calcium chloride on physical parameters and microbial count of firm flesh jackfruit bulbs during refrigerated storage. *International Journal of Current Microbiology and Applied Sciences, 6*(10), 1101–1111.
Danza, A., Conte, A., Mastromatteo, M., & Nobile, M. A. L. (2015). A new example of nanotechnology applied to minimally processed fruit: The case of fresh-cut melon. *Journal of Food Processing & Technology, 6,* 4.
Guerreiro, A. C., Gago, C. O., Faleiro, M. L., Miguel, M. G. C., & Antunes, M. D. C. (2017). The effect of edible coatings on the nutritional quality of 'Bravo de Esmolfe' fresh-cut apple through shelf-life. *LWT-Food Science and Technology, 75,* 210–219.
Heinrich, V., Zunabovic, M., Bergmair, J., Kneifel, W., & Jeager, H. (2015). Post-packaging application of pulsed light for microbial decontamination of solid foods: A review. *Innovative Food Science and Emerging Technologies, 30,* 145–156.
Lamikanra, O., & Watson, M. A. (2004). Effect of calcium treatment temperature on fresh-cut cantaloupe melon during storage. *Journal of Food Science, 69*(6), 468–472.
Lima, M. M., Tribuzi, G., De Souza, J. A., Laurindo, J. B., & Carciofi, B. A. M. (2016). Vacuum impregnation and drying of calcium-fortified pineapple snack. *LWT-Food Science and Technology, 72,* 501–509.
Mangaraj, S., Goswami, T. K., & Mahajan, P. V. (2009). Applications of plastic films for modified atmosphere packaging of fruits and vegetables: A review. *Food Engineering Reviews, 1,* 133–158.
Martin-Diana, A. B., Rico, D., Frias, J., Mulcahy, J., Henehan, G. T. M., & Barry-Ryan, C. (2007). Effect of calcium lactate on quality, safety and nutritional senescence parameters of minimally processed vegetables. *Acta horticulturae,* 687.
Porta, R., Marquez, G. R., Mariniello, L., Sorrentino, A., Giosafatto, C. V. L., Esposito, M., et al. (2013). Edible coating as packaging strategy to extend the shelf-life of fresh-cut fruits and vegetables. *Journal of Biotechnology and Biomaterials, 3,* 4.
Raybaudi-Massilia, R., Vásquez, F., Reyes, A. M., Graciela Troncone, G., & Tapia, M. S. (2015). Novel edible coating of fresh-cut fruits: Application to prevent calcium and vitamin D deficiencies in children. *Journal of Scientific Research and Reports, 6*(2), 142–156.

Roca, B. S., Fortuny, R. S., Welti-Chanes, J., & Martín-Belloso, O. (2016). Combined effect of pulsed light, edible coating and malic acid dipping to improve fresh-cut mango safety and quality. *Food Control, 66,* 190–197.

Silvana Albertini, S., Reyes, A. E. L., Juliana Moreno Trigo, J. M., Sarriés, G. A., & Spoto, M. H. F. (2016). Effects of chemical treatments on fresh-cut papaya. *Food Chemistry, 190,* 1182–1189.

Valencia, C., Gago, S. A., Río, D., & Palou, L. (2011). Antimicrobial edible films and coatings for fresh and minimally processed fruits and vegetables: A review. *Critical Reviews in Food Science and Nutrition, 51,* 872–900.

Xu-Shena, Zhanga, M., Devahastind, S., & Guoe, Z. (2019). Effects of pressurized argon and nitrogen treatments in combination with modified atmosphere on quality characteristics of fresh-cut potatoes. *Postharvest Biology and Technology, 149,* 159–165.

Zhao, Y., Park, S.-i., Leonard, S. W., & Traber, M. G. (2005). Vitamin E and mineral fortification in fresh-cut apples (Fuji) using vacuum impregnation. *Nutrition & Food Science, 35*(6), 393–402.

7

Modified and controlled atmosphere packaging

Luciana de Siqueira Oliveira, Kaliana Sitonio Eça, Andréa Cardoso de Aquino, Larissa Morais Ribeiro da Silva
Department of Food Engineering, Federal University of Ceará, Fortaleza, Brazil

1 Introduction

Minimally processed fruits and vegetables have gained great popularity world-wide due to the increasing awareness of consumers about its health-benefit, sensorial and nutritional quality, besides the advantages of its convenience while still maintaining freshness, nutrition and flavor which have allowed its significant economic importance in the fruit and vegetable processing industry. However, the processing operations applied to fresh-cut produce, such as peeling, dicing, slicing, and shredding cause damage (surface browning, off-flavor, softening) that are responsible for shortening its shelf-life.

The maintenance of fresh-cut fruit and vegetable quality through its preservation and shelf-life extension has been a major concern over these products because the tissue wounding during preparation accelerates respiration and transpiration rates; this makes fresh-cut produce susceptible to increased quality deterioration and physiological stress, decreasing the shelf-life during storage. Therefore, it is necessary that technologies for the preservation and maintenance of the quality of fresh products are employed in order to obtain safe products with high taste and nutritional quality along with long shelf-life.

Modified and controlled atmosphere packaging are traditional physical technologies for fresh-cut fruits and vegetable preservation. Modified atmosphere packaging (MAP) has commonly and successfully been used for extending the shelf-life of fresh-cut fruit and vegetables. This effective technology modifies the gas composition surrounding the products, or replaces it with inactive gases, before sealing in containers or packages, helping to

Fresh-cut Fruits and Vegetables. https://doi.org/10.1016/B978-0-12-816184-5.00007-0

increase shelf-life and maintain quality due to decreasing the fresh product's respiration rate (Ma, Zhang, Bhandari, & Gao, 2017; Yousuf, Qadria, & Srivastava, 2018). However, during storage of fresh products in MAP, an alteration of gas composition because of reactions between components of the atmosphere and the product may be observed; this may also occur by the leakage of gases into or out of the package, whereas the atmosphere within controlled atmosphere (CA), although initially modified is maintained invariant during storage (Yang & Whang, 2014).

In this chapter, modified and controlled atmosphere packaging applied to fresh-cut fruit and vegetables is discussed as a promising technology, which may allow preservation and a better maintenance of minimally processed produce quality and safety. In addition, the effects of these shelf-life extension physical technologies on quality of fresh-cut produce are also presented and assayed from physicochemical properties, nutritional, and microbiological aspects of the fresh-cut produce.

2 Modified atmosphere packaging

Fresh-cut fruits and vegetables quality is evaluated based on sensorial (appearance, color, flavor, and texture), nutritional (vitamins, phytochemicals, minerals) and safety aspects (absence of defects and microorganisms). Changes of these properties can strongly influence the shelf-life of the product and consumers' acceptance. Browning, decaying, softening, loss of moisture, lignin formation, chlorophyll degradation, and fermentation are some causes of loss of quality of some fresh-cut produce (Finnegan & O'Beirne, 2015). They are the result of compound interactions (polyphenol oxidase with the phenols causing browning), enzymatic degradation (pectins catalyzed by pectin methylesterase and polygalacturonase resulting in loss of firmness), chemical degradation (transformation volatile oils, aliphatic esters, and decrease of physicochemical compounds), or microbial contamination. These modifications could be easily observed in a wide variety of fruits and vegetables like apple, pear, mango, watermelon, potato, lettuce, eggplant, onion, and green chili (Amaro, Beaulieu, Grimm, Stein, & Almeida, 2012; Arnal & Del Río, 2003; Chitravathi, Chauhan, & Raju, 2015; Costa et al., 2011).

Many types of food preservation technologies have been applied to control these disorders, and they are classified as physical-based preservation, chemical-based preservation, and bio preservation technology, according to their action mechanism. Modified atmosphere packaging (MAP) is a physical-based preservation technology that promotes the product's shelf-life extension by the adjustment of gas composition. In other words, the gas composition surrounding the product, in containers or packages, can be modified or replaced with inactive gases before sealing (Mangaraj, Goswami, & Mahajan, 2009).

The determination of the gas composition and concentration (MAP systems) involves understanding the interplay between gas and film (gas permeability in packages), the integrated knowledge of product characteristics (respiration and transpiration rate, for example), and atmospheric conditions for the given product (Sandhya, 2010; Sousa-Gallagher & Mahajan, 2013). The result of these combined parameters will be establishment of an equilibrium-modified atmosphere and the shelf-life of the product will probably increase (Reinas, Oliveira, Pereira, Mahajan, & Poças, 2016). To achieve this benefit, it is essential to consider factors as respiration and transpiration rates, gas permeability properties in the package, storage conditions (time, temperature, and relative humidity), product weight, and heat and mass transfer to determine the MAP properly (Chitravathi et al., 2015; González-Buesa, Ferrer-Mairal, Oria, & Salvador, 2009; Sahoo, Bal, Pal, & Sahoo, 2015).

2.1 Conventional gases (O_2, CO_2, and N_2)

The effects of the application of MAP in fruit quality have been extensively investigated since the 1980s, and, nowadays, in the market there is a wide variety of fresh-cut produce available using this technology (Gorny, 1997). It has been considered widely for fresh-cut products application because of proven efficiency for extension of shelf-life. However, the combination of MAP with storage under refrigeration is essential to maintain freshness, quality, and nutritional aspects, and to ensure microbial safety of fresh-cut fruits and vegetables (Sandhya, 2010).

The most frequently used gases in MAP for fresh-cut products are CO_2, O_2, and N_2, and they are considered a suitable choice because of their capacity to preserve the overall postharvest quality of fruits and vegetables. The reason to use these convectional gases (CO_2, O_2, and N_2) is that they are directly involved in the fresh-cut fruits and vegetables' metabolism once these produces keep respiring (consuming O_2/producing CO_2) after harvest; therefore, it is well proven that MAP will affect enzymatic activity and microbial growth preventing the loss quality (Al-Ati & Hotchkiss, 2003).

There have been MAP systems proposed to control the negative effects caused by processing, as well as preventing or retarding the ripening associated with biochemical and physiological changes, by different action mechanisms (Sandhya, 2010).

Low and high concentration of oxygen can control microbial growth in many ways. High oxygen MAP is used when the aim is to inhibit the anaerobes growth; however, it can promote the growth of aerobic bacteria and molds, depending on the concentration at which it is applied. On the other hand, the use of low levels of oxygen (<5%) can impede aerobic bacterial growth, and consequently, favor the growth of anaerobes, and also can control respiration rate, and slow down senescence for concentrations <10% (Jacxsens, Devlieghere, Van der Steen, & Debevere, 2001). The concentration of oxygen in MAP should be designed correctly, as this gas is essential to maintain aerobic respiration and avoid anaerobic respiration. If this MAP system is not carefully thought out, the resulting anaerobic respiration could cause the accumulation of compounds (ethanol, acetaldehyde, and organic acids) and the deterioration of sensory aspects (Kontominas, 2014).

MAP with high oxygen concentration, between 50% and 90%, can inhibit or control enzymatic activity, mainly enzymatic browning, and the growth of certain microorganisms; yeast, for example (Jacxsens et al., 2001; Lopez-Gálvez et al., 2015). Furthermore, depending on the concentration, it can also prevent moisture loss, changes in color, and the decay of fresh-cut fruits and vegetables as a result of the effects cited before. To ensure the effect of MAP, low temperature is required and package should have appropriate oxygen permeability.

Some studies have shown relevant results for high-oxygen MAP when compared to control samples and low-oxygen MAP on shelf-life of fresh-cut fruits and vegetables. Jacxsens et al. (2001) could double overall shelf-life of grated celeriac, shredded chicory endives, and mushroom slices, under refrigeration (4°C). Although use of high-oxygen MAP compared with low-oxygen MAP (3% of O_2 plus N_2) showed advantages in terms of browning, the accumulation of ethylene and development of russet spotting in lettuce was observed (Lopez-Gálvez et al., 2015). Another important characteristic that can be preserved by use of high-oxygen MAP are some aroma volatiles, like acetate esters (Amaro et al., 2012) from melon for 12 days, under refrigeration (5°C). In this study, is was also observed that inhibition of wall enzymes activities resulted in maintenance of physical characters of cell membrane, as well as low degradation of phenolic compounds when high-oxygen was combined with ascorbic acid.

Considering that the main objective of MAP application is to obtain high-quality fruit for a longer period of storage and extension of marketing life, many different gas compositions can be used; the choice will depend on the produce's peculiarities. Two methods could be used: keeping the product under anaerobic conditions or exposing it to conditions that enhance respiration (Arnal & Del Río, 2003). If the main point is to get maximum results for appearance, nitrogen treatment is indicated, whereas for keeping flavor and aromatic compounds, carbon dioxide treatment has shown good results, for example (Bibi, Khattak, & Mehmood, 2007).

Nitrogen is an inert gas used to replace, partially or totally, the oxygen content of the atmosphere that would be consumed by the product in the respiration process. It can be also combined with high levels of oxygen to act against the growth of aerobic microorganisms (Kontominas, 2014). Ahmed, Yousef, and Sarrwy (2011) verified that the use of high nitrogen MAP (100% of N_2) on persimmon combined with refrigeration promotes an extension of the storage time and also at room temperature good results were achieved. Furthermore, it highlighted the effect of packaging thickness and composition on gas permeability, and their results suggested that the highest thickness of Low Density Polyethylene film (7 μm) is a suitable option for nitrogen MAP.

In the vegetables' metabolism, the rate of respiration is considered a good parameter to understand their storage shelf-life. For MAP systems, the reduction of O_2 partial pressure and/or the increase of CO_2 partial pressure would result in a deceleration of the metabolism, decreasing the rate of respiration, as well as less sugar oxidation, which can maintain the product quality and extend shelf-life (Costa et al., 2011; Kontominas, 2014; Sen, Mishra, & Srivastav, 2012). However, when the respiratory quotient (ratio of CO_2 produced by O_2 consumed) is higher than 1.3, it can be an indicator of an anaerobic respiration happening and the result of this is an acidic metabolic process, which can be associated with the loss of freshness (Conesa, Verlinden, Artes-Hernandez, Nicolai, & Artes, 2007).

Among the MAP conventional gases (O_2, CO_2, and N_2), CO_2 is consider the principal in terms of antimicrobial effect, because of its broader antimicrobial action (bacteriostatic and fungistatic). Considering that fungi are the most active group of microorganism that deteriorate vegetables and fruit, CO_2 MAP can be a good choice. This group of fungi includes *Alternaria, Botrytis, Fusarium, Aspergillus, Penicillium, Rhizopus, Monilina,* and *Sclerotinia.* The combination of reduced storage temperatures and CO_2

concentrations (>20%) also has an effect on the reduction of growth of some bacteria, such as *Erwinia* and *Xanthomonas*, however this effect is limited or does not extend to lactic acid bacteria, *Clostridium perfringens, and Clostridium botulinum* (Kontominas, 2014). An appropriate concentration of CO_2 can also control the rate of fresh produce respiration and the rate of chemical oxidation through the reduction of the oxygen concentration within the package (Kontominas, 2014).

Another alternative that has been studied for MAP is the use of inactive or noble gases, such as argon (Ar), helium (He), and nitrous oxide (N_2O). Although they have the capacity to control negative effects of metabolic mechanisms after processing without directly affect the plant tissues, they may increase the diffusivity of other gases (O_2, C_2H_4, and CO_2) that could influence the mitochondria activity (Gorny & Agar, 1998; Sowa & Towill, 1991).

2.2 Unconventional gases

Recently, the use of unconventional atmospheres has gained considerable interest in fruits and fruits products conservation, being involved in several studies in the literature using different gases, like argon (Ar)-, nitrous oxide (N_2O), helium (He), and superatmospheric oxygen (O_2) (Table 1) in modified atmosphere packaging (MAP).

Table 1 Studies with modified atmosphere unconventional gases

Food product	Atmosphere	Reference
Fresh-cut apples	Ar and N_2O	Cortellino, Gobbi, Bianchi, and Rizzolo (2015)
Arugula (*Eruca vesicaria* Mill.)	He, O_2, and N_2	Inestroza-Lizardo, Silveira, and Escalona (2016)
Minimally processed kiwifruit	Ar and N_2O	Rocculi, Romani, and Dalla Rosa (2005)
Fresh-cut watercress	N_2 and Ar	Pinela et al. (2016)
Sardine (*Sardina pilchardus*) fillets	Ar and N_2O	Pinheiro et al. (2019)
Fresh-cut watercress	Ar, He, and N_2O	Silveira, Araneda, Hinojosa, and Escalona (2014)
Strawberries	Superatmospheric oxygen	Van de Velde et al. (2019)
Fresh-cut potatoes	Ar	Shen, Zhang, Devahastin, and Guo (2019)
Sweet cherries	Ar	Yang, Zhang, Wang, and Zhao (2019)
Fresh-cut cucumber	Air, N_2, and Ar	Olawuyi, Park, Lee, and Lee (2019)

When properly used, this technology may preserve and extend the quality of food better than conventional atmosphere, allowing a longer period for commercialization (Pinela et al., 2016), presenting benefits to producers and consumers, with a view to reducing postharvest losses of fruits, preserving bioactive and sensorial constituents, and reducing economic losses. Their commercial use for packaging foods needs approval from the regulatory authorities.

Some unconventional gases can exert an effect on the metabolic activity of various vegetable products through unknown mechanisms.

Ar is a major component of the atmosphere inside packaging and can be considered a biochemically active gas, presenting enhanced solubility in water, and appearing to interfere with enzymatic oxygen receptor sites, reducing metabolic activity of vegetable (Char et al., 2012), improving the quality of fresh produce. This atmosphere has also been associated with microbial growth reduction (Jamie & Saltveit, 2002), which has a direct influence on the final quality of the fresh product.

Pinela et al. (2016) studying watercress samples storage in different atmospheres observed that Ar-enriched MAP was the best option to preserve the DPPH scavenging activity and b-carotene blanching inhibition capacity, as well as to increase the total phenolic content. In this research, Ar-enriched MAP was the most suitable choice to preserve the overall postharvest quality of fresh-cut watercress.

Pinheiro et al. (2019) studied the effect of modified atmosphere packaging (MAP) with unconventional gas mixtures on the parameters of sardine fillets during refrigerated storage and found the Ar atmosphere resulted in the best MAP condition adopted. Similar behavior was found by Olawuyi et al. (2019) for fresh-cut cucumber, with tissue respiration, physiological changes, and chlorophyll degradation retarding.

The use of mixed unconventional, or with conventional gases, may provide a better food conservation than when the gases are used separately. Shen et al. (2019) studied the use of argon and nitrogen atmospheres and concluded the combination of pressurized argon or mixed gas treatment with modified atmosphere is an effective method to maintain the quality of fresh-cut potatoes during refrigerated storage. A similar result was obtained for Yang et al. (2019) for sweet cherries.

The growing demand to decrease postharvest use of chemicals and the need for more sustainable technologies has led to the development of improved MAP storage methods (Falagán & Terry, 2018). Research concerning the use of helium (He), argon

(Ar) or xenon (Xe), neon (Ne), nitrous oxide (N_2O), sulfur dioxide, or superatmospheric oxygen (O_2) can be considered more and more common in literature, suggesting the potential use of these gases in food preservation, yet considered widely used commercially.

The use of He atmospheres increase O_2 diffusion, decreasing the concentration gradient between the inside and outside of the cell, which maintains ultra-low O_2 concentrations, minimizing the risk of fermentation (Day, 1998).

N_2O has a chemical structure similar to that of CO_2, providing advantageous physical properties (Gouble, Fath, & Soudain, 1995), being able to decrease the respiratory rate of fruits, and thus the release of ethylene and maturation.

Silveira et al. (2014) studied the quality of watercress storage using different conditions and found the use of unconventional modified atmospheres, especially He and N_2O atmospheres, increased the antioxidant activity of watercress at the end of the storage period. Nevertheless, the authors describe no clear effect of unconventional gases on the color parameters, polyphenol contents, and sensory parameters of fresh-cut watercress.

Superatmospheric oxygen can inhibit the microbial growth and enzymatic discoloration (Ayala-Zavala, Wang, Wand, & Gonzãlez-Aguilar, 2007) and can be an option for food conservation.

Unconventional gases has been successfully used to preserve different food products, like fresh-cut vegetables and fish, although their commercial use requires further research concerning commercial value and the possibility of different gas combinations.

3 Controlled atmosphere packaging

Controlled atmosphere is a similar technique to modified atmosphere, but there is a more accurate and precise control of the concentrations of the gases in the environment where the fruits are stored, which must be monitored and corrected throughout the storage period. Thus, the controlled atmosphere requires hermetically sealed storage environment, O_2 and CO_2 analyzers, as well as gas removal and injection mechanisms. The balance between O_2 and CO_2 is critical, and an optimal ratio is required for each product (Brackmann, Neuwald, Steffens, Sestari, & Giehl, 2005).

The development of controlled atmosphere (CA) technology can be broken into three eras. In the first era, beginning in the 1920s, CA storage involved maintaining a constant temperature and atmosphere with the introduction of mechanical refrigeration, airtight rooms and using only the product's respiration to lower O_2 and increase CO_2, and ventilation to add O_2 and remove CO_2. In the second era, the lime, activated carbon, and N_2 injection systems were used to remove CO_2, and open-flame generators, catalytic generators, and N_2 injection systems to remove O_2. The third era, called the dynamic controlled atmosphere (DCA) era, involves 2 phases. In the first phase, the dynamic nature of the CA conditions is determined from empirical data. In the second phase, biosensing of the product physiology is employed to reset the CA conditions to reflect the changing metabolic condition of the stored product (Prange, 2018).

The use of controlled atmospheres (CA) is one effective technique that has been widely and successfully used in fresh and fresh-cut produce. The controlled atmosphere technology applied in fresh-cut produce provides decreasing of the respiration rate, microbial growth, chlorophyll break down, and enzymatic browning, as well as improving the texture due to the action of CO_2 on hydrolytic cell membrane enzymes and reducing some physiological disorders induced by ethylene (C_2H_4) (Eskin & Robinson, 2000; Yousuf et al., 2018).

However, these same authors state undesirable effects in the quality of fresh products storage in CA, such as development of off-flavors due to accumulation of ethanol, acetaldehyde, and other volatiles, besides the increase of softening, development of physiological disorders, such as brown stain, internal browning, and surface pitting (Nakasone & Paull, 1998).

Poubol and Izumi (2005) showed that atmosphere with 60 kPa O_2 reduced the respiration rate of fresh-cut "Carabao" mango when stored at 5°C or 13°C for 42 h, but the browning was accelerated under these same conditions; while in fresh-cut "Nam Dokmai" mango, the high O_2 level did not affect the respiration rate, browning, and appearance. It has also been shown that, for both mango cultivars, the high O_2 level did not change their texture and ascorbic acid content; moreover, lactic acid bacteria and mold counts were below the detection level during storage at 5°C and 13°C. However, atmosphere 60 kPa O_2 stimulated the growth of mesophilic aerobic bacteria and yeasts on "Carabao" and "Nam Dokmai" fresh-cut mangos, respectively at 13°C, suggesting this atmosphere condition is not suitable for fresh-cut mango. Antoniolli, Benedetti, Sigrist,

& Silveira (2007) showed a mild browning in fresh-cut pineapple in controlled atmosphere (5% O_2 and 15% CO_2) during storage besides a slightly reduction of microbial growth.

Dong et al. (2015), suggested fresh-cut burdock storage in controlled atmosphere as a promising method to prevent browning and maintaining quality of this fresh-cut product. These authors showed that the burdock slices treated with CO_2 for 4, 6, and 8h exhibited better visual quality during 8days storage, compared with the ones treated with air. CO_2 treatment for 6h on the fresh-cut burdock slices reduced the respiration rate, lowered the activity of polyphenol oxidase (PPO) enzyme associated with darkening of the tissues, an aspect often causing rejection by consumers (Peixoto, 1999). The maintenance of color in yellow onion was studied using controlled atmosphere and it was observed that prepared diced yellow onion did not always expand enzymatic browning during storage. However, discoloration resulting from yellowing has been reported. Keeping O_2 at 2kPa, while CO_2 increased from 0 to 15kPa, the b* values were reduced. The CO_2 effect was also observed after cooking. Amaro et al. (2012) found that higher O_2 concentration is conducive to the preservation of aroma volatiles like acetate esters in fresh-cut cantaloupe and honeydew melons stored at 5°C for 12days.

Different effects of controlled atmosphere on fresh-cut products' bioactive compounds content have been observed. Wright and Kader (1997) revealed a reduction in the loss of vitamin C in fresh-cut strawberries and persimmons during postcutting life when submitted to 8days at atmosphere controlled with 2% O_2, air+12% CO_2, or 2% O_2+12% CO_2 at 0°C. While Agar et al. (1999) observed that concentrations of 5%, 10%, or 20% CO_2 caused degradation of vitamin C in fresh-cut kiwifruit slices. These authors reported that increase of vitamin C losses in response to higher than 10% CO_2 atmosphere might be a stimulating effect on acid ascorbic (AA) oxidation and inhibition of dehydroascorbic acid (DHA) reduction to AA. In vegetables, Moretti, Araújo, and Mattos (2003), studying different oxygen and carbon dioxide rate combinations to extend shelf-life revealed that fresh-cut collard greens stored under 3% O_2 and 4% CO_2 showed more vitamin C content than samples stored under 5% O_2 and 5% CO_2 (25%) and air (56%) besides more total chlorophyll content, 24% and 45%, respectively. Fresh-cut collard greens stored under 3% O_2 and 4% CO_2 presented around 44% more organic acids than control, however, total soluble solids content were not significantly affected.

References

Ahmed, D. M., Yousef, R. M., & Sarrwy, S. M. A. (2011). Modified atmosphere packaging for maintaining quality and shelf life extension of persimmon fruits. *Asian Journal of Agricultural Sciences, 3*(4), 308–316.

Al-Ati, T., & Hotchkiss, J. H. (2003). The role of packaging film permselectivity in modified atmosphere packaging. *Journal of Agricultural and Food Chemistry, 51*, 4133–4138.

Amaro, A. L., Beaulieu, J. C., Grimm, C. C., Stein, R. E., & Almeida, D. P. F. (2012). Effect of oxygen on aroma volatiles and quality of fresh-cut cantaloupe and honeydew melons. *Food Chemistry, 130*, 49–57.

Antoniolli, L. R., Benedetti, B. C., Sigrist, J. M. M., & Silveira, N. F. A. (2007). Quality evaluation of fresh-cut "Pérola" pineapple stored in controlled atmosphere. *Ciência e Tecnologia de Alimentos Camp, 27*, 530–534.

Arnal, L., & Del Río, M. A. (2003). Removing astringency by carbon dioxide and nitrogen-enriched atmospheres in persimmon fruit cv. Rojo brillante. *Journal of Food Science, 68*(4), 1516–1518.

Ayala-Zavala, J. F., Wang, S. Y., Wand, C. Y., & Gonzãlez-Aguilar, G. A. (2007). High oxygen treatment increases antioxidant capacity and postharverst life of strawberry fruit. *Food Technology and Biotechnology, 45*, 169–173.

Bibi, N., Khattak, A. B., & Mehmood, Z. (2007). Quality improvement and shelf life extension of persimmon fruit (*Diospyros kaki*). *Journal of Food Engineering, 79*, 1359–1363.

Brackmann, A., Neuwald, D. A., Steffens, C. A., Sestari, I., & Giehl, R. F. H. (2005). Conservada artificialmente. *Revista Cultivar Hortaliças e Frutas*. 30.

Char, C., Silveira, A. C., Inestroza-Lizardo, C., Hinojosa, A., Machuca, A., & Escalona, V. H. (2012). Effect of noble gas-enriched atmospheres on the overall quality of ready-to-eat arugula salads. *Postharvest Biology and Technology, 73*, 50–55.

Chitravathi, K., Chauhan, O. P., & Raju, P. S. (2015). Influence of modified atmosphere packaging on shelf-life of green chillies (*Capsicum annuum* L.). *Food Packaging and Shelf Life, 4*, 1–9.

Conesa, A., Verlinden, B. E., Artes-Hernandez, F., Nicolai, B., & Artes, F. (2007). Respiration rates of fresh-cut bell peppers under super atmospheric and low oxygen with or without high carbon dioxide. *Postharvest Biology and Technology, 45*, 81–88.

Cortellino, G., Gobbi, S., Bianchi, G., & Rizzolo, A. (2015). Modified atmosphere packaging for shelf life extension of fresh-cut apples. *Trends in Food Science and Technology, 46*(2), 320–330.

Costa, C., Lucera, A., Conte, A., Mastromatte, M., Speranza, B., Antonacci, A., et al. (2011). Effects of passive and active modified atmosphere packaging conditions on ready-to-eat table grape. *Journal of Food Engineering, 102*, 115–121.

Day, B. P. F. (1998). Novel MAP. A brand new approach. *Food Manufacture, 73*, 22–24.

Dong, T., Shi, J., Jiang, C.-Z., Feng, Y., Cao, Y., & Wang, Q. (2015). A short-term carbon dioxide treatment inhibits the browning of fresh-cut burdock. *Postharvest Biology and Technology, 110*, 96–102.

Eskin, M., & Robinson, D. S. (2000). *Food shelf life stability: Chemical, biochemical, and microbiological changes* (1st ed.). CRC Press, Taylor and Francis group.

Falagán, N., & Terry, L. A. (2018). Recent advances in controlled and modified atmosphere of fresh produce. *Johnson Matthey Technology Review, 62*(1), 107–117.

Finnegan, E., & O'Beirne, D. (2015). Characterising deterioration patterns in fresh-cut fruit using principal component analysis. II: Effects of ripeness stage, seasonality, processing and packaging. *Postharvest Biology and Technology, 100*, 91–98.

González-Buesa, J., Ferrer-Mairal, A., Oria, R., & Salvador, M. L. (2009). A mathematical model for packaging with micro perforated films of fresh-cut fruits and vegetables. *Journal of Food Engineering, 95*, 158–165.

Gorny, J. R. (1997). Modified atmosphere packaging in fresh-cut revolution. *Perishables Handling Newsletter, 90*, 4–5.

Gorny, J., & Agar, I. (1998). Are argon-enriched atmospheres beneficial? *Reviews in Food Science and Nutrition, 28*, 1–30.

Gouble, B., Fath, D., & Soudain, P. (1995). Nitrous oxide inhibition of ethylene production in ripening and senescing climacteric fruits. *Postharvest Biology and Technology, 5*, 311–321.

Inestroza-Lizardo, C., Silveira, A. C., & Escalona, V. H. (2016). Metabolic activity, microbial growth and sensory quality of arugula leaves (*Eruca vesicaria* Mill.) stored under non-conventional modified atmosphere packaging. *Scientia Horticulturae, 209*, 79–85.

Jacxsens, L., Devlieghere, F., Van der Steen, C., & Debevere, J. (2001). Effect of high oxygen modified atmosphere packaging on microbial growth and sensorial qualities of fresh-cut produce. *International Journal of Food Microbiology, 71*, 197–210.

Jamie, P., & Saltveit, M. E. (2002). Postharvest changes in broccoli and lettuce during storage in argon, helium, and nitrogen atmospheres containing 2% oxygen. *Postharvest Biology and Technology, 26*, 113–116.

Kontominas, M. G. (2014). Modified atmosphere packaging of foods. *Encyclopedia of Food Microbiology, 2*, 1012–1016.

Lopez-Gálvez, F., Ragaert, P., Haque, M. A., Eriksson, M., van Labeke, M. C., & Devlieghere, F. (2015). High oxygen atmospheres can induce russet spotting development in minimally processed iceberg lettuce. *Postharvest Biology and Technology, 100*, 168–175.

Ma, L., Zhang, M., Bhandari, B., & Gao, Z. (2017). Recent developments in novel shelf life extension technologies of fresh-cut fruits and vegetables. *Trends in Food Science and Technology, 64*, 23–38.

Mangaraj, S., Goswami, T. K., & Mahajan, P. V. (2009). Applications of plastic films for modified atmosphere packaging of fruits and vegetables: A review. *Food Engineering Reviews, 1*, 133–158.

Moretti, C. L., Araújo, A. L., & Mattos, L. M. (2003). Evaluation of different oxygen, carbon dioxide and nitrogen combinations employed to extend the shelf-life of fresh-cut collard greens. *Horticultura Brasileira, 21*, 676–680.

Nakasone, H. K., & Paull, R. E. (1998). *Tropical fruits.* Wallingford: CAB International.

Olawuyi, I. F., Park, J. J., Lee, J. J., & Lee, W. Y. (2019). Combined effect of chitosan coating and modified atmosphere packaging on fresh-cut cucumber. *Food Science & Nutrition, 7*(3), 1043–1052.

Peixoto, P. H. P. (1999). Aluminum effects on lipid peroxidation and on the activities of enzymes of oxidative metabolism in sorghum. *Revista Brasileira de Fisiologia Vegetal, 11*(3), 137–143.

Pinela, J., Barreira, J. C., Barros, L., Antonio, A. L., Carvalho, A. M., Oliveira, M. B. P., et al. (2016). Postharvest quality changes in fresh-cut watercress stored under conventional and inert gas-enriched modified atmosphere packaging. *Postharvest Biology and Technology, 112*, 55–63.

Pinheiro, A. C. D. A. S., Urbinati, E., Tappi, S., Picone, G., Patrignani, F., Lanciotti, R., et al. (2019). The impact of gas mixtures of Argon and Nitrous oxide (N_2O) on

quality parameters of sardine (*Sardina pilchardus*) fillets during refrigerated storage. *Food Research International, 115*, 268–275.
Poubol, J., & Izumi, H. (2005). Physiology and microbiological quality of fresh-cut mango cubes as affected by high-O_2 controlled atmospheres. *Journal of Food Science, 70*(6), 286–291.
Prange, R. K. (2018). Dynamic controlled atmosphere (DCA) storage of fruits and vegetables. In *Reference module in food science* (pp. 1–10): Elsevier.
Reinas, I., Oliveira, J., Pereira, J., Mahajan, P., & Poças, F. (2016). A quantitative approach to assess the contribution of seals to the permeability of water vapour and oxygen in thermosealed packages. *Food Packaging and Shelf Life, 7*, 34–40.
Rocculi, P., Romani, S., & Dalla Rosa, M. (2005). Effect of MAP with argon and nitrous oxide on quality maintenance of minimally processed kiwifruit. *Postharvest Biology and Technology, 35*(3), 319–328.
Sahoo, N. R., Bal, L. M., Pal, U. S., & Sahoo, D. (2015). Effect of packaging conditions on quality and shelf-life of fresh pointed gourd (*Trichosanthes dioica* Roxb.) during storage. *Food Packaging and Shelf Life, 5*, 56–62.
Sandhya. (2010). Modified atmosphere packaging of fresh produce: Current status and future needs. *LWT-Food Science and Technology, 43*, 381–392.
Sen, C., Mishra, H. N., & Srivastav, P. P. (2012). Modified atmosphere packaging and active packaging of banana (*Musa spp.*): A review on control of ripening and extension of shelf life. *Journal of Stored Products and Postharvest Research, 3* (9), 122–132.
Shen, X., Zhang, M., Devahastin, S., & Guo, Z. (2019). Effects of pressurized argon and nitrogen treatments in combination with modified atmosphere on quality characteristics of fresh-cut potatoes. *Postharvest Biology and Technology, 149*, 159–165.
Silveira, A. C., Araneda, C., Hinojosa, A., & Escalona, V. H. (2014). Effect of non-conventional modified atmosphere packaging on fresh cut watercress (*Nasturtium officinale* R. Br.) quality. *Postharvest Biology and Technology, 92*, 114–120.
Sousa-Gallagher, M. J., & Mahajan, P. V. (2013). Integrative mathematical modeling for MAP design of fresh-produce: Theoretical analysis and experimental validation. *Food Control, 29*, 444–450.
Sowa, S., & Towill, L. (1991). Effects of nitrous oxide on mitochondrial and cell respiration and growth in *Distichlis spicata* suspension cultures. *Plant Cell, Tissue and Organ Culture, 27*, 197–201.
Van de Velde, F., Méndez-Galarraga, M. P., Grace, M. H., Fenoglio, C., Lila, M. A., & Pirovani, M. É. (2019). Changes due to high oxygen and high carbon dioxide atmospheres on the general quality and the polyphenolic profile of strawberries. *Postharvest Biology and Technology, 148*, 49–57.
Wright, K. P., & Kader, A. (1997). Effect of slicing and controlled-atmosphere storage on the ascorbate content and quality of strawberries and persimmons. *Postharvest Biology and Technology, 10*(1), 39–48.
Yang, X., & Whang, H. (2014). Controlled atmosphere. *Encyclopedia of Food Microbiology*, 1006–1011.
Yang, Q., Zhang, X., Wang, F., & Zhao, Q. (2019). Effect of pressurized argon combined with controlled atmosphere on the postharvest quality and browning of sweet cherries. *Postharvest Biology and Technology, 147*, 59–67.
Yousuf, B., Qadria, O. S., & Srivastava, A. K. (2018). Recent developments in shelf-life extension of fresh-cut fruits and vegetables by application of different edible coatings: a review. *LWT-Food Science and Technology, 89*, 198–209.

Further reading

Bender, R. J., Brecht, J. K., & Sargent, S. A. (2000). Mango tolerance to reduced oxygen levels in controlled atmosphere storage. *Journal of the American Society for Horticultural Science, 125,* 707–713.

Kader, A. A., Zagory, D., & Kerbel, E. L. (1989). Modified atmosphere packaging of fruits and vegetables. *Critical Reviews in Food Science and Nutrition, 28,* 1–30.

Teixeira, G. H. A., Santos, L. O., Cunha Júnior, L. C., & Durigan, J. F. (2018). Effect of carbon dioxide (CO2) and oxygen (O2) levels on quality of "Palmer" mangoes under controlled atmosphere storage. *Journal of Food Science and Technology, 55*(1), 145–156.

8

Natural additives with antimicrobial and flavoring potential for fresh-cut produce

Shalini Gaur Rudra, Gajanan Gundewadi, Ram Roshan Sharma
Division of Food Science and Postharvest Technology, ICAR-Indian Agricultural Research Institute, New Delhi, India

1 Introduction

Fruits and vegetables are considered as protective foods due to their health protecting properties other than providing nutrition (Boeing et al., 2012; WHO, 2003; Barry-Ryan & O'Beirne, 1999; Francis et al., 2012; Lamikanra, 2002).

However, FCFV products are highly prone to microbial contamination and spoilage due to wounding in unit operations like peeling and size reduction to prepare them for consumption out-of-box. Handling of fresh produce and their modification leads to mechanical injuries to plant tissues and promotes biochemical changes and degradation through microbial growth on surface leading to food safety and quality issues (Ayala-Zavala, del-Toro-Sanchez, Alvarez-Parrilla, & Gonzalez-Aguilar, 2008; Gonzalez-Aguilar, Ayala-Zavala, Olivas, De La Rosa, & Álvarez-Parrilla, 2010; Rico, Mart´ın-Diana, Barat, & Barry-Ryan, 2007). Furthermore, fresh-cut produce is more susceptible to spoilage and can facilitate rapid growth of spoilage microorganisms as well as microorganisms of public health significance (Qadri, Yousuf, & rivastava, 2016; Zambrano-Zaragoza, Quintanar-Guerrero, Real, Pinon-Segundo, & Zambrano-Zaragoza, 2017). While conventional food-processing methods extend the shelf-life of fruit and vegetables, the minimal processing to which fresh-cut fruit and vegetables are subjected renders produce highly perishable, requiring chilled storage to ensure a reasonable shelf-life (Barrett & Garcia, 2002). Microorganisms are natural contaminants of fresh produce and minimally processed fresh-cut products, and contamination arises from a number of sources, including postharvest handling and processing

Fresh-cut Fruits and Vegetables. https://doi.org/10.1016/B978-0-12-816184-5.00008-2

(Beuchat, 1996). Due to the nature of the treatments applied to this type of product, a favorable environment and time for proliferation of spoilage organisms and microorganisms of public health significance are created (Ahvenainen, 1996; Francis, Thomas, & O'Beirne, 1999).

Microbial contamination of fresh-cut produce may pose threats to human health by causing diseases like diarrhea, abdominal cramps, vomiting, and even death. The past few decades have witnessed a dramatic increase in food-borne illness outbreaks mainly caused by consumption of fresh and minimally processed fruits because of *Escherichia coli O157:H7* and *Salmonella* (Faber et al., 2003; Qadri, Yousuf, & Srivastava, 2015). Unless effective steps for microbial control are followed, these FCFVs may act as potential vehicles for transmitting infectious diseases. Hence, safety in the fresh-cut fruit industry is of paramount importance. Food safety thus constitutes a growing concern for all stakeholders in the FCFV chain like producers, transporters, retailers, consumers, and the regulatory agencies (Qadri et al., 2015; Saftner, Abbott, Bhagwat, & Vinyard, 2005). The main aim of the fresh-cut technology is to provide consumers with convenient, safe, and fresh-like product, having desirable nutritional and sensory qualities. Therefore, critical steps to ensure the safety of these FCFVs, like validated procedures for preparation, packing, and disinfection, need to be in place and should be under close monitoring (Abadias, Canamas, Asensio, Anguera, & Vinas, 2006; Qadri et al., 2015).

2 Factors affecting the quality and spoilage of fresh-cut fruits and vegetables

Numerous factors limit the quality and shelf-life of fresh-cut produce. Among the internal factors, considerations of type of fruit and their genotype, morphology, physiology, biochemical defense mechanisms, and stress-induced senescence programs are required. The external factors include environmental conditions such as storage temperature, humidity, load in refrigerator/cold store, sharpness of cutting-knife, and chemical treatments (Hodges & Toivonen, 2008).

2.1 Physical factors

Fruits and vegetables generally have high moisture content (90%–95%), but on processing into ready to consume or use form, there is a tendency to lose internal water content as compared to the natural, intact form, and this rapid loss culminates into shrinkage, loss of texture, wilted appearance, and finally, discoloration. This not only reduces physiological weight, but leads to loss of nutrients through leaching of juices.

2.2 Biochemical changes

Fruits and vegetables are generally rich in polyphenols and other mineral elements. When they are cut open, the nearby cells and cell membranes are damaged or ruptured, the oxidase enzymes are suddenly activated and start acting on the substrate leading to formation of brown pigments and triggering other oxidative reactions. The most common example is enzymatic browning in apples, pears, unripe jackfruit, brinjal, etc. Phenolic compounds come into contact with polyphenoloxidase enzyme in presence of oxygen to form the chromogenic quinone polymer. Lipoxygenase is another enzyme that catalyzes the oxidation of polyunsaturated fatty acids containing a cis,cis-1,4-pentadiene unit to form hydroperoxide derivatives, which undergo scission and dismutation reactions, resulting in the development of off-flavors. Siroli et al. (2015a, 2015b) reported application of citral and hexanal+2- (*E*)-hexenal for greater control of enzymatic activity in fresh-cut apples. Coating of sliced mango with opuntia mucilage-rosemary oil has shown reduced peroxidase enzyme activity (Alikhani, 2014). Furthermore, use of encapsulated star anise essential oil in xanthan gum coating for fresh-cut Chinese yam is reported to have antioxygenic and antimicrobial properties, significantly reducing the L^* value, browning index, and polyphenol oxidase activity (Zhang, Gu, Lu, Yuan, & Sun, 2019).

2.3 Microbial spoilage

The microbial load on FCFVs is largely determined by their origin, various agricultural practices, harvesting methods, conditions, preparation procedures, type of package and storage. During distribution and storage of FCFVs, high in-package humidity and cut surfaces provide favorable conditions for growth of spoilage microorganisms. At low temperature storage, the major bacterial populations belonging to Pseudomonadaceae (*Pseudomonas fluorescens*), Enterobacteriaceae (*Rahnella aquatilis* and *Erwinia herbicola*), and lactic acid bacteria (*Leuconostoc mesenteroides*) are often reported on the surface of commodities (Ragaert, Devlieghere, & Debevere, 2007). In addition, several yeast species of genera *Candida*, *Cryptococcus*, *Rhodotorula*,

Pichia, *Trichosporon*, and *Torulaspora* have been reported during storage (Siroli, Patrignani, Serrazanetti, Gardini, & Lanciotti, 2015; Siroli, Patrignani, Serrazanetti, Tabanelli, et al., 2015a, 2015b). The slightly acidic to neutral pH of fruits and vegetables tends to favor the growth of bacteria and yeast as compared to molds (Barth, Hankinson, Zhuang, & Breidt, 2009). The excessive proliferation of these pathogens occurs due to the congenial conditions resulting in spoilage of minimally processed fruits (Siroli, Patrignani, Serrazanetti, Gardini, et al., 2015; Siroli, Patrignani, Serrazanetti, Tabanelli, et al., 2015a, 2015b).

Between 1996 and 2004, the US Food and Drug Administration (FDA) reported nearly 14 food-borne outbreaks, causing 859 illness cases, due to the consumption of tomatoes or lettuce (Patrignani, Siroli, Serrazanetti, Gardini, & Lanciotti, 2015). Psychotropic pathogen *Listeria monocytogenes* and mesophilic pathogen such as *Salmonella* and *E. coli* O157:H7 are the most common minimally processed food-borne pathogens that are associated with human illness (Siroli, Patrignani, Serrazanetti, Gardini, et al., 2015; Siroli, Patrignani, Serrazanetti, Tabanelli, et al., 2015a, 2015b; Sivapalasingam, Friedman, Cohen, & Tauxe, 2004). *E. coli* O157:H7 and *Salmonella* are two human pathogens that made major news headlines in the United States in 2006 and 2007 due to outbreaks linked with the consumption of prepared green vegetables and other produce (Sivapalasingam et al., 2004). Generally, *E. coli* O157:H7 infection has been linked with consumption of sprouts and lettuce while salmonellosis is linked with tomatoes, cantaloupe, seed sprouts, and orange and apple juice (Martin-Belloso, 2007).

Mesophilic and psychrotrophic microorganisms cause major microbiological concerns regarding FCFVs. In most cases, the profile of microorganisms present on fresh-cut fruits and vegetables is similar to that in the source field (Qadri et al., 2015). Thus, the preharvest environment serves as the origin of microbes responsible for microbial contamination or decay. Nguyen-The and Carlin (1994) have reported mesophilic bacterial counts on different fresh-cut vegetables (fresh-cut honeydew melon, spinach, broccoli florets, chopped bell peppers, lettuce, Japanese radish shreds, etc.) in the range of 10^3–10^9 CFU/g, whereas densities on minimally processed products, evaluated soon after processing, ranged from 10^3 to 10^6 CFU/g. Approximately 80%–90% of the mesophilic microflora of produce comprised of *Erwinia* spp., *Enterobacter* spp., and *Pseudomonas* spp. 50%–90% of the pseudomonads were identified as *P. fluorescens*. Lactic acid bacteria (LAB) such as *Leuconostoc mesenteroides* and *Lactobacillus* spp. and several species of yeasts and molds are also commonly reported in fresh-cut fruits and vegetables (Qadri et al., 2015). Presence of pathogens like *Salmonella* spp., *E. coli O157:H7*, and *Listeria monocytogenes* have been linked to disease outbreaks following consumption of contaminated produce (Leverentz et al., 2006).

3 Approaches to preserve quality of fresh-cut fruits and vegetables

Fresh-cut products are wounded tissues, which deteriorate more rapidly than intact fruits and vegetables. Researchers have been involved in developing new technologies that can improve the quality and safety of fresh-cut products with the aim to meet consumer expectations (Yousuf & Srivastava, 2017). Different approaches including surface treatments by means of edible coating, modified atmosphere packaging (MAP) to maintain appropriate gas concentration surrounding the cut surface, controlled atmosphere to delay softening of fruit, and chemical treatments such as calcium dips, low temperature storage, and gamma irradiation have been investigated to reduce detrimental changes such as microbial growth, desiccation, and discoloration (Yousuf & Srivastava, 2017).

Chemical approaches like chlorine, iodine, quaternary ammonium compounds, and ozone hydrogen peroxide have been found traditionally used and found effective in reducing rate of decay, but consumers are concerned about their residues in the product; particularly so in fresh-cut produce, which are consumed for their health benefits. Therefore, alternative methods for controlling fresh-cut fruit decay are being sought. Application of natural additives is one of the major emerging technologies for reducing quality loss and assuring safety of fresh-cut fruits and vegetables (Ayala-Zavala, González-Aguilar, & Del-Toro-Sánchez, 2009; Lanciotti et al., 2004).

Owing to rising interest in natural antimicrobial compounds for preserving fresh-cut produce quality, numerous studies on the antimicrobial activity of a wide range of natural compounds have been conducted (Ayala-Zavala et al., 2008; Burt, 2004). The food industry has paid attention to this perception and is actively working towards the replacement of synthetic preservatives and adoption of natural alternatives to maintain safety and quality of fresh-cut produce. Continuing search for effective natural antimicrobial compounds to preserve quality of fresh-cut produce without being detrimental to its sensorial attributes has yielded promising results. Safety and aroma appeal of fresh-cut fruits and vegetables could even be improved using the right combination of flavors inherently found in the antimicrobial essential oils (EOs) and in the target fresh-cut produce (Ayala-Zavala et al., 2009).

3.1 Essential oils as antimicrobial agents

Natural additives are compounds or groups of compounds that are already used empirically by the population for edible purposes. Fungi, seaweeds, and algae are also interesting sources of natural additives. These natural compounds have historically been known for their antimicrobial purposes, but in recent years, they have gained more interest from the food industry for direct application or in synergy with other natural or chemical additives. Among natural additives, essential oils (EOs), being complex mixtures of volatile compounds characterized by a strong sensorial impact, are produced by many plants as secondary metabolites. Also called volatile oils, they may be obtained from all the organs of the plant, i.e., flowers, buds, seeds, leaves, roots, wood, stems, twigs, fruits, or bark, and are stored in secretory cells, cavities, canals, epidermis cells, or glandular trichomes (Bakkali, Averbeck, Averbeck, & Idaomar, 2008; Hyldgaard, Mygind, & Meyer, 2012). They are used in perfumes, cosmetics, bath products, scents, incense, household cleaning products, as well as in food and drink flavorings (Antunes & Cavacob, 2010; Sivakumar & Bautista-Baños, 2014).

Antunes and Cavaco (2010) have referred to EO as concentrated, hydrophobic liquid containing a mixture of antimicrobial volatile aroma compounds commonly derived from plant tissues. EOs are "essential" in the common sense that they carry a distinctive scent of the source plant, and are generally extracted by steam distillation and other processes like maceration, cold pressing, or solvent extraction. These volatile EOs are environmentally safe and are known as "reduced risk" pesticides. Hence they have earned the generally considered as safe (GRAS) status for both human health and environment. Essential oils are composed of an array of chemical compounds of which specific active compounds exhibit antimicrobial activity against several postharvest pathogens at miniscule doses without compromising the environment or consumer's well-being (Sivakumar & Bautista-Baños, 2014). However, apart from the major bioactive components identified for antimicrobial activity, minor components present in whole essential oils also play critical roles and often have synergistic effects. Some studies have concluded that whole EO extracts have a greater antibacterial activity than their major constituents in isolation.

The aromatic constituents of known EOs are mostly comprised of short hydrocarbon chains, complemented with oxygen, nitrogen, and sulfur atoms attached at various points of the chain (Braca, Siciliano, D'Arrigo, & Germano, 2008). The inherent aroma and antimicrobial activity of EOs is related commonly to the chemical configuration of the components, their proportions, and to their interactions, which in turn affect their bioactive properties (Fisher & Phillips, 2008). Such mixtures of different-scented molecules with highly reactive atoms give different functional properties to EOs, which could be considered for many food applications. Antioxidant activity is also one of the most intensively studied properties in EO research, because reactive oxygen species damage various biological substances and subsequently causes many diseases, including cancer, liver disease, Parkinson's and Alzheimer's disease, aging, arthritis, inflammation, diabetes, atherosclerosis, and AIDS (Becerril, Gomes-Lus, Goni, Lopez, & Nerin, 2007).

Several investigations have focused on the search for natural antimicrobials able to enhance the quality and safety of the minimally processed fruits (Table 1). The *in vitro* antimicrobial activity of oregano (*Origanum vulgare*) and thyme (*Thymus vulgaris*) EOs and their main components (carvacrol and thymol) against a huge variety of Gram-positive, Gram-negative bacteria, yeasts, and molds are well documented (Burt, 2004; Lanciotti et al., 2004; Viuda-Martos, Ruiz-Navajas, Fernandez-Lopez, & Perez-Alvarez, 2008).

In previous years, use of chemical preservatives in edible coatings for this purpose was prevalent. García et al. (2001) reported edible coating of strawberry with starch incorporated with potassium sorbate and citric acid inhibited the growth of mesophilic aerobes, mold, and yeast counts during storage.

Valero et al. (2006) developed an active packaging with addition of thymol or eugenols for table grapes stored for 56 days under MAP. The results revealed lower microbial spoilage in table grapes as compared to control. Initial load of microbes on table grapes was 2.3 ± 0.15 and 2.9 ± 0.17 log CFU/g for total mesophilic aerobic and yeast and mold counts, respectively. The treated samples showed significant reduction of yeast and mold (1.70–2.40 log CFU/g) and for mesophilic aerobic (2.20–2.40 log CFU/g) counts as compared to control berries with yeast and mold (4.2 ± 0.32 log CFU/g) and mesophilic aerobic (4.2 ± 0.18 log CFU/g) populations after 28 and 56 days. Rojas-Grau et al. (2007) investigated the effect of lemongrass, oregano oil, and vanillin incorporated into apple puree-alginate edible coatings, on the shelf-life of fresh-cut "Fuji" apples. Edible coatings containing EOs were found effective in inhibiting the growth of psychrophilic aerobes, yeasts, and molds. The 0.1% *w*/w of oregano oil-coating treatment

Table 1 Overview of antimicrobial activity of different essential oils (EO) or their components on fresh-cut fruits and vegetables

Commodity	Treatment description	Results	Reference
Fruits			
Table grapes	Thymol or eugenol + MAP	Reduced microbial spoilage	Valero et al. (2006)
Fresh-cut "Fuji" apples	Lemongrass, oregano oil, and vanillin incorporated into apple puree-alginate edible coatings	Inhibited the growth of psychrophilic aerobes, yeasts, and molds and also shown strongest antimicrobial activity against *Listeria innocua*	Rojas-Grau et al. (2007)
Fresh-cut fruits and vegetables	Citral (25–125 ppm) and Citron essential oil (300, 600, 900 ppm) in sugar syrup	Inhibited the growth of gram-positive pathogen *Listeria monocytogenes*	Belletti, Lanciotti, Patrignani, and Gardini (2008)
Fresh-cut pineapple slices	Mint oil (0.5%, 1.0%, 1.5% *w*/w)	Reduced the microflora of *Escherichia coli* and *Salmonella enteritidis* and also maintained better postharvest quality parameters in fresh-cut pineapple	Bitencourt et al. (2014)
Sweet cherries	Eugenol, thymol, menthol, eucalyptol at1000 mL in gas used for MAP	Reduction of natural microflora	Serrano, Martínez-Romero, Castillo, Guillén, and Valero (2005)
Fresh sliced apples	Hexanal, hexyl acetate, E(2) Hexenalat 20–250 ppm	Decontamination of *Salmonella enteritidis*, *Escherichia coli*	Lanciotti et al. (2003)
Sliced mango	Treated with opuntia mucilage-rosemary oil (Mu+RO), 2 g rosemary oil microencapsul (ROM), and 2 g (ROM) plus (Mu +RO) placed in plastic trays, and stored at 6°C	Reduced loss of color, decreasing activity POD enzyme, and also inhibited the decay incidence	Alikhani (2014)
Fresh-cut papaya	Psyllium gum (0.5%, 1%, and 1.5%), sunflower oil, and 1% calcium chloride	1% psyllium gum coating was found to be effective in enhancing the shelf life and quality of papaya	Yousuf and Srivastava (2015)
Sliced apple and grape berry	Cocoa based coating	Maintained better texture and reduced microbial spoilage	Glicerina et al. (2019)
Strawberries	Starch + potassium sorbate + citric acid	Inhibited the growth of mesophilic aerobes, mold, and yeast counts	García, Martino, and Zaritzky (2001)

Table 1 Overview of antimicrobial activity of different essential oils (EO) or their components on fresh-cut fruits and vegetables—Cont'd

Commodity	Treatment description	Results	Reference
Fresh-cut apple	Alginate and gellan + lemongrass, oregano oil, and vanillin	4 log reduction of *Listeria innocua* in fresh-cut apple	Rojas-Grau et al., 2007
Fresh-Cut Pears	*Litsea cubeba* essential oil (50, 100, and 250 ppm)	It maintained better postharvest quality and enhanced the shelf life of fruits	
Vegetables			
Lamb's lettuce	Oregano and thyme EOs 250 mg/L	Inhibited the spoilage microflora	Siroli, Patrignani, Serrazanetti, Gardini, et al. (2015) and Siroli, Patrignani, Serrazanetti, Tabanelli, et al. (2015a, 2015b)
Melon	Alginate coating incorporated with malic acid 2.5 (*w/v*) + cinnamon 0.7 (v/v) + palmarosa 0.7 (*v/v*) + lemongrass 0.7 (v/v)	It reduced the microbial spoilage up to 3.1 log CFU/g after 30 days of storage	Raybaudi-Massilia, Mosqueda-Melgar, and Martín-Belloso (2008)
Fresh-cut lettuce	Phytoncide EO (pine leaves)	Reduced enzymatic browning in fresh-cut lettuce	Kim, Kim, Chung, and Moon (2014)
Fresh-cut kajari melons	Honey (0, 50, 100, 150 mL/L of water) followed by coating with soy protein isolate (50 g/L of water)	150 mL honey/L of water +50 g SPI/L was maintained better quality and lower total plate count (5 log CFU/g as compared to 9 log CFU/g in control)	Yousuf and Srivastava (2017)
Iceberg lettuce	Oregano oil (25–75 mg/L) and chlorine washing	Decontamination of *Salmonella enteritidis*	Gunduz, Gonul, and Karapınar (2010)
Lettuce	Oregano and rosemary EO	Reduced the microflora *Listeria monocytogenes*, *Yersinia enterocolitica*, and *Aeromonas hydrophila*	De Azeredo et al. (2011)
Fresh-cut lettuce	Thymol, carvacrol, trans-cinnamaldehyde, eugenol, and vanillin	Reduced the spoilage of *Escherichia coli*	Yuan, Teo, and Yuk (2019)
Lettuce	Oregano oil (0.05%) nanoemulsion	Control of L. *monocytogens*, *E coli* 0157:H7 and *S. typhimurium*	Bhargava, Conti, da Rocha, and Zhang (2015)

reduced yeast and mold counts by 4.1 log CFU/g in cut apples after 21 days of storage.

Belletti et al. (2008) investigated the effect of citral (25–125 ppm) and citron essential oil (300, 600, 900 ppm) blended in sugar syrup and used for dipping of ready-to-eat fruit salad (apple, pineapple, melon, kiwifruit, orange, and grape). As compared to citral, the citron essential oil has significantly inhibited the growth of gram-positive pathogen *Listeria monocytogenes by* 3.7 log CFU/g reduction and reduced the survival potential of gram-negative species *Salmonella enteritidis* and *E. coli*. Edible coating of fresh-cut pineapple slices with mint oil (0.5%, 1.0%, 1.5% *w*/w) was found effective in reducing the *E. coli* and *Salmonella enteritidis* and also maintained better postharvest quality parameters in fresh-cut pineapple (Bitencourt et al., 2014).

In general, EO nanoemulsions have been used in minimally processed fruits and vegetables as either washing disinfectant or with incorporation into edible coatings. A recent review by Prakash, Baskaran, Paramsivam, and Vadivel (2018) has covered various aspects of use of EO based nanoemulsions to improve the quality of minimally processed fruits and vegetables. Salvia-Trujillo, Rojas-Graü, Soliva-Fortuny, and Martín-Belloso (2015) investigated the efficacy of lemongrass essential oil (0.1%, 0.5% or 1% *v*/v) encapsulated nanoemulsion coating of fresh-cut Fuji apple fruits. The results revealed that 0.5% and 1% formulation completely inhibited the natural microflora *E. coli* of fresh-cut apple. Further, in another study nanoemulsion containing 0.05% oregano oil coating on the surface of the lettuce reduces the microbial population 3.44, 3.05 and 2.31 log CFU/g in *L. monocytogens*, *E. coli* O157:H7 and *S. typhimurium*, respectively was observed as compared to control (only water dip) (Bhargava et al., 2015). Likewise, washing with carvacrol or eugenol nanoemulsion significantly reduced *E. coli* and *S. enterica* up to 4.5 log CFU/g on spinach leaves (Ruengvisesh, Loquercio, Castell-Perez, & Taylor, 2015). Herein, the authors did not observe any difference in results based on the method of application i.e., spray and immersion.

Landry, Micheli, McClements, and McLandsborough (2015) evaluated the efficacy of carvacrol nanoemulsion against *E. coli* O157:H7 and *Salmonella enteritidis* in minimally processed sprouts, such as alfalfa, mung bean, radish, and broccoli. Soaking of these sprouts in 4000 ppm carvacrol nanoemulsion for 60 min, inactivated 2–3 log CFU/g of *E. coli* and *S. enteritidis* on radish seeds. However, the same treatment was found ineffective for decontamination of broccoli. Kim et al. (2013) investigated the efficacy of lemongrass oil emulsified carnauba wax nanoemulsions on microbial spoilage of plums. The 3% lemongrass EO showed the reduction of >2.8 log CFU/g of *E. coli* O157:H7 and *Salmonella typhimurium*. Furthermore, the coating did not negatively impact the perceived fracturability, flavor, or glossiness of plums. Sessa, Ferraria, and Donsìb (2015) reported an extension of shelf-life up to 7 days upon use of modified chitosan, lemon EO nanoemulsion on rucola leaf as compared to control samples.

However, limited research has been conducted to investigate the antimicrobial efficacy of these essential oils in ready to use vegetables. Gutierrez, Bourke, Lonchamp, and Barry-Ryan (2009) investigated the efficacy of oregano oil on natural microflora of ready-to-eat lettuce and carrot. The results revealed that both oregano oil and chlorine disinfection showed similar levels of decontamination effects. At the end of a storage period of 7 days, in the case of oregano and thyme oil treatment, 6.10, 3.55, 5.81 log cfu/mL was recorded in comparison to control: 6.57, 3.39, 6.01 log cfu/mL for total viable counts, lactic acid bacteria, and Pseudomonas, respectively. Gunduz et al. (2010) investigated the effect of oregano oil and chlorine washing on inhibition of *Salmonella typhimurium* inoculated onto iceberg lettuce at 20°C. Reduction of *S. typhimurium* by washing with oregano did not exceed 1.92 logarithmic units regardless of the washing times and concentrations. The effectiveness of lettuce wash with oregano oil (75 ppm) on inactivation of *S. typhimurium* was comparable to that of 50 ppm chlorine.

Yuan et al. (2019) evaluated combined use of antimicrobial activities of thymol, carvacrol, trans-cinnamaldehyde, eugenol, and vanillin on spoilage potential of *E. coli* O157:H7. The results revealed that thymol, carvacrol, trans-cinnamaldehyde, eugenol, and vanillin significantly reduced the spoilage of fresh-cut lettuce. Thymol, carvacrol, trans-cinnamaldehyde, eugenol, and vanillin displayed minimum inhibitory concentrations (MICs) in a range of 0.63–2.5 mg/mL, whereas citral and linalool were ineffective (MIC >10 mg/mL). Siroli, Patrignani, Serrazanetti, Gardini, et al. (2015) and Siroli, Patrignani, Serrazanetti, Tabanelli, et al. (2015a, 2015b) evaluated the efficacy of oregano and thyme EOs in comparison with chlorine for decontamination in lamb's lettuce. Results revealed that the essential oil checked the growth of mesophilic aerobic bacteria and yeasts, and maintained the *L, a, b* color parameters and the volatile molecule profiles with the same efficacy as that of chlorine. A further study highlighted that both these EOs strongly inhibited the spoilage microflora in lamb's lettuce, without negatively affecting the quality and sensory properties of the products. The efficacy of these EOs has been shown to increase further when combined with other hurdles like MAP and radiation treatment.

3.1.1 Mode of action of essential oils

The antimicrobial activity of essential oils is well documented by studying *in vitro* experiments. EOs (terpenes) penetrate the cell membrane of microbes and destroy the cell membrane lipid structure. Some researchers (Ultee, Kets, & Smid, 1999) also theorize their ability to disrupt membrane permeability and affect ion gradients leading to cell death. EOs are known to cause denaturation of microbial proteins and leakage of cell metabolites, ions such as Ca^{+}, depolarization of mitochondria, pH change, collapse of proton pump, and depletion of ATP pool (Sivakumar & Bautista-Baños, 2014). Essential oils are able to affect active sites of enzymes and cellular metabolism leading to inhibition of mycelial growth and spore germination of pathogenic fungi (Juglal, Govinden, & Odhav, 2002).

3.1.2 Essential oil as additives for enhancing the flavor potential

Pleasing the palate of consumers is one of the most important keys to success in the food industry, including the fresh-cut fruit and vegetable market (Kader, 2008). Sometimes even the healthiest foods can be difficult to make popular for regular consumption if they have poor sensory properties (Park, Kodihalli, & Zhao, 2005). In a recent study, application of two cocoa-based edible coatings on quality characteristics of fresh-cut apple and whole berries of grape fruits during storage was investigated Glicerina et al. (2019). The coated fruits were packed in polyethylene terephthalate and stored for 10 days at 6°C. The edible coating maintained significantly higher firmness and microbiological safety in table grapes and apple slices as compared to control. In general, cocoa-based coatings gave a positive effect only on the sensory properties of both fruit samples during storage, showing a shelf-life very similar to control ones.

The sensorial impact of application of EOs as preservatives in fresh-cut fruits and vegetables must be considered because the specific aroma and flavor associated with each commodity is registered in human minds and any alteration in the same leads to dissociations in sensory and memory perceptions. This may even lead to rejection of produce by consumers owing to suspicions and rejection of produce due to altered aromatic profile. Some fresh-cut fruits and vegetables might be the least affected by this phenomenon; however, new combinations of flavors could be considered as biased in consumer studies (Baranauskiene, Venskutonis, & Demyttenaere, 2005).

Combination of flavor profiles traditionally used in global seasoning and flavoring mixtures could be helpful to support combinations between the antimicrobial EO and fresh-cut fruit (Rozin & Rozin, 1981). The aromatic profile of a fresh-cut fruit could be merged with that of EO or their combinations using their aroma notes as connectors. Most vegetables present herbal and spicy notes, while fruits present sweet and fresh aroma notes. Thus, aroma notes of fresh-cut produce and EOs could be suitably combined. For example, oregano, coriander, garlic, thyme, laurel, rosemary, and ginger oils, present a spicy, herbal aroma that could be combined with the aroma notes of fresh-cut vegetables. Fresh-cut fruits like apple, pear, peach, watermelon, strawberry, grapes, lemon, mandarin, orange, grapefruit, pineapple, mango, and papaya, present a sweet, fresh, flavor with characteristic pome, citrus, berry, and tropical aroma notes, depending on the produce (Kader, 2008). On the other hand, cinnamon, citrus, peppermint, and clove oils present sweet and fresh aroma notes, which can be very well combined with sweet and tropical fresh-cut fruits.

The idea of adding an extra flavor to the EO-treated product could be perceived within the philosophy of seasoning (Rozin & Rozin, 1981). Such combinations could serve to increase value and acceptability whilst the fresh-cut produce is naturally preserved. The characteristic flavor in lemon (*C. limon*) EO are due to the aldehyde citral, along with smaller amounts of linear aliphatic aldehydes (C7–C13). (+)-Limonene is by far the major component (approximately 65%) in most citrus species contributing to the typical aroma of distilled lime oil (Gahagan, 2008). Mandarin (*C. reticulata*) oil made by cold pressing the peels of mandarin has a deep, sweet, orange-like character, and finds ample use in fragrances, liqueurs, and cosmetics. Flowering herbs of peppermint (*M. piperita*) are steam distilled to obtain the colorless to pale greenish-yellow liquid with a characteristic peppermint odor. Its main components are menthol, menthone, and menthyl acetate (Gahagan, 2008). These flavors can easily complement the sweet and fresh characters of fresh-cut fruits. Use of a particular EO with a specific fruit needs considerable evaluation of both of their characteristics for attaining the desired acceptability among consumers.

4 Future thrusts

Demand for fresh-cut fruits and vegetables is increasing considerably in the metropolitan cities. However, FCFV products are highly prone to microbial contamination and spoilage due to wounding in unit operations like peeling and size reduction to prepare them for consumption out-of-box. Searching for a non-chemical approach towards minimizing this microbial spoilage is of the utmost importance. Consumer preference and trends in fresh-cut fruit and vegetable consumption need to be identified to offer specific combinations in flavor among the fresh-cut produce and the antimicrobial EO. Investigation of the correct application procedure and required doses to achieve the target antimicrobial and sensorial efficacy need to be conducted.

The nutraceutical potential and toxicity properties of EOs must be considered during their application and merit further investigation. The use of natural additives such as essential oils and plant extracts may be the best solutions for curbing existing problems. There is vast scope for exploiting large numbers of available plant based natural essential oils for replacing the synthetic surface sanitizing agent chlorine and other chemical preservatives like sodium hypochlorite. Use of novel hurdle technologies to enhance their efficacy further needs to be explored. Most of the studies emphasize superiority of thyme and oregano essential oils over others. There is vast scope for evaluation of many other essential oils for their efficacy on minimizing microbial spoilage, thus enhancing the shelf-life of fresh-cut fruits and vegetables. Use of EOS in encapsulated form presents an effective means to enhance their efficacy and performance. These EOs may be subject to either microencapsulation or nanoencapsulation in any of the base material and used as an edible coating. To ensure organoleptic acceptance, appropriate care should be taken on the selection of essential oil, based on its compatibility with the specific food matrix. However, further studies are required on the toxicity of essential oil nanoemulsion to assure its commercial exploitation in minimally processed fruits and vegetables.

References

Abadias, M., Canamas, T. P., Asensio, A., Anguera, M., & Vinas, I. (2006). Microbial quality of commercial 'Golden Delicious' apples throughout production and shelf-life in Lleida (Catalonia, Spain). *International Journal of Food Microbiology, 108*, 404–409.

Ahvenainen, R. (1996). New approaches in improving the shelf life of minimally processed fruits and vegetables. *Trends in Food Science and Technology, 7*, 179–186.

Alikhani, M. (2014). Enhancing safety and shelf life of fresh-cut mango by application of edible coatings and microencapsulation technique. *Food Science & Nutrition, 2*(3), 210–217.

Antunes, M. D. C., & Cavacob, A. M. (2010). The use of essential oils for postharvest decay control. *Flavour and Fragrance Journal, 25*(5), 351–366.

Ayala-Zavala, J. F., del-Toro-Sanchez, L., Alvarez-Parrilla, E., & Gonzalez-Aguilar, G. A. (2008). High relative humidity in-package of fresh-cut fruits and vegetables:

advantage or disadvantage considering microbiological problems and antimicrobial delivering systems. *Journal of Food Science, 73,* 7–41.

Ayala-Zavala, J. F., González-Aguilar, G. A., & Del-Toro-Sánchez, L. (2009). Enhancing safety and aroma appealing of fresh-cut fruits and vegetables using the antimicrobial and aromatic power of essential oils. *Journal of Food Science, 74*(7), 84–91.

Bakkali, F., Averbeck, S., Averbeck, D., & Idaomar, M. (2008). Biological effects of essential oils—A review. *Food and Chemical Toxicology, 46*(2), 446–475.

Baranauskiene, R., Venskutonis, P. R., & Demyttenaere, J. C. R. (2005). Sensory and instrumental evaluation of sweet marjoram (*Origanum majorana* L.) aroma. *Flavour and Fragrance Journal, 20,* 492–500.

Barrett, D. M., & Garcia, E. (2002). Preservative treatments for fresh-cut fruits and vegetables. In *Fresh-cut fruits and vegetables* (pp. 273–309). CRC Press.

Barry-Ryan, C., & O'Beirne, D. (1999). Ascorbic acid retention in shredded iceberg lettuce as affected by minimal processing. *Journal of Food Science, 64*(3), 498–500.

Barth, M., Hankinson, T. R., Zhuang, H., & Breidt, F. (2009). *Microbiological spoilage of fruits and vegetables. Compendium of the microbiological spoilage of foods and beverages*: (pp. 135–183). Springer.

Becerril, R., Gomes-Lus, R., Goni, P., Lopez, P., & Nerin, C. (2007). Combination of analytical and microbiological techniques to study the antimicrobial activity of a new active food packaging containing cinnamon or oregano against *E. coli* and *S. aureus. Analytical and Bioanalytical Chemistry, 388,* 1003–1011.

Belletti, N., Lanciotti, R., Patrignani, F., & Gardini, F. (2008). Antimicrobial efficacy of citron essential oil on spoilage and pathogenic microorganisms in fruit-based salads. *Journal of Food Science, 73*(7), 331–338.

Beuchat, L. R. (1996). Pathogenic organisms associated with fresh produce. *Journal of Food Protection, 59*(2), 204–216.

Bhargava, K., Conti, D. S., da Rocha, S. R., & Zhang, Y. (2015). Application of an oregano oil nanoemulsion to the control of foodborne bacteria on fresh lettuce. *Food Microbiology, 47,* 69–73.

Bitencourt, R. G., Possas, A. M. M., Camilloto, G. P., Cruz, R. S., Otoni, C. G., & Soares, N. D. F. F. (2014). Antimicrobial and aromatic edible coating on fresh-cut pineapple preservation. *Ciencia Rural, 44*(6), 1119–1125.

Boeing, H., Bechthold, A., Bub, A., Ellinger, S., Haller, D., Kroke, A., et al. (2012). Critical review: vegetables and fruit in the prevention of chronic diseases. *European Journal of Nutrition, 51*(6), 637–663.

Braca, A., Siciliano, T., D'Arrigo, M., & Germano, M. P. (2008). Chemical composition and antimicrobial activity of Momordica charantia seed essential oil. *Fitoterapia, 79,* 123–125.

Burt, S. (2004). Essential oils: Their antibacterial properties and potential applications in foods. Review. *Journal of Food Microbiology, 94,* 223–253.

De Azeredo, G. A., Stamford, T. L. M., Nunes, P. C., Neto, N. J. G., De Oliveira, M. E. G., & De Souza, E. L. (2011). Combined application of essential oils from *Origanum vulgare* L. and *Rosmarinus officinalis* L. to inhibit bacteria and autochthonous microflora associated with minimally processed vegetables. *Food Research International, 44*(5), 1541–1548.

Faber, J. N., Harris, L. J., Parish, M. E., Beuchat, L. R., Suslow, T. V., Gorney, J. R., et al. (2003). Microbiological safety of con- trolled and modified atmosphere packaging of fresh and fresh-cut produce. *Comprehensive Reviews in Food Science and Food Safety, 2,* 142–160.

Fisher, K., & Phillips, C. (2008). Potential antimicrobial uses of essential oils in food: Is citrus the answer. *Trends in Food Science and Technology, 19,* 156–164.

Francis, G. A., Gallone, A., Nychas, G. J., Sofos, J. N., Colelli, G., Amodio, M. L., et al. (2012). Factors affecting quality and safety of fresh-cut produce. *Critical Reviews in Food Science and Nutrition, 52*(7), 595–610.

Francis, G. A., Thomas, C., & O'Beirne, D. (1999). The microbiological safety of minimally processed vegetables. *International Journal of Food Science and Technology, 34*, 1–22.

Gahagan, M. (2008). *The essential oil database. Available from http://www.chamomile.co.uk.*

García, M. A., Martino, M. N., & Zaritzky, N. E. (2001). Composite starch-based coatings applied to strawberries (*Fragaria ananassa*). *Food/Nahrung, 45*(4), 267–272.

Glicerina, V., Tylewicz, U., Canali, G., Siroli, L., Dalla Rosa, M., Lanciotti, R., et al. (2019). Influence of two different cocoa-based coatings on quality characteristics of fresh-cut fruits during storage. *LWT-Food Science and Technology, 101*, 152–160.

Gonzalez-Aguilar, G. A., Ayala-Zavala, J. F., Olivas, G. I., De La Rosa, L. A., & Álvarez-Parrilla, E. (2010). Preserving quality of fresh-cut products using safe technologies. *Journal für Verbraucherschutz und Lebensmittelsicherheit, 5*(1), 65–72.

Gunduz, G. T., Gonul, S. A., & Karapınar, W. (2010). Efficacy of oregano oil in the inactivation of *Salmonella typhimurium* on lettuce. *Food Control, 21*, 513–517.

Gutierrez, J., Bourke, P., Lonchamp, J., & Barry-Ryan, C. (2009). Impact of plant essential oils on microbiological, organoleptic and quality markers of minimally processed vegetables. *Innovative Food Science and Emerging Technologies, 10*, 195–202.

Hodges, D. M., & Toivonen, P. M. (2008). Quality of fresh-cut fruits and vegetables as affected by exposure to abiotic stress. *Postharvest Biology and Technology, 48* (2), 155–162.

Hyldgaard, M., Mygind, T., & Meyer, R. L. (2012). Essential oils in food preservation: mode of action, synergies, and interactions with food matrix components. *Frontiers in Microbiology, 3*(1), 12.

Juglal, S., Govinden, R., & Odhav, B. (2002). Spice oils for the control of co-occurring mycotoxin-producing fungi. *Journal of Food Protection, 65*(4), 683–687.

Kader, A. A. (2008). Flavor quality of fruits and vegetables. *Journal of the Science of Food and Agriculture, 88*, 8–1863.

Kim, D.-H., Kim, H.-B., Chung, H.-S., & Moon, K.-D. (2014). Browning control of fresh-cut lettuce by phytoncide treatment. *Food Chemistry, 159*, 188–192.

Kim, I. H., Lee, H., Kim, J. E., Song, K. B., Lee, Y. S., Chung, D. S., et al. (2013). Plum coatings of lemongrass oil-incorporating carnauba wax-based nanoemulsion. *Journal of Food Science, 78*(10), 1551–1559.

Lamikanra, O. (2002). *Fresh-cut fruits and vegetables: Science, technology, and market.* CRC Press.

Lanciotti, R., Belletti, N., Patrignani, F., Gianotti, A., Gardini, F., Guerzoni, M. E., et al. (2003). Application of hexanal, (E)-2-hexenal, and hexyl acetate to improve the safety of fresh-sliced apples. *Journal of Agricultural and Food Chemistry, 51* (10), 2958–2963.

Lanciotti, R., Gianotti, A., Patrignani, F., Belletti, N., Guerzoni, M. E., & Gardini, F. (2004). Use of natural aroma compounds to improve shelf-life and safety of minimally processed fruits. *Trends in Food Science and Technology, 15*(3–4), 201–208.

Landry, K. S., Micheli, S., McClements, D. J., & McLandsborough, L. (2015). Effectiveness of a spontaneous carvacrol nanoemulsion against *Salmonella enterica Enteritidis* and *Escherichia coli* O157: H7 on contaminated broccoli and radish seeds. *Food Microbiology, 51*, 10–17.

Leverentz, B., Conway, W. S., Janisiewicz, W., Abadias, M., Kurtzman, C. P., & Camp, M. J. (2006). Biocontrol of the food-borne pathogens *Listeria monocytogenes* and *Salmonella entericaserovarpoona* on fresh-cut apples with naturally occurring bacterial and yeast antagonists. *Applied and Environmental Microbiology, 72,* 1135–1140.

Martin-Belloso, O. (2007). *Pros and cons of minimally processed foods.* London: Elsevier Science.

Nguyen-The, C., & Carlin, F. (1994). The microbiology of minimally processed fresh fruits and vegetables. *Critical Reviews in Food Science and Nutrition, 34*(4), 371–401.

Park, S., Kodihalli, I., & Zhao, Y. Y. (2005). Nutritional, sensory, and physicochemical properties of vitamin E- and mineral-fortified fresh-cut apples by use of vacuum impregnation. *Journal of Food Science, 70,* 9–593.

Patrignani, F., Siroli, L., Serrazanetti, D. I., Gardini, F., & Lanciotti, R. (2015). Innovative strategies based on the use of essential oils and their components to improve safety, shelf-life and quality of minimally processed fruits and vegetables. *Trends in Food Science and Technology, 46,* 311–319.

Prakash, A., Baskaran, R., Paramsivam, N., & Vadivel, V. (2018). Essential oil based nanoemulsions to improve the microbial quality of minimally processed fruits and vegetables: a review. *Food Research International, 111,* 509–523.

Qadri, O. S., Yousuf, B., & rivastava, A. K. (2016). Fresh-cut produce: advances in preserving quality and ensuring safety. In M. W. Siddiqi, & A. Ali (Eds.), *Postharvest management of horticultural crops: Practices for quality preservation* (pp. 265–290). Waretown, NJ: Apple Academic Press.

Qadri, O. S., Yousuf, B., & Srivastava, A. K. (2015). Fresh-cut fruits and vegetables: Critical factors influencing microbiology and novel approaches to prevent microbial risks—A review. *Cogent Food and Agriculture, 1*(1), 606–1121.

Ragaert, P., Devlieghere, F., & Debevere, J. (2007). Role of microbiological and physiological spoilage mechanisms during storage of minimally processed vegetables. *Postharvest Biology and Technology, 44,* 185–194.

Raybaudi-Massilia, R. M., Mosqueda-Melgar, J., & Martín-Belloso, O. (2008). Edible alginate-based coating as carrier of antimicrobials to improve shelf-life and safety of fresh-cut melon. *International Journal of Food Microbiology, 121*(3), 313–327.

Rico, D., Mart´ın-Diana, A. B., Barat, J. M., & Barry-Ryan, C. (2007). Extending and measuring the quality of fresh-cut fruit and vegetables: a review. *Trends in Food Science and Technology, 18,* 373–386.

Rojas-Grau, M. A., Raybaudi-Massilia, R. M., Soliva-Fortuny, R. C., Avena-Bustillos, R. J., McHugh, T. H., & Martín-Belloso, O. (2007). Apple pureeealginate edible coating as carrier of antimicrobial agents to prolong shelf-life of fresh-cut apples. *Postharvest Biology and Technology, 45,* 254–264

Rozin, E., & Rozin, P. (1981). Culinary themes and variations: traditional seasoning practices both a sense of familiarity and a source of variety. *Natural History, 90,* 6–14.

Ruengvisesh, S., Loquercio, A., Castell-Perez, E., & Taylor, T. M. (2015). Inhibition of bacterial pathogens in medium and on spinach leaf surfaces using plant-derived antimicrobials loaded in surfactant micelles. *Journal of Food Science, 80,* 2522–2529.

Saftner, R. A., Abbott, J. A., Bhagwat, A. A., & Vinyard, B. T. (2005). Quality measurement of intact and fresh-cut slices of Fuji, Granny Smith, Pink Lady, and Gold-Rush apples. *Journal of Food Science, 70,* 317–324.

Salvia-Trujillo, L., Rojas-Graü, M. A., Soliva-Fortuny, R., & Martín-Belloso, O. (2015). Use of antimicrobial nanoemulsions as edible coatings: Impact on safety and quality attributes of fresh-cut Fuji apples. *Postharvest Biology and Technology, 105,* 8–16.

Serrano, M., Martínez-Romero, D., Castillo, S., Guillén, F., & Valero, D. (2005). The use of natural antifungal compounds improves the beneficial effect of MAP in sweet cherry storage. *Innovative Food Science & Emerging Technologies, 6*(1), 115–123.

Sessa, M., Ferraria, G., & Donsìb, F. (2015). Novel edible coating containing essential oil nanoemulsions to prolong the shelf life of vegetable products. *Chemical Engineering Transactions, 43*, 55–60.

Siroli, L., Patrignani, F., Serrazanetti, D. I., Gardini, F., & Lanciotti, R. (2015). Innovative strategies based on the use of bio-control agents to improve the safety, shelf-life and quality of minimally processed fruits and vegetables. *Trends in Food Science and Technology, 46*, 302–310.

Siroli, L., Patrignani, F., Serrazanetti, D. I., Tabanelli, G., Montanari, C., Gardini, F., et al. (2015a). Lactic acid bacteria and natural antimicrobials to improve the safety and shelf-life of minimally processed sliced apples and lamb's lettuce. *Food Microbiology, 47*, 74–84.

Siroli, L., Patrignani, F., Serrazanetti, D. I., Tabanelli, G., Montanari, C., Tappi, S., et al. (2015b). Potential of natural antimicrobials for the production of minimally processed fresh-cut apples. *Journal of Food Processing and Technology, 6*(2), 1–9.

Sivakumar, D., & Bautista-Baños, S. (2014). A review on the use of essential oils for postharvest decay control and maintenance of fruit quality during storage. *Crop Protection, 64*, 27–37.

Sivapalasingam, S., Friedman, C. R., Cohen, L., & Tauxe, R. V. (2004). Fresh produce: a growing cause of outbreaks of foodborne illness in the United States, 1973 through 1997. *Journal of Food Protection, 67*, 2342–2353.

Ultee, A., Kets, E. P. W., & Smid, E. J. (1999). Mechanisms of action of carvacrol on the food-borne pathogen *Bacillus cereus. Applied and Environmental Microbiology, 65*(10), 4606–4610.

Valero, D., Valverde, J. M., Martínez-Romero, D., Guillén, F., Castillo, S., & Serrano, M. (2006). The combination of modified atmosphere packaging with eugenol or thymol to maintain quality, safety and functional properties of table grapes. *Postharvest Biology and Technology, 41*(3), 317–327.

Viuda-Martos, M., Ruiz-Navajas, Y., Fernandez-Lopez, J., & Perez-Alvarez, J. (2008). Antifungal activity of lemon (*Citrus lemon* L.), mandarin (*Citrus reticulata* L.), grapefruit (*Citrus paradisi* L.) and orange (*Citrus sinensis* L.) essential oils. *Food Control, 19*, 1130–1138.

World Health Organization (2003). *The world health report 2003: Shaping the future.* World Health Organization. WHO Library Cataloguing-in-Publication Data.

Yousuf, B., & Srivastava, A. K. (2015). Psyllium (Plantago) gum as an effective edible coating to improve quality and shelf life of fresh-cut papaya (*Carica papaya*). *International Journal of Biological, Biomolecular, Agricultural, Food and Biotechnological Engineering, 9*, 702–707.

Yousuf, B., & Srivastava, A. K. (2017). A novel approach for quality maintenance and shelf life extension of fresh-cut Kajari melon: Effect of treatments with honey and soy protein isolate. *LWT-Food Science and Technology, 79*, 568–578.

Yuan, W., Teo, C. H. M., & Yuk, H. G. (2019). Combined antibacterial activities of essential oil compounds against *Escherichia coli* O157: H7 and their application potential on fresh-cut lettuce. *Food Control, 96*, 112–118.

Zambrano-Zaragoza, M. L., Quintanar-Guerrero, D., Real, A. D., Pinon-Segundo, E., & Zambrano-Zaragoza, J. F. (2017). The release kinetics of β-carotene nanocapsules/xanthan gum coating and quality changes in fresh-cut melon (cantaloupe). *Carbohydrate Polymers, 157*, 1874–1882.

Zhang, G., Gu, L., Lu, Z., Yuan, C., & Sun, Y. (2019). Browning control of fresh-cut Chinese yam by edible coatings enriched with an inclusion complex containing star anise essential oil. *RSC Advances, 9*(9), 5002–5008.

9

Fortification in fresh and fresh-cut horticultural products

Alka Joshi, Uma Prajapati, Shruti Sethi, Bindvi Arora, Ram Roshan Sharma

Division of Food Science and Postharvest Technology, ICAR-Indian Agricultural Research Institute, New Delhi, India

1 Food production in India

Due to improved varieties, cultivation practices, and improved mechanization, the food platter of India has been converted from the condition of food scarcity to self-sufficiency in 50 years. India also enjoys several superlatives in food production. It is one of the world's largest food producers after China and the United States. It is the largest producer of milk (~165 million tonnes), pulses (~25.23 million tonnes), tea, and spices in the world. India is the second largest producer of horticultural crops (310 million tonnes) and the third largest producer of food grains (285 million tonnes). In spite of such abundance, as per 2018 estimates, India ranks at 103 positions in Global Hunger Index on the basis of percentage of stunting, wasting, child mortality rate and overall malnutrition status. A large population (1.36 billion), poverty, and lack of awareness are some of the major factors for this thought-provoking situation. Most developing countries are facing problems of nutrient deficiencies particularly for vital minerals. In this direction, the World Bank identified 36 countries in the world including India, Africa, Indonesia, Bangladesh, etc. for the dire consequences for mobility, mortality, productivity, and economic growth (Word Bank Report, 2009).

2 What may be the strategy?

For combating nutritional deficiencies challenges food fortification has a proven history in various countries such as iodine fortification in salt (mandatory in Switzerland and the United

Fresh-cut Fruits and Vegetables. https://doi.org/10.1016/B978-0-12-816184-5.00009-4

States), folic acid fortification in flour, vitamin B_1 fortification in flour (mandatory in Canada), etc. (Chopra, 1974; Clarke, 1995). Recently, mineral fortification has been targeted for various staple crops, especially cereal crops, for targeted mineral delivery to the undernourished (Frano, De Moura, Boy, Lonnerdal, & Burri, 2008). Similar approaches are required for addressing the issues related to mineral deficiencies particularly iron, zinc, calcium etc. in growing population groups (specifically children and teen agers) using a suitable carrier and fortificant/s.

It is a well-established fact that through conventional plant breeding, agronomic management, or genetic engineering, the problem of nutrient deficiencies can be mitigated in a sustainable manner. However, these approaches are time-consuming, they require high skill for making cross-selection, can be challenging in terms of obtaining regulatory approvals, and are affected by agronomic traits, environmental factors, variability in micronutrient concentrations, bioavailability issues, etc. (Nestel, Bouis, Meenakshi, et al., 2006). Controls over vulnerable population, consumption pattern, and availability also create difficulty in addressing these challenging issues through biofortification approaches. Keeping these difficulties in view, food fortification seems a practically feasible, controlled, and well-adopted approach to meet these challenges. Jain and Pandey (2012) have highlighted various factors on which the success of fortification depends:

1. effective fortificant and carrier (food) system;
2. impact of fortification color, texture, flavor, storability of food products;
3. consumption of carrier in staple diet form;
4. centralized processing, especially if market-driven;
5. frequent and continuous consumption of fortified foods;
6. economic feasibility;
7. political and industrial support;
8. consumer acceptance;
9. economic sustainability;
10. validation; and
11. nutrient interactions, processing losses, and bioavailability of the fortificant(s).

All the above-reported factors are well within human control as consumers, food industry personnel, or policy makers. Therefore, among all the existing methodologies, food fortification seems a practically feasible, consumer-oriented, target- and market-driven approach.

3 Background of food fortification

The history of nutrient supplementation can be traced back to the year 400 BC when a Persian physician, Melanpus, suggested adding iron fillings to wine to increase soldiers' potency. In the early 1920s, salt fortification was one of the popular technologies where iodization of salt was done in Switzerland and the United States to eliminate the higher incidence of goiter, reduced mental ability, and short stature (Burgi, Supersaxo, & Selz, 1990). Further, it was between the First and Second World War (1924–44) that the nutrient supplementation became an established measure to correct or prevent nutritional deficiencies in human population or to restore nutrients lost during food processing. In the early 1940s, cereals and cereal products were fortified with thiamine, niacin, and riboflavin, milk was fortified with vitamin D, and margarine was fortified with vitamin A. Folic acid fortification has become popular in countries like the United States and Canada (Lindsay, de Benoist, Dary, & Hurrell, 2006). The concept of double fortification has also fetching the market such as double fortification of salt, vitamin D and calcium fortification in milk products, vitamin E and A fortification in fats, etc. All over the world, various food commodities are fortified in order to reduce the prevalence of diseases and disorders like xerophthalmia (vitamin A deficiency), goiter (iodine deficiency), rickets (vitamin D deficiency), and anemia (iron deficiency) (Mozaffarian, Rosenberg, & Uauy, 2018).

4 What is fortification/enrichment?

According to the Codex's "General Principles for the Addition of Essential Nutrients to Foods," the term "fortification" or "enrichment" is associated with the addition of one or several essential nutritional elements to a food product, regardless of whether it is or not habitually contained in foods, toward the prevention or correction of proven deficiencies in one or more nutrients, either for the entire population or certain specific groups (Codex Alimentarius Commission, 1987).

The FAO also defined food fortification as the addition of one or more essential nutrients to a food, whether or not it is normally contained in the food, for the purpose of preventing or correcting a demonstrated deficiency of one or more nutrients in the population or specific population groups (FAO/WHO, 1994). In this case, the amount of nutrients added may be higher than that present before processing.

Enrichment has been used interchangeably with fortification (FAO/WHO, 1994); however, a slight difference exists between fortification and enrichment. Fortification is target-specific, using a targeted dose for targeted population to offer targeted benefits, while enrichment is not target-specific, but is for the well-being of the general population. However, similar to the FAO/WHO approach, these two terms are also used interchangeably in commercial practice.

A similar term, "restoration," means the process in which the amount of nutrients added is almost equal to the content naturally present in the food before processing. It generally compensates for the nutrient content which is lost during the various steps of processing and storage. However, taking into consideration the fact that the aim in all the cases is to improve the nutritional value of foods, the term "nutrification" was suggested, which includes both enrichment and fortification:

$$\text{Nutrification} = \text{Fortification} + \text{Enrichment}$$

5 Types of fortification

5.1 Mass fortification

The fortification of foods that are widely consumed by the general population is almost always mandatory.

5.2 Targeted fortification

To fortify foods designed for specific population subgroups, such as complementary foods for young children or rations for displaced populations can be either mandatory or voluntary, depending on the public health significance of the problem it is seeking to address.

5.3 Market-driven fortification

Allowing food manufacturers voluntarily to fortify foods available in the market place is always voluntary, but governed by regulatory limits.

6 Fortification of horticultural produce

The concept of horticultural produce fortification is new but quite appealing and realistic. A horticultural matrix in the form of fresh-cut or minimally processed can act as an ideal carrier for fortification, which is the foremost condition for success of any fortification program. Fresh-cut fruit and vegetables without any alteration or any further processing can become part of a consumer's daily diet. It is evident from the background of food fortification that till now, the major success stories of fortification are about fortification of salt, cereals, and milk and dairy products. These may be due to cost-effectiveness and the staple consumption pattern of salt and cereal products by the population.

A bumper rise in horticultural production, attention to underutilized and low-value produce as well as high-value horticultural crops, and a significant rise in per capita income have shifted the paradigm of food consumption by the consumer from cereal crops to horticultural crops. Not only fresh produce, but also minimally processed fruits and vegetables, are attracting customers. Considering these points, horticultural produce as a whole, including fresh-cut and minimally processed fruits and vegetables, can be considered as an ideal matrix for food fortification. Several other points make horticultural produce (fresh-cut) a relevant matrix of food fortification, provided low-value crops are also targeted for fortification, such as:

1. recommended to be consumed by all age groups;
2. options for all economic class of consumers;
3. RDI of minerals can be achieved by consuming fresh-cut produce;
4. thermo-resistant as well as thermo-labile—nutrients can be delivered through fresh-cut or minimally processed fortified fruits and vegetables;
5. inherent porosity of horticultural matrix facilitates fast imbibition and adherence of fortificants;
6. not only nutrient enhancement but shelf-life extension and probiotics enrichment can be achieved through the same fortification protocols;
7. some horticultural produce has the ability to add bulk to the diet, such as potatoes, jackfruit, etc.; and
8. horticultural commodities (except GLVs for oxalate) are relatively free of, or have significantly lower levels of, antinutritional factors compared to cereal crops, which further helps in better absorption of vital nutrients (bioavailable) in the human body.

7 What may the fortificants be?

The fortificant should be the GRAS substance (as per FDA recommendations). Additionally, it should affect the sensory and processing quality of the matrix only minimally. Thirdly, the storability of produce, whether fresh-cut or whole, should not be negatively affected by the addition of fortificants. They should be more bioavailable and cost-effective. By definition, bioavailability is the amount of the nutrient that is accessible for utilization in normal physiological functions, metabolism, and storage, which can be enhanced or inhibited by the presence of other food components and food-processing techniques (Frano et al., 2008). Generally, the availability of any nutrient in a biological system is bioavailability.

The criteria of fortificant selection may be its safety, availability, bioavailability, cost effectiveness, neutral taste and aroma, or its effect on the produce's appearance, overall acceptability, and handling, etc. Many different fortificants are available in the market. Table 1 provides the details of some commonly used

Table 1 Suitable fortificants for targeted nutrient

S. No.	Targeted nutrient	Possible fortificants	Remarks
1.	Iron	Ferrous sulfate (100%) Ferrous fumarate (100%) Ferric pyrophosphate (21%–74%) and Ferric orthophosphate (25%–31%) Encapsulated ferrous sulfate NaFeEDTA, ferrous gluconate (89%) and ferrous succinate (98%) Encapsulated ferrous sulfate Encapsulated ferrous fumarate	Provoking oxidation (except iron-EDTA forms) and browning reactions, lower bioavailability and discoloration are common problems with iron fortificants. Encapsulated forms due to lower bioavailability are recommended to be added in higher (~two times) proportion than the normal iron fortificants. However, encapsulated forms due to their lesser impact on the sensory attributes of carrier foods are generally preferred
2.	Folic acid	Pure and simple derivative form	The daily requirement of folic acid is very low (0.1–0.4 mg/day of folic acid) but its deficiency in infants leads to a birth defect known as neural tube defect (Spina bifida)
3.	Zinc	The FDA lists five zinc compounds as GRAS for foods: zinc chloride, zinc gluconate, zinc oxide, zinc stearate and zinc sulfate	Zinc with lactate and citrate ligands are gaining popularity for their enhanced functional bioavailability
4.	Vitamin-A	Retinyl acetate Retinyl acetate (vitamin A acetate) ß carotene: provitamin A	Water solubility is a significant challenge; however, the mentioned three are soluble in warm water, warm water with milky appearance, and partially soluble in the presence of BHT:emulsifier, respectively

Values in parentheses show the bioavailability of iron fortificants.

Based on Shelke, K., & Feder, D. R. D. (2006). Iron and zinc fortification National health surveys claim almost three in four Americans don't meet requirements for zinc and iron. Food Processing. *Available from: https://www.foodprocessing.com/articles/2006/019 (Accessed 24 February 2006); Jain, S. C., & Pandey, A. (2012). Improvement of nutritional quality of bakery products through fortification/enrichment.* Processed Food Industry, 16 *(2), 18–25.*

fortificants for iron, zinc, folic acid and vitamin A. These are the major micronutrients which are targeted by maximum fortification programs in India and abroad.

8 Nutrient interactions

When targeting any nutrient, its interaction with other nutrients should be kept in mind (Nair, 2012). For example, some inhibitors like phytates or oxalates combine with minerals like Fe, Ca, and Zn, and render them insoluble/unavailable. This shows that for fortification appropriate variety, the inherent antinutrients factor is the foremost important criterion since these factors restrain effective nutrient delivery in the human body.

9 Methods of fortification for fresh-cut produce

Here osmotic treatment, vacuum impregnation, ultrasound assisted vacuum impregnation, ohmic-assisted vacuum impregnation, pulsed electric field-assisted vacuum impregnation, and dipping treatment are described as potential methods for fortification of horticulture matrices.

9.1 Osmotic treatment

Osmotic treatment (OT) is considered to be a feasible technology for developing fruit and vegetable matrices to which functional ingredients can be successfully added to provide novel functional product categories and new commercial opportunities (Alzamora et al., 2005). OT is a nonthermal treatment, the aim of which is to modify the composition of food material by partially removing water and impregnating it with solutes, without affecting the material's structural integrity (Spiess & Behsnilian, 1998). OT has two advantages: it will partially dehydrate the sample, and it enables the introduction of controlled quantities of solutes into food. When samples are immersed in a hypertonic media, the osmotic solute is transported from the solution to the food material and water flows out of the product counter-currently. In OT, quality can be improved not only by removing water with the minimal thermal stress, but also by impregnating the solutes and modifying the structure (Torreggiani & Bertolo, 2001). Under atmospheric conditions, previous authors developed antioxidant-rich aonla candy, known as Pusa nutra aonla candy (Kaur & Sharma, 2015), papaya candy (Joshi & Sagar, 2016), etc. To fasten the osmotic imbibition process, atmospheric impregnation can be replaced by vacuum-assisted impregnation or vacuum impregnation (VI).

9.2 Vacuum impregnation

In recent years, vacuum impregnation has emerged as a useful tool for incorporating physiologically active compounds into the porous structure of fruits and vegetables without disturbing their cellular structure (Anino, Salvatori, & Alzamora, 2006; Barrera, Betoret, & Fito, 2004; Betoret et al., 2003; Gras, Vidal, Betoret, Chiralt, & Fito, 2003; Martinez-Monzo, Martinez-Navarrete, Chiralt, & Fito, 1998). Porosity (intercellular air spaces) is common in parenchymatous tissue of horticultural matrices and has been estimated to be 20%–25% of the total volume in apple, 15% in peach, 37%–45% in mushroom, and 1% in potato (Alzamora et al., 2005). When a porous tissue is immersed in a physiologically active component (PAC) concentrated solution under vacuum conditions, air is extracted from the pores. When the atmospheric pressure is restored, the impregnation solution penetrates the intercellular spaces by capillary action and by the pressure gradients (i.e., the hydrodynamic mechanism, HDM) that are imposed on the system, helping incorporation of active components (Fito, 1994). In short, in this technique, air escapes or is sucked from the pores of the native matrix through a vacuum, and when such vacated tissues are exposed to a nutrient(s) solution mixture due to the concentration gradient, the solution moves towards the vacated pores.

This shows how in vacuum impregnation, porosity plays a major role. Any process that enhances porosity will reduce the process time, such as blanching, sonication, ohmic heating, etc. Generally, in all horticulture produce, porosity can be enhanced by blanching. Blanching treatment often produces profound structural alterations (swelling of cell walls, disruption of membranes, etc.), which affects mass transport phenomena, resulting in the extensive uptake of solute inside the cells (Alzamora et al., 2005). It has been reported by Joshi et al. (2019) that after blanching, due to increased porosity, the rate of infusion of bioactives was increased 1.5 to 1.8-fold. Therefore, porosity measurement is an important process parameter to control the infusion process under vacuum or atmospheric pressure. Sahin and Sumnu (2006) have described various methods of porosity measurement such as the direct method, the optical method, the density method, the gas pycnometer method (volume fraction of air), and the porosimeter (liquid extrusion from the pores or liquid intrusion into the pores).

9.2.1 *Vacuum impregnation process*

The vacuum impregnation system has been described by Joshi et al. (2016) for mineral fortification in raw potato discs followed by microwaving; it was fabricated by Kar and Lad (2014) in the Division of Food Science & PHT, ICAR-IARI, Pusa, New Delhi (Fig. 1). A vacuum was created in a closed chamber, or a desiccator in which the raw commodity can be placed in a single layer on a perforated base or sieve. The large surface area of the commodity results in more effective exposure of the matrix to the vacuum. After the desirable vacuum level for a set time is achieved, the raw commodity should immediately be exposed to the fortificants solution mixture of a particular concentration for a predefined restoration time or infusion period at atmospheric pressure.

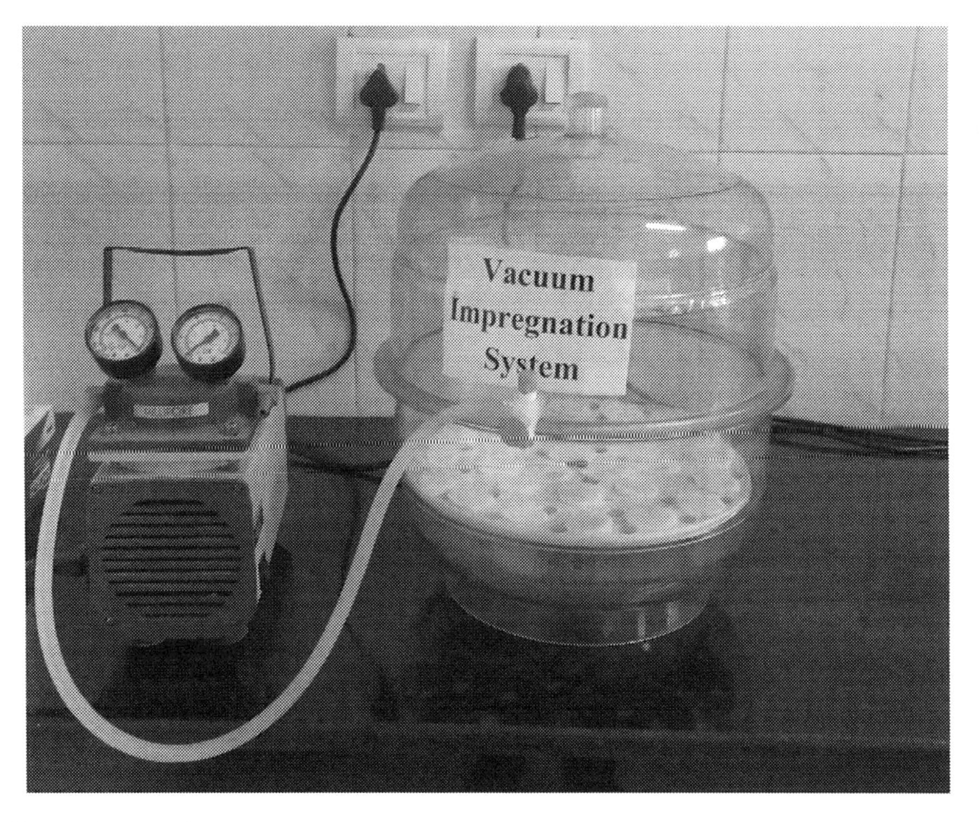

Fig. 1 Vacuum impregnation system.

9.2.2 Ultrasound assisted vacuum impregnation

Ultrasound is an effective technology in food application which is based on the principle of cavitation force where gas- or vapor-filled bubbles are formed and collapsed in the liquid medium due to pressure fluctuation generated by the passage of ultrasound waves. Due to this cavitation action, the porosity of cells is enhanced. Thus, ultrasonicated vacuum impregnation could be a promising technique to fortify fresh food products. The cumulative results of cavitation, heating effect, and sponge effect due to ultrasonication can increase the mass diffusivity and mass transfer phenomena, which make interchange of matters easier by creating microscopic channels (Yao, 2016).

Yılmaza and Bilek (2018) developed vacuum-ultrasonic equipment and analyzed the effects of different ultrasound powers ranging from 96 to 198 W at 35 kHz under vacuum and restoration period on fresh-cut apple during impregnation of calcium and black carrot phenolics. The results suggested that vacuum impregnation did not rupture cellular integrity at 130 W ultrasound and increased calcium content (13.8%), total phenolics (11.8%), total flavonoids (17.3%), total anthocyanins (24.6%), and antioxidant capacities (23.6%) of apple discs compared to nonultrasound assisted vacuum impregnation. In addition, it was found that black carrot infused apple slices by the ultrasound-assisted vacuum impregnation technique also inhibited the total population of psychrophilic and mesophilic microorganisms during a storage period of 18 days.

9.2.3 Ohmic Heating-assisted Vacuum Impregnation

Ohmic heating (OH) is a thermal process in which the food material, which serves as an electrical resistor, is heated by an electric current (Salengke & Sastry, 2007). Ohmic heating enhanced permeability of cell boundaries therefore can assist VI treatment to impregnate nutrient into the food matrix. Moreno et al. (2016) conducted a study on folic acid incorporation using ohmic heating. In this study, apple slices were enriched with folic acid by using vacuum impregnation along with ohmic heating technique at 30°C, 40°C, and 50°C and air drying at 50°C, 60°C, and 70°C. The results showed that vacuum impregnated apple slices at 40°C and vacuum impregnation along with ohmic heating at 50°C represented higher content of folic acid due to electropermeabilization effect. This effect induces folic acid retention and increases its concentration by reducing the water content of apple slices, whereas air-dried apple slices at 60°C showed higher retention of folic acid.

9.2.4 Pulsed electric field-assisted vacuum impregnation

Pulsed electric field (PEF) is a nonthermal technology that enhances mass transport by increasing permeability of cell membrane with less degradation of nutritional compounds (Janositz, Noack, & Knorr, 2011; Lebovka, Praporscic, & Vorobiev, 2004). PEF mainly works on the principle of electroporation of cell membranes where short pulses of high voltage electric fields cause increased permeability of cell membrane (Ignat, Manzocco, Brunton, Nicoli, & Lyng, 2015; Maskooki & Eshtiaghi, 2012). It is used as a pretreatment for extraction, drying, freezing, and salting because of its accelerating effects on mass transfer. Due to its efficiency in mass transfer, it is clubbed with vacuum impregnation for fortification of various food commodities. Mashkour, Yahya Maghsoudlou, and Aalami (2018) clubbed vacuum impregnation and pulsed electric field for fortifying whole potato with ferric pyrophosphate. In this, whole potatoes were pretreated at 394 V/cm with 36 pulses followed by vacuum treatment at 3.5 kPa for 37 min., and maintained for 39 min. restoration time in atmospheric pressure. The results suggested that iron content of vacuum impregnated and pulsed electric field with vacuum impregnation increased iron content about 126% and 457%, respectively, in comparison to unprocessed potatoes. This is the highest achievable iron fortification with the lowest physicochemical changes in comparison to any other processing techniques.

9.3 Application of vacuum impregnation technique

9.3.1 Probiotics

Generally, the term "probiotic food" is related to dairy products but, through vacuum impregnation, it is feasible to infuse probiotics cultures in a horticulture matrix. Probiotics are live microorganisms that improve gut health. Most frequently, bacteria used in probiotics belong to *Lactobacillus* and *Bifidobacterium*, *Streptococcus thermophiles*, and *Saccharomyces boulardii* species. Probiotics can play a role in nutritional management, which includes immunomodulation, infection control, eradication of multidrug-resistant microbes, cholesterol, blood pressure lowering, improving lactose tolerance, antimutagenic/anticarcinogenic activity, etc. (Rastall, Fuller, Gaskins, & Gibson, 2000; Saarela, Lähteenmäki, Crittenden, Salminen, & Mattila-Sandholm, 2002). Impregnation studies were conducted in 'Granny Smith' apples with different microorganisms, such as *Saccharomyces cerevisiae*, *Lactobacillus acidophilus*, and *Phoma glomerata*. Here the cut cylinders of the apple were impregnated with an isotonic solution of sucrose containing microorganisms at one vacuum pulse of 2 min and at five different absolute pressures (75, 125, 225, 325, 425 mmHg) at 25°C. The results suggested that the lower the absolute pressure of the vacuum pulse applied, the higher the incorporation of microorganisms by the hydrodynamic model (Rodrı´guez, 1998). A similar study was also conducted in guava where guava pieces was impregnated with 10^7 CFU/g. counts of *Bifidobacterium* spp., which later decreased to three log cycles after 12 days of storage at 5°C when no anaerobic condition was provided by the packaging (Ortiz, Salvatori, & Alzamora, 2002).

9.3.2 Minerals

Mineral fortification is one of the most popular methods of food fortification. In particular, calcium, zinc, and iron are incorporated in food matrices. Calcium is mainly required for good bone and teeth health, as well as helping to reduce lead absorption in the blood. Zinc is an essential component for our immune system, and responsible for the proper functioning of almost 300 enzymes in the human body. Recent analysis of dietary patterns has revealed that approximately 48% of the world's population is at risk of zinc deficiency, in spite of the low value of its recommended dietary allowance.

Calcium impregnation is done by using VI techniques in apple tissue, where apple cylinders were immersed in an isotonic glucose aqueous solution containing 5.24% (w/w) Ca^{2+} salts (5266 ppm Ca^{2+}). In vacuum impregnation, a pressure of 30 mmHg was applied to the system for 10 min; atmospheric pressure was then restored and maintained for 10 min. For atmospheric impregnation (AI), fruit samples were taken out of the solution at different immersion times (0, 2, 6, 10, and 22 h).Ca^{2+} content significantly increased along with AI treatment, reaching 1300 ppm in samples treated for 6 h and 3100 ppm after 22 h. Calcium incorporation in 200 g of fruit would satisfy about 41%–62% of the adequate intake, AI_{Ca} (1000 mg/day) (Alzamora, Anino, & Salvatori, 2001).

Calcium fortification in different vegetables like eggplant, oyster mushroom, and carrot has been performed using VI. VI (37.5 mmHg for 10 min and immersion at atmospheric pressure for 10 min) was made in sucrose isotonic aqueous solutions containing calcium lactate (Gras et al., 2003). Melon was also utilized for impregnation of zinc and calcium (Tapia, Schulz, Gómez, López-Malo, & Welti-Chanes, 2003). de Lima et al. (2016) used vacuum impregnation and the drying method to produce calcium-enriched pineapple snacks. Pineapple slices were fortified with calcium (using 1% $CaCl_2$ solution) by three different methods: by keeping slices at atmospheric pressure for 20 min; by applying vacuum (13.33 kPa) to a container containing pineapple slices immersed in calcium impregnating solution for 10 min. followed by impregnation at atmospheric pressure for 10 min.; and by applying vacuum impregnation followed by vacuum pulses (6.70 kPa) after removing the samples from the impregnating solution. The samples were then dried using convective heat and freeze drying. Results suggested that samples that were vacuum impregnated retained 91% higher Ca^{+2} concentration than samples impregnated at atmospheric pressure. The combination of vacuum impregnation and drying is the best practice to produce calcium-fortified pineapples with better textural and nutritional properties.

Gras, Vidal-Brotóns, and Vásquez-Forttes (2011) conducted research on iceberg lettuce to fortify its leaves with calcium using the vacuum impregnation technique. For this study, the whole leaf was differentiated into three distinct regions (apical, medium, and basal) on the basis of distribution of the vascular system. For vacuum impregnation, pressure of 500 mbar was maintained for 10 min., and samples remained immersed into the solution (5.4 g Ca/L) for 10 min. at atmospheric pressure. The results suggested that fortified iceberg lettuce leaves contained 169 mg of calcium per 250 g of matrix.

9.3.3 Ascorbic acid

Ascorbic acid is one of the essential components of food commodities. Ascorbic acid (AA), as a natural antioxidant, protects against oxidative stress and acts as a free radical scavenger, which otherwise leads to tissue damage from lipid peroxidation and causes severe degenerative diseases such as cancer and heart disease (Bates, 1997). According to the available biochemical, clinical, and epidemiological evidence, the current recommended daily allowance (RDA) for AA is 100 mg/day for adults in order to reduce risk of heart disease, stroke, and cancer in healthy individuals (Naidu, 2003). Due to several health benefits, this ascorbic acid has been incorporated in fresh commodities to utilize its benefits. Hironaka et al. (2011) conducted a study on fresh potato tubers where whole potatoes were immersed in 10% ascorbic acid solution and a vacuum pressure of 70 cmHg was applied for 0–60 min. followed by atmospheric restoration of 3 h in a vacuum impregnation solution. Results suggested that ascorbic acid concentration in whole potato tubers increased with an increase in vacuum time, i.e., max 150 mg/100 g FW. A study on steam cooking also suggested that 100 g of the steam-cooked VI potatoes for 25 min. could also provide 90%–100% of the RDA of AA to adults. The storage study showed that VI-treated potatoes had high AA concentrations, i.e., 50 mg/100 g FW in storage period for 14 days at low temperature.

9.3.4 Phenolics

Joshi et al. (2016), Joshi et al. (2019), and Joshi and Sagar (2016) successfully enriched black carrot anthocyanins and beetroot betanins in raw potato and raw papaya matrices followed by microwave drying and osmotic drying, respectively. Tapi, Tylewicz, Romani, and Rosa (2016) incorporated green tea extract (GTE) in minimally processed apples by using vacuum impregnation in order to achieve a nutritionally fortified product. Green tea extract is a rich source of flavonoids, having strong antioxidant properties, and has been studied worldwide for in vitro and in vivo trials. Apples were impregnated in a vacuum at an ambient temperature, and a pressure of 200 mbar for 5 min was maintained before restoring atmospheric pressure. A relaxation time of 5 min was also applied. Apples were incorporated with isotonic sucrose solutions containing 1% GTE and/or 1% ascorbic acid (AA). Results suggested that impregnation was satisfactory in minimally processed apples, with a strong increase of the antioxidant compounds' content and activity. Other quality parameters were slightly affected and color was also influenced in VI-treated apples. However, the addition of 1% of AA preserved better color and antioxidant properties during storage, due to reducing the oxidative phenomena. The infusion of green tea catechins and ascorbic acid seems to be a promising tool to obtain functionally fortified fruit produce. Similarly, Bellary and Rastogi (2012) also impregnated curcuminoids extracted from turmeric in coconut slices.

In summary, the important reports of vacuum impregnation are as follows:

1. Alzamora et al. (2005) reported that porous fruits and vegetables like apples, peaches, and mushrooms are the ideal matrix for impregnation of biologically active compounds like minerals, phenolics, and probiotics.
2. Mineral and probiotics infusion in apple and melon tissues has been done by Alzamora et al. (2005) and Betoret et al. (2003), using the vacuum impregnation technique.

3. Shelf-life extension through calcium impregnation in fresh, whole, and minimally processed fruits and vegetables has already been reported by vacuum (Martın-Dianaa et al., 2007).
4. Calcium fortification in various vegetables is an alternative in developing functional foods by applying vacuum impregnation (Gras et al., 2003).
5. Fortification of a potato matrix by natural plant extractives like grape seed extracts (GSE), green tea extracts (GTE), or butylatedhydroxytoluene (BHT) and chemical antioxidants has been done (Rababah, 2012).
6. Zinc enrichment in whole potato tubers by vacuum impregnation has recently been done by Erihemu Hironaka, Koaze, Oda, and Shinanda (2015).
7. Vitamin A fortification (ß carotene, in vivo precursor of vitamin A) has been fortified in frozen products and the method of development has been reported in United States Patent No. 3,423,213 (http://www.google.com/patents/US3423213).

10 Dipping treatment

Dipping treatments with sanitizers, browning inhibiters, elicitors, etc. are commonly practiced in fresh horticultural produce. However, nutritional quality enhancement can also be achieved through dipping treatment. In comparison with the impregnation process, it is less time-consuming (1–5 min) and restricted to surface phenomena. Dipping can be done on whole, peeled, shredded, and sliced commodities. For example, strawberry dipping is done at the whole stage, while potatoes are peeled and then dipped, apples are either dipped as whole or in fresh-cut processing they are cored, sliced, and then dipped. Dipping treatments favor the dispersion of the solution on the surface of the commodity. A larger surface area results in more effective dipping results. One of the major advantages of dipping is the removal of cellular exudates, which affects the postharvest quality of commodity; this is immediately taken care of by the dipping solvent (Soliva-Fortuny & Martin-Belloso, 2003).

Dipping time, frequency, solute composition, temperature, and concentration are the process variables of the dipping treatment. Various researchers have tried the dipping treatment with calcium salt for shelf-life extension in which calcium enrichment was an added advantage. Calcium chloride dips in fresh-cut cantaloupe (Luna-Guzman, Cantwell, & Barrett, 1999), calcium lactate dips on fresh-cut lettuce and carrots (Martin-Diana et al., 2005), and calcium dips on whole strawberry

(Suutarinen, Anakainen, & Autio, 1998) are some examples of this added advantage. Warm temperature (40°C–60°C) (Rico et al., 2007) and acidic pH (Wiley, 1994) have been found to be the major influencing factors of dipping treatment in calcium salt solution dipping experiments.

11 Labeling and claims

Until now, no mandatory fortification has been decided for fresh-cut horticultural produce like salt, cereals, and vegetable oils. However, for mandatory fortification, a declaration on the label stating "fortified with…" (the blank to be filled in with the name of the nutrient and the quantity of the nutrient added, as finalized by the Authority) shall be given. It is also advised that the percentage RDA fulfilled by a serving of the fortified product should also be mentioned on the label.

12 Overdoses impact

Doses of fortificant should be decided firstly on the basis of the RDA of a particular nutrient; secondly, one should keep in mind the concept of RACC (Recommended Amount Customarily Consumed); thirdly, the fortified food is not the sole source of a targeted nutrient. Therefore, to avoid chances of overdoses, nutrient doses calculation must be done with proper attention, keeping consumption pattern, frequency of consumption, and targeted population in mind. For example, iodized salt contains 0.002%–0.004% iodine, supplied as either potassium iodide or potassium iodate. In the United States, iodine is added as potassium iodide in table salt at slightly higher levels (0.006%–0.01% potassium iodide, equivalent to 0.0046%–0.0077% iodine) since for iodine, consumers depend mainly on salt. All these levels are sufficient to provide the RDI of iodine which is approximately 150 μg for adult. An overdose of any nutrient is also harmful specifically if it is fat-soluble. Keeping the recommended intake of 2000 Kcal for sedentary workers in mind, Table 2 shows the adverse effects of overdoses of some reported nutrients (in food or supplements).

13 Conclusion

- The success of fortification efforts depends on serious consideration being given to appropriate carrier (variety) and fortificant selection, bioavailability of nutrients, nutrient interaction, nutrition labeling, appropriate fortificant doses calculation, packaging design, maximum retention of fortificant until consumption, etc.

Table 2 Effect of overdoses of nutrients.

Vitamins

- Overdoses of folic acid mislead the results of vitamin B_{12} deficiency (at intake level >400 μg/day)
- Since vitamin A is fat soluble, its high doses can accumulate in the body and lead to joint pain (at intake level, many times higher than RDI, i.e., 600 μg)
- Riboflavin overdoses may change the urine color while niacin excess leads to liver malfunction (at intake level >1.6 and >150 mg/day, respectively)
- Pantothenic acid excess leads to nausea; pyridoxine in excess leads to hardness in limbs and muscles (at intake level >1200 and >100 mg/day, respectively)
- High intake of vitamin B_{12} (at intake level >3000 μg/day) may cause ill effects to eyes
- At the time of writing, no ill effects of ascorbic acid have been reported
- The highly insoluble nature of vitamin D in excess (at intake level >50 μg/day) leads to coma and death
- Excess of tocopherol (at intake level >1000 mg/day) leads to internal blood clotting
- Excess of vitamin K leads to liver cell damage and anemia

Minerals

- The higher dose health impact of calcium has rarely been reported, but at intake level >1500 mg/day may cause stomach problems similar to magnesium at intake level >400 mg/day, phosphorus at intake level >250 mg/day, and potassium at intake level >3500 mg/day
- Chromium and selenium at intake level >200 μg/day are considered as toxicants similar to copper at intake level >10 mg/day
- No side effects of iodine and sodium have been observed yet, since sodium by its sources is a self-limiting mineral
- At intake level >25 mg/day, may cause constipation and blackening of stool. Since iron deficiency is highly prevalent globally, its impact on excess consumption is rarely noticed (exception: medication during pregnancy)
- Excess consumption of manganese hinders iron bioavailability similar to molybdenum, which leads to copper deficiencies
- Zinc at intake level >25 mg/day may cause anemia and copper deficiency, and intake level >40 mg/day is considered highly toxicant

Based on *https://www.lenntech.com/recommended-daily-intake.htm.*

- Fresh-cut fruits can be targeted for heat-labile nutrients (bioactives, antioxidants, vitamins) as well as heat-stable nutrients, whereas for thermosensible nutrients such as minerals, fresh-cut vegetables would be a more appropriate carrier.
- The improvement of nutritional quality of horticultural produce, particularly fresh-cut and minimally processed, using the tool of fortification/enrichment, can go a long way in improving the health of all age groups from children to the geriatric population, provided the carrier horticulture matrices reach and are well consumed by everyone.

- Selection of low-value horticultural produce like potato, papaya, onion, and cabbage can act as an efficient carrier in which targeted mineral delivery can be expected through GRAS fortificants, which minimally affects the appearance, texture, flavor, and storability of raw commodities in a highly controlled, cost-effective, and practically feasible manner.

References

Alzamora, S. M., Anino, S., & Salvatori, D. M. (2001). Impregnation techniques for the incorporation of calcium to apple matrix. In *2001 IFT annual meeting, Institute of Food Technologists, New Orleans, USA.*

Alzamora, S. M., Salvatori, D., Tapia, M. S., López-Malo, A., Welti-Chanes, J., & Fito, P. (2005). Novel functional foods from vegetable matrices impregnated with biologically active compounds. *Journal of Food Engineering, 67*(1–2), 205–214.

Anino, S. V., Salvatori, D. M., & Alzamora, S. M. (2006). Changes in calcium level and mechanical properties of apple tissue due to impregnation with calcium salts. *Food Research International, 39*(2), 154–164.

Barrera, C., Betoret, N., & Fito, P. (2004). Ca^{2+} and Fe^{2+} influence on the osmotic dehydration kinetics of apple slices (var. Granny Smith). *Journal of Food Engineering, 65*(1), 9–14.

Bates, C. (1997). Bioavailability of vitamin C. *European Journal of Clinical Nutrition, 51*(Suppl. 1), S28–S33.

Bellary, A. N., & Rastogi, N. K. (2012). Effect of hypotonic and hypertonic solutions on impregnation of curcuminoids in coconut slices. *Innovative Food Science and Emerging Technologies, 16*(2012), 33–40.

Betoret, N., Puente, L., MJ, D. ´a., MJ, P. ´n., MJ, G. ´a., Gras, M. L., et al. (2003). Development of probioticenriched dried fruits by vacuum impregnation. *Journal of Food Engineering, 56*(2–3), 273–277.

Burgi, H., Supersaxo, Z., & Selz, B. (1990). Iodine deficiency diseases in Switzerland one hundred years after Theodor Kocher's survey: a historical review with some new goitre prevalence data. *Acta Endocrinologica, 123*, 577–590. https://doi.org/10.1530/acta.0.1230577.

Chopra, J. G. (1974). Enrichment and fortification of foods in Latin America. *American Journal of Public Health, 64*, 19–26.

Clarke, R. (1995). *Micronutrient fortification of foods: Technology and quality control.* (CONFORT 2). Available at www.fao.org. Accessed 1 May 2012.

Codex Alimentarius Commission (1987). *General principles for the addition of essential nutrients to foods CAC/GL 09-1987 (amended 1989, 1991). Rome, Joint FAO/WHO Food Standards Programme.* Codex Alimentarius Commission. http://www.codexalimentarius.net/download/standards/299/CXG_009e.pdf. Accessed 1 October 2018.

de Lima, M. M., Tribuzi, G., de Souza, J. A. R., de Souza, I. G., Laurindo, J. B., & Carciofi, B. A. M. (2016). Vacuum impregnation and drying of calcium—fortified pineapple snacks. *LWT- Food Science and Technology, 72*, 501–509.

Erihemu Hironaka, K., Koaze, H., Oda, Y., & Shinanda, K. (2015). Zinc enrichment of whole potato tuber by vacuum impregnation. *Journal of Food Science and Technology, 52*(4), 2352–2358.

FAO/WHO (1994). Methods of analysis and sampling. Joint FAO/WHO Food Standards Programme Codex Alimentarius Commission (Vol. 13, 2nd ed.).

Annexure-IV Micronutrient Fortification of food: Technology and quality control. In *FAO Technical meeting on food fortification: Technology and quality control, Rome, Italy,* November 20–23, 1995.

Fito, P. (1994). Modelling of vacuum osmotic dehydration of foods. *Journal of Food Engineering, 23,* 313–328.

Frano, M. R. L., De Moura, F. F., Boy, E., Lonnerdal, B., & Burri, B. J. (2008). Bioavailability of iron, zinc and provitamin A carotenoids in biofortified staple crops. *Nutrition Reviews, 72*(5), 289–307.

Gras, M. L., Vidal, D., Betoret, N., Chiralt, A., & Fito, P. (2003). Calcium fortification of vegetables by vacuum impregnation interactions with cellular matrix. *Journal of Food Engineering, 56,* 279–284.

Gras, M. L., Vidal-Brotóns, D., & Vásquez-Forttes, F. A. (2011). Production of 4th range iceberg lettuce enriched with calcium. Evaluation of some quality parameters. *Procedia Food Science, 1,* 1534–1539.

Hironaka, K., Kikuchi, M., Koaze, H., Sato, T., Kojima, M., Yamamoto, K., et al. (2011). Ascorbic acid enrichment of whole potato tuber by vacuum-impregnation. *Food Chemistry, 127,* 1114–1118.

Ignat, A., Manzocco, L., Brunton, N. P., Nicoli, M. C., & Lyng, J. G. (2015). The effect of pulsed electric field pretreatments prior to deep-fat frying on quality aspects of potato fries innovative. *Innovative Food Science & Emerging Technologies, 29,* 65–69.

Jain, S. C., & Pandey, A. (2012). Improvement of nutritional quality of bakery products through fortification/enrichment. *Processed Food Industry, 16*(2), 18–25.

Janositz, A., Noack, A.-K., & Knorr, D. (2011). Pulsed electric fields and their impact on the diffusion characteristics of potato slices. *LWT- Food Science and Technology, 44,* 1939–1945.

Joshi, A., Kar, A., Rudra, S. G., Sagar, V. R., Varghese, E., Lad, M., et al. (2016). Vacuum impregnation: a promising way for mineral fortification in potato porous matrix (potato chips). *Journal of Food Science and Technology, 53*(12), 4348–4353.

Joshi, A., Kaur, C., Rudra, S. G., Sagar, V. R., Raigond, P., & Singh, B. (2016). Anthocyanins rich purple potato chips: an innovative fusion product with enhanced functionality. In *Proceedings of International Conference on Recent Advances in Food Processing and Biotechnology, BHU, Varanasi (India).*

Joshi, A., & Sagar, V. R. (2016). Anthocyanins rich papaya candy (tutty fruity): an innovative fusion product with enhanced functionality. *International Journal of Advanced Technology in Engineering and Science, 04*(05), 417–420.

Joshi, A., Sethi, S., Sagar, V. R., Tomar, B. S., Raigond, P., & Singh, B. (2019). Betanins rich purple potato chips: an innovative fusion product with enhanced functionality. *Journal of Agricultural Engineering and Food Technology (JAFET), 6*(1), 1–3. p-ISSN 2350-0085, e-ISSN 2350-0263.

Kar, A., & Lad, M. (2014). *VI assembly fabrication done at FS & PHT division of ICAR-IARI.* New Delhi: Pusa.

Kaur, C., & Sharma, R. R. (2015). *Development of Pusa Nutra Aonla Candy. Extension folder, published by FS & PHT division of ICAR-IARI.* Pusa: New Delhi.

Lebovka, N. I., Praporscic, I., & Vorobiev, E. (2004). Effect of moderate thermal and pulsed electric field treatments on textural properties of carrots, potatoes and apples. *Innovative Food Science & Emerging Technologies, 5,* 9–16.

Lindsay, A., de Benoist, B., Dary, O., & Hurrell, R. (2006). *Guidelines on Food Fortification with Micronutrients.* WHO Library: Geneva. http://www.who.int/entity/nutrition/publications/guide_food_fortification_micronutrients.pdf. Accessed 1 October 2018.

Luna-Guzman, I., Cantwell, M., & Barrett, D. M. (1999). Fresh-cut cantaloupe: effects of $CaCl_2$ dips and heat treatments on firmness and metabolic activity. *Postharvest Biology and Technology, 17,* 201e213.

Martin-Diana, A. B., Rico, D., Barry-Ryan, C., Frias, J. M., Mulcahy, J., & Henehan, G. T. M. (2005). Comparison of calcium lactate with chlorine as a washing treatment for fresh-cut lettuce and carrots: quality and nutritional parameters. *Journal of Science and Food Agriculture, 85*, 2260e2268.
Martın-Dianaa, A. B., Ricoa, D., Frıasa, J. M., Baratb, J. M., Henehana, G. T. M., & Barry-Ryana, C. (2007). Calcium for extending the shelf life of fresh whole and minimally processed fruits and vegetables: a review. *Trends in Food Science and Technology, 18*, 210–218.
Martinez-Monzo, J., Martinez-Navarrete, N., Chiralt, A., & Fito, P. (1998). Mechanical and structural changes in apple (Var. Granny Smith) due to vacuum impregnation with cryoprotectants. *Journal of Food Science and Technology, 63*(3), 499–503.
Mashkour, M., Yahya Maghsoudlou, M. K., & Aalami, M. (2018). Iron fortification of whole potato using vacuum impregnation technique with a pulsed electric field pretreatment. *Potato Research, 61*, 375–389.
Maskooki, A., & Eshtiaghi, M. N. (2012). Impact of pulsed electric field on cell disintegration and mass transfer in sugar beet. *Food and Bioproducts Processing, 90*, 377–384.
Moreno, J., Espinoza, C., Simpson, R., Petzold, G., Nuñez, H., & Gianelli, M. P. (2016). Application of ohmic heating/vacuum impregnation treatments and air drying to develop an apple snack enriched in folic acid. *Innovative Food Science and Emerging Technologies, 33*, 381–386.
Mozaffarian, D., Rosenberg, I., & Uauy, R. (2018). History of modern nutrition science—implications for current research, dietary guidelines, and food policy. *BMJ, 361*(k2392), 1–6. https://doi.org/10.1136/bmj.k2392.
Naidu, K. A. (2003). Vitamin C in human health and disease is still a mystery? An overview. *Nutrition Journal, 2*, 7–16.
Nair, K. M. (2012). *Fortification of Bakery Products.* NIN (ICMR), PPT available at. http://www.ilsi-india.org/FortificationFoods/FortificationofBakeryProducts. Accessed 22 January 2012.
Nestel, P., Bouis, H. E., Meenakshi, J. V., et al. (2006). Biofortification of staple food crops. *Journal of Nutrition, 136*, 1064–1067.
Ortiz, C., Salvatori, D. M., & Alzamora, S. M. (2002). Combined blanching and vacuum impregnation for enhancing calcium fortification in mushroom tissue. In 2002 IFT annual meeting. Anaheim, CA: Institute of Food Technologists.
Rababah, et al. (2012). Fortification of potato chips with natural plant extracts to enhance their sensory properties and storage stability. *Journal of the American Oil Chemists' Society, 89*, 1419–1425.
Rastall, R. A., Fuller, R., Gaskins, H. R., & Gibson, G. R. (2000). Colonic functional foods. In G. R. Gibson & C. M. Williams (Eds.), *Functional foods.* Cambridge: Woodhead Publishing Limited.
Rico, D., Martin-Diana, A. B., Henehan, G. T. M., Frias, J., Barat, J. M., & Barry-Ryan, C. (2007). Improvement in texture using calciumlactate and heat-shock treatments for stored ready-to-eat carrots. *Journal of Food Engineering, 79*, 1196e1206.
Rodrı´guez, M. I. (1998). *Estudio de la penetracio´n de microorganismos en frutas mediante el modelo Hidrodina´mico (HDM).* Thesis, Instituto de Ciencia y Tecnologı ´a de Alimentos, Universidad Central de Venezuela.
Saarela, M., Lähteenmäki, L., Crittenden, R., Salminen, S., & Mattila-Sandholm, T. (2002). Gut bacteria and health foods—the European perspective. *International Journal of Food Microbiology, 78*, 99–117.
Sahin, S., & Sumnu, S. G. (2006). *Physical properties of foods.* New York: Springer Science + Business Media, LLC.

Salengke, S., & Sastry, S. K. (2007). Experimental investigation of ohmic heating of solid—liquid mixtures under worst-case heating scenarios. *Journal of Food Engineering, 83*, 324–326.

Soliva-Fortuny, C., & Martin-Belloso, O. (2003). New advances inextending the shelflife of fresh-cut fruits: a review. *Trends in Food Science and Technology, 14*, 341e353.

Spiess, W. E. L., & Behsnilian, D. (1998). Osmotic treatments in food processing. Current stage and future needs. In *Drying, A* (pp. 47–56). Thessaloniki: Ziti Editions.

Suutarinen, J., Anakainen, L., & Autio, K. (1998). Comparison of Comparison of light microscopy and spatially resolved Fourier transform infrared (FT-IR) microscopy in the examination of cell wall components of strawberries. *LWT-Food Science and Technology, 31*, 595e601.

Tapi, S., Tylewicz, U., Romani, S., & Rosa, M. D. (2016). Study on the quality and stability of minimally processed apples impregnated with green tea polyphenols during storage. *Innovative Food Science & Emerging Technologies.* https://doi.org/10.1016/j.ifset.2016.12.007.

Tapia, M. S., Schulz, E., Gómez, V., López-Malo, A., & Welti-Chanes, J. (2003). A new approach to vacuum impregnation and functional foods: Melon impregnated with calcium and zinc. In 2003 IFT annual meeting technical program abstracts (60D-2), Chicago, USA.

Torreggiani, D., & Bertolo, G. (2001). Osmotic pre-treatments in fruit processing: chemical, physical and structural effects. *Journal of Food Engineering, 49* (2–3), 247–253.

Wiley, R. C. (1994). *Minimally processed refrigerated fruits and vegetables.* New York: Chapman & Hall.

Word Bank Report. (2009). *World Bank Report on Malnutrition in India (India Undernourished Children: A Call for Reform and Action).* http://web.worldbank.org. Accessed 13 December 2015.

Yao, Y. (2016). Enhancement of mass transfer by ultrasound: application to adsorbent regeneration and food drying/dehydration. *Ultrasonics Sonochemistry, 31*, 512–531. https://doi.org/10.1016/J.ULTSONCH.2016.01.039.

Yılmaza, F. M., & Bilek, S. E. (2018). Ultrasound-assisted vacuum impregnation on the fortification of fresh-cut apple with calcium and black carrot phenolics. *Ultrasonics Sonochemistry.* https://doi.org/10.1016/j.ultsonch.2018.07.007.

Further reading

Alzamora, S. M., Tapia, M. S., Leúnda, A., Guerrero, S. N., Rojas, A. M., Gerschenson, L. N., et al. (2000). Relevant results on minimal preservation of fruits in the context of the multinational project XI.3 of CYTED, an Ibero-American R&D cooperative program. In J. Lozano, M. C. Añón, & G. B. Cánovas (Eds.), *Trends in food engineering.* Gathersburg, MD: Aspen Publishers.

Anino, S., Salvatori, D., & Alzamora, S. M. (2003). Apple matrix with calcium incorporated by impregnation processes. In P. Fito, A. Mulet, A. Chiralt, & A. Andrés (Eds.), *Ingeniería de alimentos: Nuevasfronterasen el siglo XXI, Proceedings del III congreso Iberoamericano de ingeniería de alimentos (CIBIA 2001), Vol. 4, Servicio de Publicaciones de la Universidad Politécnica de Valencia, Valencia, España.*

González-Centeno, M. R., Knoerzer, K., Sabarez, H., Simal, S., Rosselló, A., & Femenia, C. (2014). Effect of acoustic frequency and power density on the aqueous ultrasonic-assisted extraction of grape pomace (Vitisvinifera L.)—a response surface approach. *Ultrasonics Sonochemistry, 21*, 2176–2184. https://doi.org/10.1016/J.ULTSONCH.2014.01.021.

Lenntech (2015). *Recommended daily intake of vitamins and minerals.* http://www.lenntech.com/recommended-daily-intake.htm. Accessed 4 March 2015.

Shelke, K., & Feder, D. R. D. (2006). Iron and zinc fortification National health surveys claim almost three in four Americans don't meet requirements for zinc and iron. *Food Processing*, Available from: https://www.foodprocessing.com/articles/2006/019 (Accessed 24 February 2006).

10

Probiotics in fresh-cut produce

J.E. Dávila-Aviña, A. Ríos-López, A. Aguayo-Acosta, L.Y. Solís-Soto

Department of Microbiology, Faculty Biological Sciences, Autonomous University of Nuevo Leon, San Nicolas de los Garza, Mexico

1 Introduction

Presently, food minimally processed ready for consumption, as well as functional food, are terms that are starting to become more and more popular. This is mostly due to changes in the life style and eating patterns of modern society around the daily schedule (Aguiar, Geraldi, Cazarin, & Junior, 2019). Efforts from international public health organizations trying to raise awareness about the relationship between a thoughtful diet and good health have been reported (Fulkerson, 2018). In this context, the market for minimally processed products, particularly the consumption of FCFVs, as well as the demand for food that not only brings a nutritious value but also has a beneficial impact in physiological functions (also known as functional food) has increased considerably (Weaver et al., 2014).

Consumers have shown great acceptance and special preference for products labeled as healthy choices (do Espírito Santo, Perego, Converti, & Oliveira, 2011). In this category, food products containing probiotics, which are considered to keep a healthy digestive system, are experiencing a very rapid growth (Shori, 2015). Etymologically, the term probiotic is a combination of Greek and Latin that means "pro-life" (Hossain, Sadekuzzaman, & Ha, 2017). This term is applied to alive nonpathogenic organisms (moistly certain strains of bacteria and yeast), which, after consumption in adequate concentrations, have the capacity of inducing beneficial effects in the health of the consumer (FAO/WHO, 2001). These are available mostly in the form of dietary supplements and in food products.

There are several reports about the benefits of consuming probiotics, such as an increase in the absorption of minerals including calcium and magnesium, prevention and treatment of infectious diarrheas associated to antibiotic ingestion, reduction

Fresh-cut Fruits and Vegetables. https://doi.org/10.1016/B978-0-12-816184-5.00010-0

in the symptoms of intestinal inflammation, regulation of immune system, and lactose intolerance reduction, among others (De Melo Pereira, De Oliveira Coelho, Júnior, Thomaz-Soccol, & Soccol, 2018; Roberfroid, 2000; Vijaya Kumar, Vijayendra, & Reddy, 2015). These beneficial effects might be due to direct or indirect regulation of intestinal microbiota or an immunological response (Schrezenmeir & De Vrese, 2001). For probiotics to apply their beneficial effects, it is necessary to have a minimum number of viable microorganisms (Denkova & Krastanov, 2012). In spite of not having a consensus about the right dose, it is recommended to have minimally 10^6 CFU/g of viable cells in a product, thus to provide a dose of 10^8–10^9 of viable cells per day (Min, Bunt, Mason, & Hussain, 2018). However, not all probiotic organisms have the same effect, and it is necessary to understand that their consumption, or the ingestion of any other functional food, should never replace a healthy diet or a medical treatment.

The beneficial effects associated with probiotic bacteria have been reported, however, it is important to consider that the amount of data available supported with clinical assays is still limited (Jankovic, Sybesma, Phothirath, Ananta, & Mercenier, 2010). It is critical to emphasize that any over the counter probiotic product (functional food) should never be labeled as a product intended to diagnose, cure, or prevent human diseases (Sanders et al., 2010). The FDA also referred to "probiotics" as live biotherapeutic products (LBP) used in the prevention, treatment, or cure of a human disease without being a vaccine, where these microorganisms have been subject of clinical assays and strict quality controls, thus assuring their therapeutic effect (O'toole & Paoli, 2017).

Probiotic products are mostly composed of microorganisms from the genera *Bifidobacteria* and *Lactobacillus* (Champagne, Ross, Saarela, Hansen, & Charalampopoulos, 2011). They belong to the group of lactic acid bacteria whose beneficial effects in human have been well stablished (Arboleya et al., 2012). A great number of bacterial species from both genera are amply recognized and have GRAS status (Generally Recognized as Safe) by international regulatory agencies (Awaisheh, 2012). A general way to divide probiotic products is into dietary supplements and probiotic food, the latter further classified as lactic products, fermented nonlactic products, and nonfermented food products (products with added probiotic strains, which this chapter centers on) (Awaisheh, 2012).

In the case of FCFVs, Martins et al. (2013) report the strains *Lactobacillus acidophilus, Lb. casei, Lb. plantarum, Lb. rhamnosus*, and *Bifidobacterium lactis* as the most utilized. Additionally, there are more commercial strains being used in food products, i.e., *B. animalis, B. adolescentis, B. bifidum, B. breve, B. infantis, B. longum, B. essencis, B. lactis, Lactobacillus* GG, *Lb. acidophilus, Lb. casei, Lb. bulgaricus, Lb. delbrueckii* spp. *bulgaricus, Lb. fermentum, Lb. helveticus, Lb. lactis, Lb. plantarum, Lb. paracasei, Lb. reuteri, Lb. rhamnosus, Lb. salivarius, Bacillus lactis, Enterococcus faecium, Streptococcus salivarius, Streptococcus thermophilus,* and *Saccharomyces boulardii* (Roy & Kumar, 2018; Soccol et al., 2010).

All the strains used must comply with certain parameters to be implemented efficiently as probiotics in food products (De Melo Pereira et al., 2018). Some of these parameters include having a precise identification of the strain, lacking virulence factors, inability to produce toxic metabolites (Sornplang & Piyadeatsoontorn, 2016), not presenting known transferable antibiotic resistance genes, stability and viability in the food product, tolerance to shelf storage conditions (regarding oxygen and temperature variables), good sensory properties (Mishra & Mishra, 2012), and after they been ingested, to show survival in the gastric tract in acidic conditions, in presence of digestive enzymes and intestinal bile salts (Tuomola, Crittenden, Playne, Isolauri, & Salminen, 2001). Once the probiotics reach the intestine tract, they must be able to adhere to and colonize it so they can provide their beneficial effects (Molin, 2001). Some probiotics strains have shown to have an antagonistic reaction (through competition and/or through the production of bacteriocins against pathogenic agents present in food products) (Ljungh & Wadstrom, 2006).

The term probiotic has become very popular in the last few years. There are many scientific and extension reports that associate the consumption of probiotic products with beneficial effects in human health, particularly their impact on digestive system regulation, hence the recommendation for their regular use along with a healthy daily diet (Duggan, Gannon, & Walker, 2002). Even though these products are readily available, traditionally, the most common probiotic vehicle has been fermented lactic products, as they are considered the ideal vector to incorporate probiotic bacteria (Turkmen, Akal, & Özer, 2019). Unfortunately, this fact limits their consumption from people that demand low cholesterol products, or who are lactose intolerant, or that just have vegetarian habits (Bansal, Mangal, Sharma, & Gupta, 2016). As a result, an increase in the interest of the food industry and scientific community to develop alternative options to the lactic probiotic products has been observed (Figueroa-González, Quijano, Ramírez, & Cruz-Guerrero, 2011).

Following this trend, minimally processed products such FCFVs could be a suitable vector for the inclusion of probiotics with all their beneficial attributes, such as vitamin content, minerals, antioxidants compounds, water, fibers, and some other components considered prebiotic that could have a symbiotic effect in the consumer (Lebaka, Wee, Narala, & Joshi, 2018). FCFVs as vehicles of probiotics could be a great alternative because of the increasing demand for nonlactic probiotic food—a very promising concept, but still with some challenges to overcome (Vinderola, Burns, & Reinheimer, 2017).

2 Fresh-cut produce enriched with probiotic

Currently, within nonlactic probiotic products from vegetal origin, fruit or vegetable beverages are a very popular developed concept with commercial products well established in the market of functional beverages (Granato, Branco, Nazzaro, Cruz, & Faria, 2010). In the case of FCFVs, there are very few scientific studies with concrete enough results to develop a functional product for the market; out of the total amount of commercial probiotic strains available, only some have been evaluated in FCFVs. A very promising area to develop, yet some challenges to overcome.

Certain strains of *Lactibacillus* and *Bifidobacterium* such as *Lb. rhamnosus, Lb. casei, Lb. paracasei, Lb. plantarum, Lb. acidophilus, Lb. delbrueckii* subsp. *bulgaricus, Lb. brevis, Lb. johnsonii, Lb. fermentum*, and Bifidobacteria—for instance *B. infantis, B. adolescentis, B. bifidum, B. longum, B. breve, B. animalis* subsp. *animalis, B. animalis* subsp. *lactis*—are considered as important probiotics (Fijan, 2014); unfortunately, only a few of them have been evaluated on FCFVs (Table 1).

Scientific studies report some examples of application of lactic acid bacteria on fresh-cut fruits and vegetables matrices such as apples, carrot slices, cantaloupe melons, lettuce, papaya, fruit salads, pear, pineapple, and mixed vegetable salads (Table 1).

Most of these studies have focused on the incorporation of probiotics in FCFVs from different perspectives where microbiologic, chemical, and sensory parameters were analyzed, as well as their incorporation on food vectors (Table 1).

3 Viability of probiotics in fresh-cut produce

Finding the right probiotic strain at the right dose is of paramount importance to develop a probiotic food product; its efficiency will depend on the number of viable cells at the moment

Table 1 Studies regarding the use of fresh-cut fruit and vegetables matrices carrying probiotic bacteria

Fresh-cut product	Microorganism strain	References
Apple slices (var *Granny Smith*)	*Lactobacillus paracasei* LL13	Akman, Uysal, Ozkaya, Tornuk, and Durak (2019)
Apple cubes (*Malus* sp. var. *Granny Smith*)	*Lactobacillus casei* NRRL B-442	Rodrigues, Silva, Mulet, Cárcel, and Fernandes (2018)
Apple slices (cv. *Braeburn*)	*Lactobacillus rhamnosus* GG; LGG	Rößle, Auty, Brunton, Gormley, and Butler (2010) and Rößle, Brunton, Gormley, Ross, and Butler (2010)
"Granny Smith" apples and melon pieces (*Cucumis melo*, var. *cantalupensis*)	*Lactobacillus plantarum* C19	Speranza et al. (2018)
"Golden Delicious" apple wedges	*Lactobacillus rhamnosus* GG	Alegre, Viñas, Usall, Anguera, and Abadias (2011)
Commercially sliced apples and minimally processed lamb's lettuce	*Lactobacillus* paracasei M3B6, *Lb. casei* V4B4, and *Lb. plantarum* CIT3 and V7B3 strains	Siroli et al. (2015)
Carrot slices	*Lactobacillus acidophilus* La-14	Shigematsu et al. (2018)
Cantaloupe melons (*Cucumis melo* var. *cantalupensis*)	*Lactobacillus plantarum* B2 and *Lb. fermentum* PBCC11.5	Russo et al. (2015)
Iceberg lettuce leaf cuts	*Lactobacillus plantarum* ATCC 11454	Trias, Bañeras, Badosa, and Montesinos (2008)
Melons, yellow type (*Cucumis melo* L.)	*Lactobacillus rhamnosus* HN001	De Oliveira, Leite Junior, Lopes Martins, Furtado Martins, and Mota Ramos (2014)
Maradol papayas (*Carica papaya* L.) and Fuji apples (*Malus domestica* Borkh) cylinders	*Bifidobacterium lactis* Bb-12	Tapia et al. (2007)
Minimally processed fruit salads (apple, banana, guava, mango, papaya, and pineapple)	*Lactobacillus rhamnosus* HN001	Martins, Ramos, Martins, and Leite Junior (2016)
Pear wedges (*Pyrus communis* L. cv. Conference)	*Lactobacillus rhamnosus* GG, *Lb. acidophilus* LA-5	Iglesias, Abadias, Anguera, Sabata, and Vi (2017); Iglesias, Echeverría, Viñas, Lopez, and Abadias (2018)
Pineapple pieces	*Lactobacillus plantarum* B2 and *Lb. fermentum* PBCC11.5	Russo et al. (2014)
Ready-to-use vegetables (salads)	*Lactobacillus casei* IMPC LC34, IMPC LC41, *Lb. plantarum* LP4	Vescovo, Torriani, Orsi, Macchiarolo, and Scolari (1996)
Ready-to-use mixed salad vegetables	*Lactobacillus casei* IMPC LC34	Torriani, Orsi, and Vescovo (1997)

of its consumption (Ranadheera, Baines, & Adams, 2010). The actual food product general conditions at the time of its distribution are factors that strongly will influence the stability, survival, and viability of the probiotic microorganisms, and consequently, their efficiency (Tripathi & Giri, 2014).

Their success will be tied not only to their bacterial physiology, but also to the food product particular chemistry, including pH, sugar content, acidity, water activity, nutrients availability, particular biopathways of each FCFV, and the application of treatments to preserve shelf-life (natural or chemical) after or during washing (Min et al., 2018).

Temperature during storage is another important factor to consider. To increase the viability of the probiotic microorganisms they must be in storage at a temperature of 4–5°C, without affecting the product sensory attributes by a slow fermentation (Rodgers, 2008). In the case of FCFVs, this condition seems to be met, since their recommended storage condition is at 4–7°C (Bansal, Siddiqui, & Rahman, 2015).

The oxygen content during packaging is another variable to consider since molecular oxygen is very harmful to probiotic microorganisms due to the anaerobic nature of most of them (Watanabe, Van Der Veen, & Abee, 2012). In order to develop functional food products in this area, a clear understanding of the relationships between the probiotic microorganisms and the substrates presents in the FCFVs where they are intended to be incorporated is necessary, to unlock their full potential.

Russo et al. (2015) evaluated the viability of *Lb. fermentum* PBCC11.5 and *Lb. plantarum* B2 strains in fresh-cut cantaloupe stored at 4°C for 11 days. Both strains showed high viability patterns. The strains of *Lb. fermentum* and *Lb. plantarum* had concentrations of 3×10^8 and 7.8×10^7 CFU/g, respectively, at the end of the shelf-life. Using these strains, Russo et al. (2014) monitored their viability on artificially inoculated pineapple pieces throughout storage. *Lb. fermentum* PBCC11.5 and *Lb. plantarum* B2 initial population on minimally processed pineapples was approximately 8.4 ± 0.42 $\log_{10}$ CFU/g. Their results showed that the concentration of both strains ranged between 6.3 and 7.3 log CFU/g, respectively, after 8 days of refrigerated storage; this without affecting the final quality of the fresh-cut pineapple.

Martins et al. (2016) proposed fruit salad as probiotic (*Lb. rhamnosus* HN001) carriers. Counts after bacteria inoculation showed 8.5 log CFU/g, and at 5 days counts were 8.49 log CFU/g approximately; this showing that time did not affect the viability of *Lb. rhamnosus* in fruit salads. A similar result was observed by Torriani et al. (1997), who used *Lb. casei* IMPC

LC34 in ready-to-use mixed salad vegetables and found that after 6 days of storage at 8°C the product contained 8.24 log CFU/g of this bacterium.

Rodrigues et al. (2018) developed an apple slice snack as a probiotic carrier, where they incorporated *Lb. casei* NRRL B-442 with an inoculum of 8.5 ± 0.5 log CFU/g on the apple slices and dried under different temperatures with or without application of ultrasound. After 100 min they found that at 40°C and 60°C without ultrasound, the concentration of probiotics was reduced to 6.3 ± 0.3 log CFU/g. On the other hand, the process carried out applying ultrasound at 10°C retained $87 \pm 2\%$ of the initial probiotic count.

Fresh-cut apple wedges (cultivar Braeburn) were used as potential carrier for the commercial strain *Lb. rhamnosus* GG (LGG) by Rößle, Auty, et al. (2010). Initial inoculation level had a CFU/g count of 10^8. This count remained stable throughout the 10-day storage period at 2–4°C. In a different study, *L. rhamnosus* was evaluated for its ability to survive in apple wedges (cultivar *Braeburn*) at 2–4°C for a storage period of 14 days. At the beginning of the storage, all samples had a *L. rhamnosus* count between 8.8 and 8.9 log CFU/g and the probiotic microorganism remained stable ($>10^7$–10^8 CFU/g) during the whole shelf-life storage of the product. A similar result was observed by Rößle, Brunton, et al. (2010), who used *L. rhamnosus* GG in minimally processed apples and found that after 10 days of storage, the product contained 7.79 Log CFU/g of this bacterium. According to Rößle, Brunton, et al. (2010), this is a similar or above recommended concentration to produce beneficial health effects in the intestinal tract and is comparable to counts of probiotic bacteria in commercially available dairy products.

4 Physicochemical and sensory parameters

As mentioned previously, most people in modern society recognize amply the importance of consuming food products with high nutritional value. Nonetheless, a deciding factor on choosing food products is still sensory attributes such as good flavor and appearance (Shori, 2016). In the case of FCFVs, sensory attributes play a fundamental role. Color, texture, smell, and packing will most likely determine the initial acceptance by the consumer (Bansal et al., 2015).

However, such attributes do not guarantee internal quality characterized by good flavor (Breslin, 2001). Although initial acquisition of such food products is based on how good they look, subsequent preference will be based on the consumer evaluation once the food product is consumed (Moser, Raffaelli, & Thilmany, 2011). The actual taste of the food products is defined by the interaction of sugars, free amino acids, organic acids, and volatile compounds, which are strongly influenced by the maturity state and metabolic processes particular to each vegetable or fruit (Velderrain-Rodríguez, Quirós-Sauceda, González-Aguilar, Siddiqui, & Ayala-Zavala, 2015). One concern when incorporating probiotic microorganism in food products is the possible modification they might have on these important sensory attributes (Torriani, Scolari, Dellaglio, & Vescovo, 1999).

The effects of probiotic bacteria sensorial and physicochemical properties on different FCFVs have been reported in some assays. In Iglesias et al. (2018) analyzed the effect of a probiotic strain, *L. rhamnosus* GG, on food-borne pathogens inoculated on fresh-cut pear. In this case, despite the antimicrobial effect, the authors observed that the presence of probiotic bacteria did not affect the soluble solid content and titratable acidity of pear after 9 days of storage at 5°C. A higher concentration of 12 volatiles compounds in *L. rhamnosus* GG-treated samples was also detected. From a sensory point of view, the positive effect of volatile compounds is mainly due to their quantitative abundance, olfactory thresholds, and to the odor descriptor, obtaining a positive response in the flavor perception of these samples by consumers, thus suggesting that pear treated with probiotics could add a good flavor to the final product.

Siroli et al. (2015), using a combination of antimicrobial compounds with lactic acid bacteria as *Lb. plantarum* CIT3 and V7B3, evaluated the organoleptic properties of sliced apples after 9 days of storage at 6°C, taking into account the usual shelf-life of this type of product. No significant differences with respect to the dipping solution were recorded for various descriptors, including smell, firmness, crispness, juiciness, and sweetness. Results showed that a high dose of probiotic did not affect most of the sensory qualities after 9 days of storage. Particularly interesting was the *Lb. plantarum* CIT3 strain because it was shown to preserve the quality of treated fresh-cut apples when used alone and up to 16 days when combined with natural antimicrobials.

Tapia et al. (2007) analyzed the physiochemical characteristics of some fruit products when coated with different formulations and probiotics. After coating apple cylinders and freshly cut papaya with gellan coating (0.025%) or alginate (2%) with sunflower oil and antioxidants, plus 2% *B. lactis* Bb12 (added to each formulation before coating the products). They found that gellan coatings showed better barrier properties against water ($P < 0.05$) than alginate coatings for papaya and apple cylinders. The water vapor resistance values were maintained 40%–50% lower (11.87 and 12.72 s/cm, for alginate and gellan with bacteria, respectively) than the coatings without bacteria. This suggests that the addition of probiotic biomass (2% *w/v*) to coatings composed of alginate or gellan sunflower oil causes an increase in the spacing between the polymer chains, due to the inclusion of bacterial cells between the chains favoring the diffusivity through the coatings and accelerating the transmission of water.

Shigematsu et al. (2018) after treatment slices of carrot coated in a solution of sodium alginate with and without *Lb. acidophilus* La-14 at 8 ± 2°C for 19 days, found a slight reduction in pH for the treatments, where the carrots covered with probiotics showed the lowest statistically significant variation, suggesting that these probiotics could reduce the metabolism of the minimally processed carrot slices. At the same time, the authors also indicated that the barrier created by the coatings of *Lb. acidophilus* contributed to the quality of the minimally processed carrots since they conserved their moisture and minimized the color changes during storage.

Recently, Akman et al. (2019) developed an apple slice snack inoculated with *Lb. paracasei* and treated it with oven and vacuum drying at 45°C, kept under storage for 28 days at 4°C. Under these conditions, physicochemical assays where conducted and showed that the incorporation of probiotics significantly increased the total phenolic content. At the end of 4-week storage, apple samples enriched with probiotics had higher antioxidant capacity values than control without probiotics, maybe due to the protective effect of probiotic bacteria on phenolic compounds. The use of probiotics also showed a nonsignificant effect on apple water activity, color change, or sensory characteristics compared to untreated slices. Most studies have reported that sensory and physicochemical attributes such as texture, color Lab, °brix, pH, titratable acidity, and flavor remain stable or with no significant differences when compared to FCFVs without the addition of probiotics. These results suggest that products of this nature would be of comparable quality to FCFVs currently available in supermarkets (Rößle, Auty, et al., 2010; Rößle, Brunton, et al., 2010; Torriani et al., 1997).

5 Bio preservative agents

Food-borne illnesses have been a worldwide public health problem for many years. Vegetables are food products that are consumed mostly raw, making easy the transmission of pathogen microorganisms (Zoellner et al., 2016). The possibility of microbial contamination on FCFVs is always latent due to the processes involved in their preparation. The surface of the fruit or vegetable itself might have an elevated moisture content and nutrient availability, which is compatible with the microbiota growth, favoring its establishment (Ayala-zavala, Del Toro-Sánchez, Alvarez-Parrilla, & González-Aguilar, 2008). Additionally, the texture in some cases favors the adhesion of certain pathogens and their subsequent development and reproduction. Looking to increase the innocuity of food products, the sanitation of the surfaces of food products after the washing and before packing is conducted by physical, chemical, or synthetic methods, all in ample use (Tapia et al., 2015). However, there is a trend among consumers to choose natural food products, minimally processed, and free of synthetic antimicrobial compounds, because of the fear that such compounds could have harmful effects on human health (Alvarez, Moreira, Roura, Ayala-Zavala, & González-Aguilar, 2015).

The need to find safe alternatives for the consumer has led to a search for natural antimicrobial compounds (Karunaratne, 2018). Natural antimicrobial compounds of bacterial origin have been evaluated to determine their potential to inhibit pathogenic growth in many food product categories, including FCFVs. Particularly, some strains of acid-lactic bacteria have gained importance as a biocontrol of food-borne illnesses, achieving this control through an antagonistic response, or based on proteins production, such as bacteriocins (Hossain et al., 2017; Woo & Ahn, 2013).

Vescovo et al. (1996) indicate that *Lb. casei* IMPC LC34, IMPC LC41 and *Lb. plantarum* LP4 have the ability to produce antimicrobial substances and grow at 4–8°C. The authors reported that *Lb. casei* IMPC LC34 was most effective in reducing total mesophilic bacteria and the coliform group in ready-to-use mixed salads during storage at 8°C for 6 days.

Iglesias et al. (2017) were interested in the use of probiotics as bio preservation agents of food-borne pathogens. Consequently, the objective of their work was to test the antagonic capacity of probiotic bacteria, *Lb. rhamnosus* GG (ATCC 53103) and *Lb. acidophilus* LA-5, against *Listeria monocytogenes* and *Salmonella enterica* subsp. *enterica* in fresh-cut pear (cv. conference) stored at different temperatures (5°C, 10°C, and 20°C), finding that *L. rhamnosus* GG reduced *L. monocytogenes* (approximately 3-log units) and *Salmonella* (approximately 2-log units) populations at all tested storage temperatures, however *Lb. acidophilus* LA-5 did not show efficacy against *L. monocytogenes* and *Salmonella*. The authors conclude that *L. rhamnosus* GG could be used to control pathogenic microorganisms, but recommended further studies that evaluate its effect on the quality of fresh-cut pear. Iglesias et al. (2018) achieved 1.8 log_{10} reduction of *L. monocytogenes* over fresh-cut pear, inoculated with *Lb. rhamnosus* GG after 9 days of storage at 5°C, but it did not have an effect on Salmonella.

Alegre et al. (2011) also found that *L. rhamnosus* GG could be a suitable probiotic for minimally processed apples, capable of reducing *L. monocytogenes* growth by 1-log units for less than 28 days of storage at 5°C and 10°C, although *Salmonella* was not affected by the probiotic microorganism. Nevertheless, they also concluded that apple wedges shelf-life should not pass 14 days to guarantee the probiotic effect. A similar result was observed by Russo et al. (2014), who used *Lb. fermentum* PBCC11.5 against *L. monocytogenes* CECT 4031 in minimally processed pineapples and found a fast reduction (2 log CFU/g) of *L. monocytogenes* after 3 days of storage at 5°C.

As outlined, the antimicrobial effect of microorganism prebiotic has been determined on a variety of FCFV products. Although the success of probiotics as bio preservative depends greatly on the type of food-borne pathogens microorganisms, it is encouraging to see they could provide an alternative to the use of synthetic antimicrobials.

6 Application alternatives

Currently there are different methods or technologies such as nonthermic techniques, natural preservatives and edible coatings to extend the shelf-life of FCFVs (González-Aguilar, Ayala-Zavala, Olivas, de la Rosa, & Álvarez-Parrilla, 2010), nonetheless, the use of packages or wrappers to preserve, protect and maintain the food product and facilitate their manipulation and shipping is essential (Abdul Khalil et al., 2018).

Packaging generally keeps FCFVs under different conditions compared to the normal air composition, thus they are classified as under controlled atmosphere (Wani, Singh, Pant, & Langowski, 2015). In the case of FCFVs, this modification usually results on an environment low in oxygen and/or high in carbon dioxide, this influencing greatly in the metabolism on the food product extending its shelf-life (Siddiqui, Chakraborty, Ayala-Zavala, & Dhua, 2011). However, it is important to be aware of their tolerance limits to avoid undesirable flavor profiles resulting from anaerobic respiration (Lin & Zhao, 2007).

Considering the efficacy of the probiotic microorganisms and their promising success if accepted in FCFVs, it is necessary to optimize their mechanism of incorporation in food products

(Soccol et al., 2014). There are reports that show that the stability and viability as well as sensory attributes on minimally processed FCFVs has been maintained or improved when combined with packaging under controlled atmosphere or using edible coatings (Quirós-Sauceda, Ayala-Zavala, Olivas, & González-Aguilar, 2014).

Tapia et al. (2007) evaluated the stability of *B. lactis* Bb12 on apple and papaya fresh cuts using gellan coating (0.025%) or alginate (2%) with sunflower oil and antioxidants, plus 2% *B. lactis* Bb12 (added to each formulation before coating the products). After 10 days at 8°C, the survival of *B. lactis* Bb-12 was determined; it was observed the survival maintained values $>10^6$ CFU/g, this indicating the probiotics stability was maintained in both food products with edible coating.

Shigematsu et al. (2018) formulated a solution of sodium alginate with the addition of a probiotic based on *Lb. acidophilus* La-14 (7.36 log CFU/g) in which they submerged carrot slices, activating gelation by immersion in a chloride calcium solution. After 19 days at 8 ± 2°C, the probiotic microorganism survival was determined, reporting 7.11 log CFU/g at the end of the treatment, they concluded that the alginate coating was an effective support for *Lb. acidophilus*. There are also indications that the probiotic can inhibit fungal contamination of the product, although more research is required to confirm these effects. The coating technique also allowed the vegetable to present probiotic potential without causing any significant physical or chemical alteration.

In a very recent study, apple slices were treated with *Lb. paracasei* LL13 and dried using two different methods, then stored for 4 weeks in refrigerated conditions. At the beginning of the storage, microbial contents of convectional and vacuum dried apples were 7.99 and 7.42 $\log_{10}$ CFU/g, respectively. Bacterial numbers of convectional and vacuum dried apples stayed above 7 and 6 $\log_{10}$ CFU/g during the storage, respectively. Akman et al. (2019) concluded that bacterial cells attached and survived on convectional dried apple surfaces better than on vacuum dried ones, and that both drying techniques are successfully applicable to produce probiotic-enriched apple slices with high bacterial stability.

Different assays have been developed to determine the potential of probiotic bacteria in several FCFVs. It has been documented that edible films/coatings can carry diverse food additives, and if viable probiotics such as *Bifidobacterias* and *Lactobacillus* are included, new possibilities for the development of fresh-cut fruit with probiotics products and their benefits are being explored, mainly in the field of fruit and vegetable preservation technology (Burgain, Gaiani, Linder, & Scher, 2011). This technology could also be an option to extend the shelf-life of fresh produce in addition to possibly acting as probiotic carriers without changing the sensory analysis, as the capacity to retard lipid oxidation in foods, and the ability to improve the flavor, texture, and protect the cell structure of the coated vegetable (Abdul Khalil et al., 2018).

More extensive research about probiotic bacteria along with other preservation technology or matrices as edible carriers would provide new proceedings for assuring a greater shelf-life of the final product.

7 Regulatory issues

Currently, there does not exist a common worldwide standard with legal status for safety evaluation for food products that contain probiotics (Przyrembel, 2001). Organizations such as the European Union differ greatly on their requirements and rules across countries. The United Nations recommends a series of specifications that must be considered for a probiotic product to be approved. These include their ability to survive the gastrointestinal tract, to colonize and/or adhere to it, and inhibition of pathogenic organisms. The label must indicate the genus and species based on international nomenclature, expected viability, recommended dose, and expiration date (Lee et al., 2018).

In North America, the regulation of probiotics relies on federal agencies such as the FDA (Food and Drugs Administration) through the FDCA (Food, Drugs and Cosmetics Act) and the DSHEA (Dietary Supplement Health and Education Act). It is important to mention that in America a norm does not exist that defines the "probiotic" for human consumption (Sanders, 2008). Nonetheless, they are classified in four categories: food ingredient, dietary supplement, medical food and biological product (Degnan, 2008). If a probiotic falls into the category of medical product and is not considered as GRAS, it is then classified as new and it is subject of all the requirements needed for a new medical product (Hoffmann et al., 2013). A product classified as a dietary supplement does not need approval before becoming commercial, it just need to disclose the effect that is expected in human health. However, the FDA must be notified if any complaint or dispute arises, so the responsible party can justify or address the issue accordingly (Vanderhoof & Young, 2008).

On the other hand, if the probiotic is classified as a food ingredient, the FDA must regulate the product as food additive or food ingredient GRAS, and it will be subject to approval before becoming commercial. If the food ingredient is not GRAS, it will be classified as a food additive and must go through a process before becoming commercial. If the FDA concludes that there are no concerns or questions about the GRAS status, then the product can go straight to market. If the probiotic falls into the category of medical food, the requirements are the same applied to food additives and GRAS (Lee et al., 2018).

Each country applies their rules and requirements differently. In France, for instance, only microorganisms with unknown background are candidates for approval. In the United Kingdom, a specific regulatory program does not exist yet. Japan was the first country worldwide to implement a regulatory system for functional food and nutraceutical products in 1991 (Lee et al., 2018; Stanton et al., 2001). The Japanese regulation classifies probiotics in different food categories and "food for specified health uses" (FOSHU). A health claim is not included in the label; if it is included, a permit must be obtained from the Health Ministry indicating that the product is considered FOSHU and its efficacy and safety have been evaluated (Awaisheh, 2012).

The European Union was second to implement a series of regulations regarding functional food. In 1995, its agency for food safety allowed the introduction of the Qualified Presumption of Safety (QPS) to certify safety evaluations in food products before becoming commercial. It defines probiotics as functional food present in food and medical products. In India, the government created the FSSA (Food Safety and Standards Authority) under the Ministry of Health and Family Welfare in 2005 to regulate food safety affairs including regulation of functional food and dietary supplements (Arora, Sharma, & Baldi, 2013; Lee et al., 2018).

8 Concluding remarks

As it is reviewed in this chapter, it is of paramount importance to select the appropriate probiotic strain that will be able to adapt to storage conditions and commercialization of the food products; this, along with specific biopathways of those food products, oxygen presence, pH, ability to stay viable and in adequate amounts ($>10^6$), moving safely through the stomach to get to the intestinal tract, so they can accomplish their goal as a functional food. It was also reviewed that the potential of these microorganisms as a biocontrol of food-borne illnesses, particularly FCFVs, where very positive results have been reported. Exploring technological alternatives that allow efficient incorporation of these microorganisms in FCFVs is an ongoing task. If supported with science and technology, including food microbiology, developing the concept of FCFVs as rich sources of nutrients, fiber, vitamins, minerals, and antioxidants combined with the ideal probiotic microorganisms would, in the words of Isolauri (2001), kill two birds with the same stone.

References

Abdul Khalil, H. P. S., Banerjee, A., Saurabh, C. K., Tye, Y. Y., Suriani, A. B., Mohamed, A., et al. (2018). Biodegradable films for fruits and vegetables packaging application: Preparation and properties. *Food Engineering Reviews, 10*, 139–153.

Aguiar, L. M., Geraldi, M. V., Cazarin, C. B. B., & Junior, M. R. M. (2019). Functional food consumption and its physiological effects. In M. R. S. Campos (Ed.), *Bioactive compounds* (pp. 205–225): Woodhead Publishing.

Akman, P. K., Uysal, E., Ozkaya, G. U., Tornuk, F., & Durak, M. Z. (2019). Development of probiotic carrier dried apples for consumption as snack food with the impregnation of Lactobacillus paracasei. *LWT, 103*, 60–68.

Alegre, I., Viñas, I., Usall, J., Anguera, M., & Abadias, M. (2011). Microbiological and physicochemical quality of fresh-cut apple enriched with the probiotic strain Lactobacillus rhamnosus GG. *Food Microbiology, 28*, 59–66.

Alvarez, M. V., Moreira, M. D. R., Roura, S. I., Ayala-Zavala, J. F., & González-Aguilar, G. A. (2015). Using natural antimicrobials to enhance the safety and quality of fresh and processed fruits and vegetables: Types of antimicrobials. In T. M. Taylor (Ed.), *Handbook of natural antimicrobials for food safety and quality* (pp. 287–313). Oxford: Woodhead Publishing.

Arboleya, S., González, S., Salazar, N., Ruas-Madiedo, P., de los Reyes-Gavilán, C. G., & Gueimonde, M. (2012). Development of probiotic products for nutritional requirements of specific human populations. *Engineering in Life Sciences, 12*, 368–376.

Arora, M., Sharma, S., & Baldi, A. (2013). Comparative insight of regulatory guidelines for probiotics in USA, India and Malaysia: A critical review. *International Journal of Biotechnology for Wellness Industries, 2*, 51–64.

Awaisheh, S. S. (2012). Probiotic food products classes, types, and processing. In - *Probiotics*. IntechOpen.

Ayala-zavala, J. F., Del Toro-Sánchez, L., Alvarez-Parrilla, E., & González-Aguilar, G. A. (2008). High relative humidity in-package of fresh-cut fruits and vegetables: Advantage or disadvantage considering microbiological problems and antimicrobial delivering systems? *Journal of Food Science, 73*, R41–R47.

Bansal, S., Mangal, M., Sharma, S. K., & Gupta, R. K. (2016). Non-dairy based probiotics: A healthy treat for intestine. *Critical Reviews in Food Science and Nutrition, 56*, 1856–1867.

Bansal, V., Siddiqui, M. W., & Rahman, M. S. (2015). Minimally processed foods: Overview. In M. W. Siddiqui, & M. S. Rahman (Eds.), *Minimally processed foods: Technologies for safety, quality, and convenience* (pp. 1–16): Springer International Publishing.

Breslin, P. (2001). Human gustation and flavour. *Flavour and Fragrance Journal, 16*, 439–456.

Burgain, J., Gaiani, C., Linder, M., & Scher, J. (2011). Encapsulation of probiotic living cells: From laboratory scale to industrial applications. *Journal of Food Engineering, 104*, 467–483.

Champagne, C. P., Ross, R. P., Saarela, M., Hansen, K. F., & Charalampopoulos, D. (Check for Year). Recommendations for the viability assessment of probiotics as concentrated cultures and in food matrices. *International Journal of Food Microbiology, 149*, 185–193.

De Melo Pereira, G. V., De Oliveira Coelho, B., Júnior, A. I. M., Thomaz-Soccol, V., & Soccol, C. R. (2018). How to select a probiotic? A review and update of methods and criteria. *Biotechnology Advances, 36*(8), 2060–2076.

De Oliveira, P. M., Leite Junior, B. R. C., Lopes Martins, M., Furtado Martins, E. M., & Mota Ramos, A. (2014). Minimally processed yellow melon enriched with probiotic bacteria. *Semina: Ciências Agrárias, 35*, 2415–2426.

Degnan, F. H. (2008). The US food and drug administration and probiotics: Regulatory categorization. *Clinical Infectious Diseases, 46*, S133–S136.

Denkova, Z., & Krastanov, A. (2012). *Development of new products: Probiotics and probiotic foods*. Intechopen https://doi.org/10.5772/47827.

do Espírito Santo, A. P., Perego, P., Converti, A., & Oliveira, M. N. (2011). Influence of food matrices on probiotic viability—A review focusing on the fruity bases. *Trends in Food Science & Technology, 22*, 377–385.

Duggan, C., Gannon, J., & Walker, W. A. (2002). Protective nutrients and functional foods for the gastrointestinal tract. *The American Journal of Clinical Nutrition, 75*, 789–808.

FAO/WHO. (2001). *Health and nutritional properties of probiotics in food including powder milk with live lactic acid bacteria. In Report from FAO/WHO expert consultation.* Paper 85 Available at http://www.fao.org/3/a-a0512e.pdf.

Figueroa-González, I., Quijano, G., Ramírez, G., & Cruz-Guerrero, A. (2011). Probiotics and prebiotics—Perspectives and challenges. *Journal of the Science of Food and Agriculture, 91*, 1341–1348.

Fijan, S. (2014). Microorganisms with claimed probiotic properties: An overview of recent literature. *International Journal of Environmental Research and Public Health, 11*, 4745–4767.

Fulkerson, J. A. (2018). Fast food in the diet: Implications and solutions for families. *Physiology & Behavior, 193*, 252–256.

González-Aguilar, G. A., Ayala-Zavala, J. F., Olivas, G. I., de la Rosa, L. A., & Álvarez-Parrilla, E. (2010). Preserving quality of fresh-cut products using safe technologies. *Journal für Verbraucherschutz und Lebensmittelsicherheit, 5*, 65–72.

Granato, D., Branco, G. F., Nazzaro, F., Cruz, A. G., & Faria, J. A. F. (2010). Functional foods and nondairy probiotic food development: Trends, concepts, and products. *Comprehensive Reviews in Food Science and Food Safety, 9*, 292–302.

Hoffmann, D. E., Fraser, C. M., Palumbo, F. B., Ravel, J., Rothenberg, K., Rowthorn, V., et al. (2013). Probiotics: Finding the right regulatory balance. *Science, 342*, 314–315.

Hossain, M. I., Sadekuzzaman, M., & Ha, S. D. (2017). Probiotics as potential alternative biocontrol agents in the agriculture and food industries: A review. *Food Research International, 100*, 63–73.

Iglesias, M. B., Abadias, M., Anguera, M., Sabata, J., & Vi, I. (2017). Antagonistic effect of probiotic bacteria against foodborne pathogens on fresh-cut pear. *LWT-Food Science and Technology, 81*, 243–249.

Iglesias, M. B., Echeverría, G., Viñas, I., Lopez, M. L., & Abadias, M. (2018). Biopreservation of fresh-cut pear using Lactobacillus rhamnosus GG and effect on quality and volatile compounds. *LWT- Food Science and Technology, 87*, 581–588.

Isolauri, E. (2001). Probiotics in human disease. *The American Journal of Clinical Nutrition, 73*, 1142S–1146S.

Jankovic, I., Sybesma, W., Phothirath, P., Ananta, E., & Mercenier, A. (2010). Application of probiotics in food products—Challenges and new approaches. *Current Opinion in Biotechnology, 21*, 175–181.

Karunaratne, A. M. (2018). A multifaceted approach to harness probiotics as antagonists on plant based foods, for enhanced benefits to be reaped at a global level. *Journal of the Science of Food and Agriculture, 98*, 5189–5196.

Lebaka, V. R., Wee, Y. J., Narala, V. R., & Joshi, V. K. (2018). Development of new probiotic foods—A case study on probiotic juices. In M. A. Grumezescu, & A. M. Holban (Eds.), *Therapeutic, probiotic, and unconventional foods* (pp. 55–78): Academic Press.

Lee, Y. K., Margolles, A., Mayo, B., Ruas-Madiedo, P., Gueimonde, M., De los Reyes-Gavilán, C., et al. (2018). Probiotic microorganisms. In K. Y. Lee, & S. Salminen (Eds.), *Handbook of probiotics and prebiotics* (2nd ed., pp. 1–139): John Wiley & Sons, Inc.

Lin, D., & Zhao, Y. (2007). Innovations in the development and application of edible coatings for fresh and minimally processed fruits and vegetables. *Comprehensive Reviews in Food Science and Food Safety, 6*, 60–75.

Ljungh, A., & Wadstrom, T. (2006). Lactic acid bacteria as probiotics. *Current Issues in Intestinal Microbiology, 7*, 73–90.

Martins, E., Ramos, A., Martins, M., & Leite Junior, B. R. (2016). Fruit salad as a new vehicle for probiotic bacteria. *Food Science and Technology, 36*, 540–548.

Martins, E. M. F., Ramos, A. M., Vanzela, E. S. L., Stringheta, P. C., De Oliveira Pinto, C. L., & Martins, J. M. (2013). Products of vegetable origin: A new alternative for the consumption of probiotic bacteria. *Food Research International, 51*, 764–770.

Min, M., Bunt, C. R., Mason, S. L., & Hussain, M. A. (2018). Non-dairy probiotic food products: An emerging group of functional foods. *Critical Reviews in Food Science and Nutrition, 9*, 1–16.

Mishra, S., & Mishra, H. (2012). Technological aspects of probiotic functional food development. *Nutrafoods, 11*, 117–130.

Molin, G. (2001). Probiotics in foods not containing milk or milk constituents, with special reference to Lactobacillus plantarum 299v. *The American Journal of Clinical Nutrition, 73*, 380s–385s.

Moser, R., Raffaelli, R., & Thilmany, D. D. (2011). Consumer preferences for fruit and vegetables with credence-based attributes: A review. *International Food and Agribusiness Management Review, 14*, 121.

O'toole, P. W., & Paoli, M. (2017). The contribution of microbial biotechnology to sustainable development goals: Microbiome therapies. *Microbial Biotechnology, 10*, 1066–1069.

Przyrembel, H. (2001). Consideration of possible legislation within existing regulatory frameworks. *The American Journal of Clinical Nutrition, 73*, 471s–475s.

Quirós-Sauceda, A. E., Ayala-Zavala, J. F., Olivas, G. I., & González-Aguilar, G. A. (2014). Edible coatings as encapsulating matrices for bioactive compounds: A review. *Journal of Food Science and Technology, 51*, 1674–1685.

Ranadheera, R. D. C. S., Baines, S. K., & Adams, M. C. (2010). Importance of food in probiotic efficacy. *Food Research International, 43*, 1–7.

Roberfroid, M. B. (2000). Prebiotics and probiotics: Are they functional foods? *The American Journal of Clinical Nutrition, 71*, 1682S–1687S.

Rodgers, S. (2008). Novel applications of live bacteria in food services: Probiotics and protective cultures. *Trends in Food Science & Technology, 19*, 188–197.

Rodrigues, S., Silva, L. C. A., Mulet, A., Cárcel, J. A., & Fernandes, F. A. N. (2018). Development of dried probiotic apple cubes incorporated with Lactobacillus casei NRRL B-442. *Journal of Functional Foods, 41*, 48–54.

Rößle, C., Auty, M. A. E., Brunton, N., Gormley, R. T., & Butler, F. (2010). Evaluation of fresh-cut apple slices enriched with probiotic bacteria. *Innovative Food Science & Emerging Technologies, 11*, 203–209.

Rößle, C., Brunton, N., Gormley, R. T., Ross, P. R., & Butler, F. (2010). Development of potentially synbiotic fresh-cut apple slices. *Journal of Functional Foods, 2*, 245–254.

Roy, P., & Kumar, V. (2018). Functional food: Probiotic as health booster. *Journal of Food, Nutrition and Population Health, 2*, 12.

Russo, P., de Chiara, M. L. V., Vernile, A., Amodio, M. L., Arena, M. P., Capozzi, V., et al. (2014). Fresh-cut pineapple as a new carrier of probiotic lactic acid bacteria. *BioMed Research International, 2014*, 1–9.

Russo, P., Peña, N., de Chiara, M. L. V., Amodio, M. L., Colelli, G., & Spano, G. (2015). Probiotic lactic acid bacteria for the production of multifunctional fresh-cut cantaloupe. *Food Research International, 77*, 762–772.

Sanders, M. E. (2008). Probiotics: Definition, sources, selection, and uses. *Clinical Infectious Diseases, 46*, S58–S61.

Sanders, M. E., Akkermans, L. M., Haller, D., Hammerman, C., Heimbach, J. T., Hörmannsperger, G., et al. (2010). Safety assessment of probiotics for human use. *Gut Microbes, 1*, 164–185.

Schrezenmeir, J., & De Vrese, M. (2001). Probiotics, prebiotics, and synbiotics—Approaching a definition. *The American Journal of Clinical Nutrition, 73*, 361s–364s.

Shigematsu, E., Dorta, C., Rodrigues, F. J., Cedran, M. F., Giannoni, J. A., Oshiiwa, M., et al. (2018). Edible coating with probiotic as a quality factor for minimally processed carrots. *Journal of Food Science and Technology, 55*, 3712–3720.

Shori, A. B. (2015). The potential applications of probiotics on dairy and non-dairy foods focusing on viability during storage. *Biocatalysis and Agricultural Biotechnology, 4*, 423–431.

Shori, A. B. (2016). Influence of food matrix on the viability of probiotic bacteria: A review based on dairy and non-dairy beverages. *Food Bioscience, 13*, 1–8.

Siddiqui, M. W., Chakraborty, I., Ayala-Zavala, J., & Dhua, R. (2011). Advances in minimal processing of fruits and vegetables: A review. *Journal of Scientific and Industrial Research, 70*, 823–834.

Siroli, L., Patrignani, F., Serrazanetti, D. I., Tabanelli, G., Montanari, C., Gardini, F., et al. (2015). Lactic acid bacteria and natural antimicrobials to improve the safety and shelf-life of minimally processed sliced apples and lamb's lettuce. *Food Microbiology, 47*, 74–84.

Soccol, C. R., De Souza Vandenberghe, L. P., Spier, M. R., Medeiros, A. B. P., Yamaguishi, C. T., De Dea Lindner, J., et al. (2010). The potential of probiotics: A review. *Food Technology and Biotechnology, 48*, 413–434.

Soccol, C., Prado, M., Garcia, L., Rodrigues, C., Medeiros, A., & Soccol, V. (2014). Current developments in probiotics. *Journal of Microbial and Biochemical Technology, 7*, 011–020.

Sornplang, P., & Piyadeatsoontorn, S. (2016). Probiotic isolates from unconventional sources: A review. *Journal of Animal Science and Technology, 58*, 26.

Speranza, B., Campaniello, D., Bevilacqua, A., Altieri, C., Sinigaglia, M., & Corbo, M. R. (2018). Viability of Lactobacillus plantarum on fresh-cut chitosan and alginate-coated apple and melon pieces. *Frontiers in Microbiology, 9*, 2538.

Stanton, C., Gardiner, G., Meehan, H., Collins, K., Fitzgerald, G., Lynch, P. B., et al. (2001). Market potential for probiotics. *The American Journal of Clinical Nutrition, 73*, 476s–483s.

Tapia, M. R., Gutierrez-Pacheco, M. M., Vazquez-Armenta, F. J., González-Aguilar, G. A., Ayala-Zavala, J. F., Rahman, M. S., et al. (2015). Washing, peeling and cutting of fresh-cut fruits and vegetables. In M. W. Siddiqui, & M. S. Rahman (Eds.), *Minimally processed foods: Technologies for safety, quality, and convenience* (pp. 57–78): Springer International Publishing.

Tapia, M. S., Rojas-Graü, M. A., Rodríguez, F. J., Ramírez, J., Carmona, A., & Martin-Belloso, O. (2007). Alginate- and gellan-based edible films for probiotic coatings on fresh-cut fruits. *Journal of Food Science, 72*, E190–E196.
Torriani, S., Orsi, C., & Vescovo, M. (1997). Potential of Lactobacillus casei, culture permeate, and lacti acid to control microorganisms in ready-to-use vegetables. *Journal of Food Protection, 60*, 1564–1567.
Torriani, S., Scolari, G., Dellaglio, F., & Vescovo, M. (1999). Biocontrol of leuconostocs in ready-to-use shredded carrots. *Annali di Microbiologia ed Enzimologia, 49*, 23–32.
Trias, R., Bañeras, L., Badosa, E., & Montesinos, E. (2008). Bioprotection of golden delicious apples and iceberg lettuce against foodborne bacterial pathogens by lactic acid bacteria. *International Journal of Food Microbiology, 123*, 50–60.
Tripathi, M. K., & Giri, S. K. (2014). Probiotic functional foods: Survival of probiotics during processing and storage. *Journal of Functional Foods, 9*, 225–241.
Tuomola, E., Crittenden, R., Playne, M., Isolauri, E., & Salminen, S. (2001). Quality assurance criteria for probiotic bacteria. *The American Journal of Clinical Nutrition, 73*, 393s–398s.
Turkmen, N., Akal, C., & Özer, B. (2019). Probiotic dairy-based beverages: A review. *Journal of Functional Foods, 53*, 62–75.
Vanderhoof, J. A., & Young, R. (2008). Probiotics in the United States. *Clinical Infectious Diseases, 46*, S67–S72.
Velderrain-Rodríguez, G. R., Quirós-Sauceda, A. E., González-Aguilar, G. A., Siddiqui, M. W., & Ayala-Zavala, J. F. (2015). Technologies in fresh-cut fruit and vegetables. In M. W. Siddiqui, & M. S. Rahman (Eds.), *Minimally processed foods: Technologies for safety, quality, and convenience* (pp. 79–103): Springer International Publishing.
Vescovo, M., Torriani, S., Orsi, C., Macchiarolo, F., & Scolari, G. (1996). Application of antimicrobial-producing lactic acid bacteria to control pathogens in ready-to-use vegetables. *Journal of Applied Bacteriology, 81*, 113–119.
Vijaya Kumar, B., Vijayendra, S. V. N., & Reddy, O. V. S. (2015). Trends in dairy and non-dairy probiotic products—A review. *Journal of Food Science and Technology, 52*, 6112–6124.
Vinderola, G., Burns, P., & Reinheimer, J. (2017). Probiotics in nondairy products. In F. Mariotti (Ed.), *Vegetarian and plant-based diets in health and disease prevention* (pp. 809–835): Academic Press.
Wani, A. A., Singh, P., Pant, A., & Langowski, H. C. (2015). Packaging methods for minimally processed foods. In M. W. Siddiqui, & M. S. Rahman (Eds.), *Minimally processed foods: Technologies for safety, quality, and convenience* (pp. 35–55): Springer International Publishing.
Watanabe, M., Van Der Veen, S., & Abee, T. (2012). Impact of respiration on resistance of Lactobacillus plantarum WCFS1 to acid stress. *Applied and Environmental Microbiology, 78*, 4062–4064.
Weaver, C. M., Dwyer, J., Fulgoni, V. L., III, King, J. C., Leveille, G. A., Macdonald, R. S., et al. (2014). Processed foods: Contributions to nutrition. *The American Journal of Clinical Nutrition, 99*, 1525–1542.
Woo, J., & Ahn, J. (2013). Probiotic-mediated competition, exclusion and displacement in biofilm formation by food-borne pathogens. *Letters in Applied Microbiology, 56*, 307–313.
Zoellner, C., Venegas, F., Churey, J. J., Dávila-Avina, J., Grohn, Y. T., García, S., et al. (2016). Microbial dynamics of indicator microorganisms on fresh tomatoes in the supply chain from Mexico to the USA. *International Journal of Food Microbiology, 238*, 202–207.

11

Preservation of fresh-cut fruits and vegetables by edible coatings

Basharat Yousuf*, Ovais Shafiq Qadri†

*Department of Post Harvest Engineering and Technology, Faculty of Agricultural Sciences, Aligarh Muslim University, Aligarh, India. †Department of Bioengineering, Integral University, Lucknow, India

1 Introduction

Foods are generally sterile from within and any contamination inside, especially microbial, starts from the surface. Nature shields food with its own unique ways until it is harvested; for instance, fruits are enclosed in peel, cereals in husk, and animals in skin. In some cases, the onset of disease and other unfavorable environmental factors may compromise the quality of foods before harvest, but otherwise the quality is maintained. The increased human intervention in agricultural practices with the help of modern technology has made us better in controlling the food quality during production. The conventional and advanced preservation methods help us take care of the food after harvest.

Fruits and vegetables are nutritionally important foods. These foods serve as a source of numerous bioactive compounds including provitamins, vitamins, minerals, antioxidants, and phytochemicals. The antioxidant, antibacterial, antifungal, antiviral, and anticarcinogenic properties of fruits and vegetables have been attributed to the presence of different phytochemicals. The presence of higher moisture content in fruits and vegetables brings them under the perishable foods category. Usually, fruits and vegetables are enclosed within a protective covering known as cuticle. The wear and tear of this protective layer results in the speedy loss of water and increased susceptibility to microbial damage, which in turn decreases the shelf-life of fruits and vegetables. The integrity of fruits and vegetables is compromised during cutting and preparation, a prerequisite in fresh-cut produce.

Fresh-cut Fruits and Vegetables. https://doi.org/10.1016/B978-0-12-816184-5.00011-2

Fresh-cut fruits and vegetables offer a convenient form of these foods to consumers, while still maintaining the fresh character. The shelf-life of fresh-cut fruit or vegetable is significantly less than its intact form. Low temperature storage of fresh-cut fruits is the widely used preservation method to prolong the shelf-life and maintain the quality. The other methods include use of different packaging materials—modified atmosphere packaging further helps in retention of quality of fresh-cut produce.

The coating of fresh-cut produce with biodegradable and biocompatible materials that are safe and edible, commonly known as edible coatings, is also one of the potential methods for quality retention and shelf-life enhancement. The cut fruits and vegetables are covered in a very thin layer of an edible material that maintains the quality and does not need to be removed before consumption. Edible coatings may help in retention of water and other volatiles within the fruit, prevention of microbial deterioration, improving the aesthetic appeal, and increasing the shelf stability of the fresh-cut fruits and vegetables (Bourtoom, 2008). Edible coatings are applied as surface treatments by dipping, brushing, or spraying and are mostly derived from natural environment friendly materials (McHugh & Senesi, 2000).

Fruits and vegetables are composed of living cells, which continue to respire until senescence. The edible coating creates a barrier between the fruit surface and atmosphere, restricting the ingress of oxygen and building up the concentration of carbon dioxide within the fruit. The cells of fruits and vegetables shift to anaerobic respiration amid reduced oxygen supply, which also hampers the ethylene (hormone responsible for accelerated ripening) production. With slowing down of ripening process in addition to reduction in water loss, the coated fruits and vegetables retain the quality in terms of texture and remain shelf stable for longer durations (Dhall, 2013).

2 Types/nature of coatings

The classification of edible coatings is usually based on the major structural component, which may be a polysaccharide, protein, or lipid. Several studies have reported the use of edible coatings that are formed by synergistically combining proteins, lipids, and polysaccharides. The properties possessed by these coatings are an amalgam of the properties of each base material used and such coatings are classified under composite coatings (Falguera, Quintero, Jiménez, Muñoz, & Ibarz, 2011). Extension of shelf-life of different fresh-cut fruits and vegetables by application of different edible coatings is given in Table 1.

Table 1 Different fresh-cut fruits preserved by applying edible coatings

Fruit	Coating	Active ingredient	Type	Reference
Melon	Chitosan	Ag-nanoparticle	Polysaccharide	Ortiz-Duarte, Pérez-Cabrera, Artés-Hernández, and Martínez-Hernández (2019)
Cantaloupe	Chitosan, pectin, alginate, gellan	–	Polysaccharide	Koh, Noranizan, Hanani, Karim, and Rosli (2017)
Apricot	Basil seed gum	*Origanum vulgare* subsp. *viride* essential oil	Polysaccharide	Hashemi, Khaneghah, Ghahfarrokhi, and Eş (2017)
Rose apple	Konjac glucomannan	Pineapple extract	Polysaccharide	Supapvanich, Prathaan, and Tepsorn (2012)
Melon	Chitosan	Trans-cinnamaldehyde	Polysaccharide	Carvalho et al. (2016)
Strawberry	Gellan	Geraniol and pomegranate extract	Polysaccharide	Tomadoni, Moreira, Pereda, and Ponce (2018)
Eggplant	Soy protein	Cysteine	Protein	Ghidelli, Mateos, Rojas-Argudo, and Pérez-Gago (2014)
Melon	Sunflower oil	Citral	Lipid	Arnon-Rips, Porat, and Poverenov (2019)
Melon	Chitosan	–	Polysaccharide	Ortiz-Duarte et al. (2019)
Goji	Sodium alginate, konjac glucomannan, and starch	Lotus leaf extract	Composite	Fan et al. (2019)
Pears	Chitosan	Rosemary extracts	Polysaccharide	Xiao, Zhu, Luo, Song, and Deng (2010)
Potato	*Opuntia dillenii* polysaccharide	–	Polysaccharide	Wu (2019)
Strawberries	Beeswax, chitosan	–	Composite	Velickova, Winkelhausen, Kuzmanova, Alves, and Moldão-Martins (2013)
Apple	Carnauba wax, cassava starch	–	Composite	Chiumarelli and Hubinger (2012)
Apple	Sodium alginate and pectin	Eugenol and citral	Polysaccharide	Guerreiro, Gago, Faleiro, Miguel, and Antunes (2017)
Kiwifruit	Rice bran oil, whey protein concentrate	–	Composite	Hassani, Garousi, and Javanmard (2012)
Oranges	Gelatin	*Aloe vera*, green tea extract	Protein	Radi, Firouzi, Akhavan, and Amiri (2017)

2.1 Polysaccharide based coatings

Polysaccharide based coatings selectively permit the transfer of gases, especially oxygen and carbon dioxide. However, the barrier offered to moisture by these coatings is usually poor owing to the hydrophilic nature of polysaccharides and may therefore be appropriate for low moisture foods. The gas barrier properties of polysaccharide-based coatings help in development of modified atmospheric conditions within the cut fruit. Since the polysaccharide polymers do not completely prevent the transfer of gases but minimize it, the aerobic respiration continues but at a very low rate and the fruit starts to respire anaerobically as well. This decrease in the respiration rate of the produce delays the senescence and increases the shelf-life. If the aerobic respiration stops completely, the fruits and vegetables respire only anaerobically and there is a build-up of different compounds. Accumulation of these compounds affects the quality attributes of the fruits and vegetables and results in undesirable changes (Yousuf, Qadri, & Srivastava, 2018).

The polysaccharide base material for edible coatings may be derived from different plants, animals, or microorganisms including marine species. Some of the polysaccharides that have been reported to be used include carboxymethyl cellulose (Ponce, Roura, del Valle, & Moreira, 2008), chitosan (Romanazzi, Nigro, Ippolito, Divenere, & Salerno, 2002), pectin (Maftoonazad, Ramaswamy, Moalemiyan, & Kushalappa, 2007), locust bean gum, guar gum, ethyl cellulose (Shrestha, Arcot, & Paterson, 2003), sorbitol and sucrose (Sobral, Menegalli, Hubinger, & Roques, 2001), cassava starch (Kechichian, Ditchfield, Veiga-Santos, & Tadini, 2010), and maize starch (Pagella, Spigno, & De Faveri, 2002), among others.

2.2 Protein based coatings

Proteins are also used to develop edible coatings for fresh-cut fruits and vegetables. Proteins are generally hydrophilic, and the coatings developed from proteins may absorb moisture and be affected by temperature and humidity. Proteins derived from animals, like casein, whey, collagen, gelatin, and keratin, are categorized under fibrous proteins and are water insoluble, whereas globular proteins of plant origin, which include wheat gluten, soy protein, peanut protein, corn-zein, cottonseed protein, may be soluble in water or in acidic, basic, or salt solutions.

The characteristics exhibited by proteins vary widely and are dependent on their nature and function. It is these molecular characteristics that determine whether a protein is suitable for film formation and the properties that are possessed by the film formed from a protein. The protein is generally applied in the form of a solution or dispersion and once applied, the solvent or carrier is allowed to evaporate, leaving a layer of protein over the fruit. The denaturation of proteins before application increases the chances of chain-to-chain interactions. Thus, the films obtained are strong, and quite resistant to transfer of gases and liquids, but the flexibility is limited (Kester & Fennema, 1986).

Different researchers have reported the use of protein-based coatings derived from corn, wheat, soybeans, peanut, and milk. Gelatin, an animal protein, has also been reported to have been used in fresh-cut fruits and vegetables.

2.3 Lipid based coatings

Lipids as coating in foods have been used for a long time. Waxing of fruits and coating of confectionary with lipid-based materials are a few examples of lipids being used as food coatings in history (Dhall, 2013). Lipids are hydrophobic and repel water. The hydrophobicity of lipids makes them excellent moisture barriers. In addition, coating of lipids also reduces the respiration rate of fruits and vegetables and improves the appearance by adding shine and gloss to the surface. The problem with lipid coatings is that they form a thick layer over the surface, are generally opaque, possess less flexibility, and because of the hydrophobic nature, may not stick to cut surfaces of fresh-cut fruits and vegetables. These factors limit the use lipid based coatings in fresh-cut produce (Olivas & Barbosa-Cánovas, 2005). The commonly used lipids for coating are waxes, acylglycerols, or fatty acids. Carnauba wax and beeswax are natural waxes that have been used to formulate coating for fresh-cut fruits and vegetables. Other lipids used in coating of fresh-cut produce include sunflower oil, rosehip oil, lemongrass essential oil, and others (Yousuf et al., 2018).

2.4 Composite coatings

The materials used to coat foods possess specific characteristics that depend mainly on the properties of the chief component. There are certain characteristics that are common in a group—for instance, hydrophobic character of lipid-based coatings—but the specific characteristics of a coating material are attributed to the main component. Some of the coating materials are better barriers to gases, others prevent the loss of moisture in a better way, but a homogenous coating does not always possess all the desirable properties. However, hydrocolloids (proteins and carbohydrates) and lipids may be integrated together to form coatings that possess better functionality. Such coatings are known as composite coatings or bilayer coatings (Lin & Zhao, 2007). Composite coatings are being studied extensively, especially with fresh-cut produce because they form potentially superior films. Composite films may be prepared for better barrier properties or improved mechanical strength. Application of these heterogeneous coatings on fresh-cut fruits and vegetables can be done either by applying the coating mixture (emulsion, suspension, dispersion, or solution) or by coating the different layers successively. The multilayer coatings are generally better in terms of barrier properties because all the layers are continuous throughout the surface of the fruit. Composite coatings present enormous possibilities in the field of edible coatings, as is evident from the intensity of studies being reported currently in the field of fresh-cut produce. Some of the combinations that have been studied include soy protein isolate and polylactic acid (Rhim, Lee, & Ng, 2007), beeswax and chitosan (Velickova et al., 2013), rice bran oil and whey protein concentrate (Hassani et al., 2012), and maize starch-chitosan-glycerin (Liu, Xie, Yu, Chen, & Li, 2009).

3 Edible coatings as a medium for active components

During fresh-cut processing, fruit tissue is injured or wounded. This wounding during preparation makes fresh-cut products more perishable than their corresponding whole uncut commodities. The physical barrier provided by the epidermis is removed during processing, which facilitates easy access to microorganisms. A number of bioactive compounds have been investigated for their effects on fresh-cut produce. Certain studies are also devoted to analyzing their mode of action, nature of activity, toxicology, and impact on sensory, biochemical, and physiological properties of the fresh-cut produce. All these measures are taken to ensure high quality and safe products to consumers. Significant research is being directed to develop effective natural preservatives. Preservatives used for fresh-cut products can be broadly categorized as antibrowning agents and antimicrobial agents. These compounds can be used directly as dip treatments to fresh-cut fruits or vegetables or they can be incorporated into the edible coatings used for these products.

Edible coatings act as effective carriers of active ingredients that can extend product shelf-life, reduce the risk of microbial growth and proliferation on foods, and retard many other unwanted changes in fresh-cut products (Pranoto, Salokhe, & Rakshit, 2005). There are a number of instances where active components have been incorporated into edible coatings and subsequently analyzed for their effectiveness in performing their functions. Some other additives may also be incorporated into the edible coatings such as texture enhancers and nutraceuticals (Rojas-Graü, Soliva-Fortuny, & Martín-Belloso, 2009). As already discussed, the basic ingredients of edible coatings are proteins, polysaccharides, and lipids; whereas, active ingredients include food-grade additives, which are classified as generally regarded as safe compounds (GRAS). The incorporation of these active compounds improves the functionality of edible coatings in terms of shelf-life, safety, and sensory qualities and compensates the limitations of the edible coatings. The role of each broad category of active components while being incorporated into coatings is discussed below.

3.1 Antimicrobial agents

Fresh-cut fruits and vegetables provide favorable conditions to different types of microorganisms for their survival, growth and proliferation. Deterioration caused by microorganisms is the major source of spoilage of fresh-cut fruits and vegetables. This microbial spoilage leads to both public health risks and economic losses (Qadri, Yousuf, & Srivastava, 2016). Fresh-cut fruits and vegetables are highly susceptible to spoilage and pathogenic bacteria growth. Fruits and vegetables may be contaminated by microorganisms from many sources, such as soil, water, animals, birds, and insects during growth at farm. Further addition of contamination may happen during various steps like harvesting, washing, cutting, packaging, and shipping. In addition, fresh-cut fruits and vegetables are never subjected to any "lethal" treatment such as thermal processing for microorganism destruction. Therefore, antimicrobial agents can be incorporated in the edible coating for shelf-life extension and to reduce microbial growth.

Incorporating antimicrobial compounds into edible coatings can provide a novel way to ensure safety and improve the shelf-life of fresh-cut produce. Edible coatings are important as an alternative to reduce the deleterious effects imposed by minimal processing on fruits and vegetables. Some of the edible coatings have inherent antimicrobial properties, such as chitosan coating. On the other hand, the antimicrobial components can also be incorporated into the coatings. For instance, essential oils possessing antimicrobial activity can be are incorporated into the edible

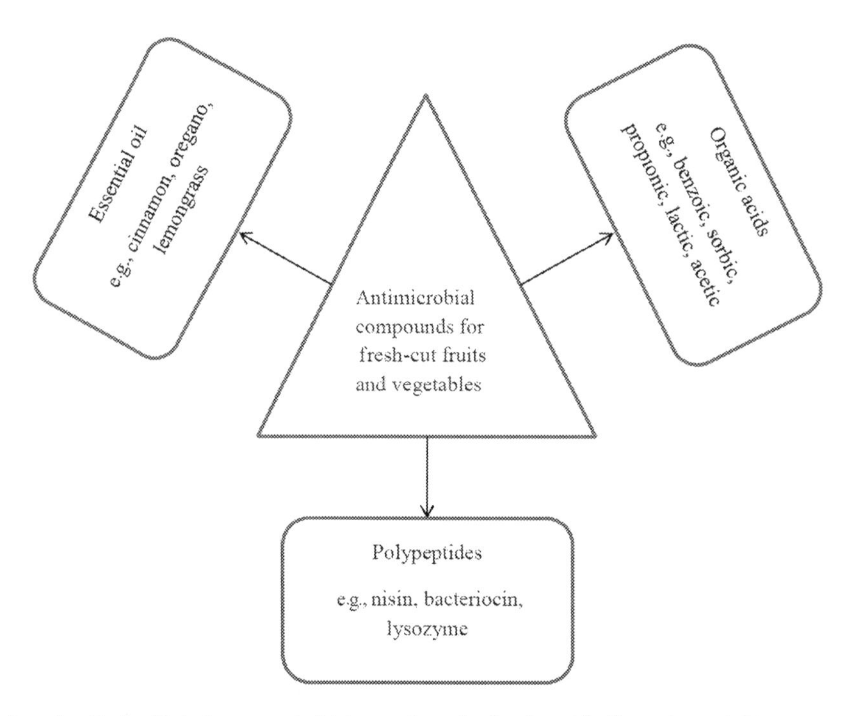

Fig. 1 Examples of antimicrobials incorporated into coatings for fresh-cut fruits and vegetables.

coatings. Some important categories of antimicrobial agents that can be incorporated into edible coatings are shown in Fig. 1. Incorporation of antimicrobial compounds into edible coatings ensures a novel way to improve the safety and shelf-life of fresh-cut fruits and vegetables. Various studies have demonstrated that the antimicrobial compounds incorporated into edible coatings are effective in suppressing the growth of pathogen and spoilage microorganisms in fresh-cut fruits and vegetables.

3.2 Antibrowning agents

Fresh-cut processing causes rapid deterioration due to tissue damage, which results in increased physiological activity and major physico-chemical changes, such as enzymatic browning. Edible coatings are gaining importance as an alternative to reduce the detrimental effects on fresh-cut products imparted by fresh-cut processing. The functional properties of coatings may be improved by the addition of food ingredients, such as antioxidants and antibrowning agents to enhance appearance, integrity, and microbial safety (Valencia-Chamorro, Pérez-Gago, del Río, & Palou, 2011).

Browning is one of the most prevalent problems in fresh-cut fruits and vegetables and is always judged as lower quality. Edible coatings act as carriers of food additives including antibrowning agents. Antibrowning agents protect fresh-cut fruits and vegetables against oxidative rancidity, degradation, and enzymatic browning (Rojas-Graü, Tapia, Rodríguez, Carmona, & Martin-Belloso, 2007). Most of the antibrowning agents for fresh-cut fruits and vegetables have been studied individually. However, antibrowning agents and edible coating can be both used on fresh-cut fruits produce. Antibrowning agents can be incorporated into the edible coatings. Coating with addition of antioxidants effectively reduces browning and retardeds problems of fresh-cut products. Natural substances that are able to act as antibrowning agents have garnered a lot of interest in recent years. Incorporation of active substances is advantageous over the direct application onto foods because edible coatings can be designed for the controlled release of such substances (Sanchís et al., 2016).

Ascorbic acid, 4-hexylresorcinol, and some sulfur-containing amino acids such as cysteine and glutathione have been widely studied, to prevent enzymatic browning and for improving shelf-life of fresh-cut fruits and vegetables. Antibrowning and antioxidant agents may sometimes perform similar functions. The incorporation of antioxidant agents such as *N*-acetylcysteine and glutathione into alginate- and gellan-based coatings helped to prevent fresh-cut apples and papayas from browning (Rojas-Grau, Tapia, & Martın-Belloso, 2008; Tapia, Rodrıguez, Rojas-Grau, & Martın-Belloso, 2005). A substantial reduction in browning of fresh-cut apples was reported when using a whey protein concentrate beeswax coating containing ascorbic acid, cysteine, or 4-hexylresorcinol (Perez-Gago, Serra, & del Rio, 2006).

4 Impact of coating on quality of fresh-cut fruits and vegetables

The primary function of coating on fresh-cut produce surface is shelf-life extension. The quality of the fruit—which includes microbiological quality, nutritional quality, and organoleptic quality—is to be preserved, if not improved, by the coating applied. The coated fresh-cut fruit or vegetable may encounter different changes depending on the type of coating applied. There are different aspects that may be influenced by the application of coating including nutritional, sensory, and safety. The effect of edible coatings on quality and shelf-life of fresh-cut products has been extensively studied.

4.1 Effect of coating on water vapor permeability and moisture/weight loss

After harvesting, water lost by fresh produce through transpiration could not be replaced, rendering the produce unmarketable, because a little water loss may have a large impact on quality. Excessive water loss results in unappealing appearance and reduced crispness. Fresh-cut products are more susceptible to water loss than the whole and/or intact fruits. Water loss is one of the major problems associated with fresh-cut produce and can be controlled with edible wax coatings (Debeaufort, Quezada-Gallo, & Voilley, 1998).

Ghavidel, Davoodi, Asl, Tanoori, and Sheykholeslami (2013) coated fresh-cut apples with different coatings including whey protein, soy protein isolate, carrageenan, and alginate coatings. The analysis showed significant differences of water content between coated and uncoated samples. Their results indicated that samples with coating of alginate had the highest water retention while soyprotein coated samples showed lowest water retention. Khan, Cakmak, Tavman, Schutyser, and Schroën (2014) applied solid lipid based coatings on apple slices by means of electrospraying and showed that the coatings significantly reduced the moisture loss of coated apple slices, and flux reduction was a function of the amount of coating applied.

Weight loss in fruits and vegetables is usually due to loss of water by transpiration, which consequently determines freshness and shelf-life. Nongtaodum and Jangchud (2009) found that chitosan coating reduced the weight loss of fresh-cut mango. During storage, they observed that the weight loss was significantly higher in the case of uncoated samples than the coated ones. Mohamed, Aboul-Anean, and Amal (2013) used edible coating in extending the shelf-life of minimally processed prickly pear and showed that there was significant weight loss in uncoated samples compared to those of guar or xanthan coated samples.

4.2 Effect of coating on texture

Food texture is one of the fundamental features determining product acceptability. Detrimental changes in texture is one of the main causes of quality loss in fresh-cut products and texture can seldom be maintained for a long period of time, even under optimal storage conditions. Generally, plant tissue loses firmness due to various unit operations employed for preparation of fresh-cut produce. Softening of tissue is a serious problem limiting the shelf-life of fresh-cut products. However, fresh-cut fruits and vegetables have to maintain firm and desirable textures as consumers relate texture with freshness and wholesomeness.

Treatment with calcium can be used to maintain firmness of fresh-cut produce. It has been reported that dipping of fresh-cut products in calcium chloride solutions (0.5%–1%) can be effective in maintaining tissue firmness. Vilas-Boas and Kader (2006) showed that calcium compounds, applied in the form of calcium dips, can help to maintain the firmness of fresh-cut apple, pear, strawberry, and banana. Martin-Diana et al. (2006) studied the textural and micro-structural changes in fresh-cut lettuce over a storage period of 12 days. Treatment with calcium lactate was found to maintain sensorial and textural properties of fresh-cut lettuce. Sometimes, use of calcium chloride as a firming agent may impart bitterness due to a residual amount of calcium chloride remaining on the surface of the product after the dip treatment. However, this bitter taste may be suppressed by combining calcium with larger ions such as lactate, gluconate, or glycerol phosphate (Lawless, Rapacki, Horne, & Hayes, 2003).

4.3 Effect of coating on color

Browning is one of the main problems occurring in fresh-cut produce. Action of Polyphenol oxidase on polyphenols is believed to be the major cause of browning reactions. Polyphenol oxidases catalyze two reactions after interacting with polyphenolic substrates present in the fruit in presence of oxygen (Toivonen & Brummell, 2008). One is hydroxylation of monophenols to diphenols, which is relatively slow and results in colorless products. The second reaction is oxidation of diphenols to quinones, which are colored compounds, and the reaction is relatively rapid. Follow-up reactions of quinones results in browning of tissue due to formation and accumulation of the brown or black pigment called melanin. In addition to polyphenol oxidase, peroxidase is also known to be involved in enzymatic browning appearing along the cut surface of fresh-cut produce.

Control of browning of the cut surfaces is one of the important issues in fresh-cut processing. Enzymatic browning in fresh-cut fruits and vegetables can be retarded by the use of various anti-browning agents. Organic acids such as ascorbic acid, citric acid and oxalic acid, often found in most fresh fruits and vegetables, are used to control excessive browning (González-Aguilar, Ruiz-Cruz, Cruz-Valenzuela, Rodrıguez-Félix, & Wang, 2004). Suttirak and Manurakchinakorn (2011) discussed the potential application of these organic acids for browning inhibition in fresh-cut fruits and vegetables. Based on mechanism of inhibition, antibrowning agents have been categorized into various groups such as acidulants, reducing agents, chelating agents, complexing agents, and enzyme inhibitors (Altunkaya & Gökmen, 2009). Kim, Kim, Chung, and Moon (2014) used phytoncide (an oil derived from pine leaves) treatment for browning control of fresh-cut lettuce. Changes in the browning characteristics of cut lettuce treated with phytoncide were investigated for 12 days at 4°C. Their results suggest that phytoncide treatment could be used as an effective method for control of enzymatic browning of fresh-cut lettuce.

4.4 Effect of coating on respiration rate and ethylene production

Fruits and vegetables are living organisms and require oxygen during respiration, and in turn release carbon dioxide and ethylene. Respiration rate of fresh-cut produce is generally elevated as a result of wounding, affecting shelf-life. When plant tissues are injured during processing operations of fresh-cut products, ethylene can accumulate in the package. This leads to detrimental effects on quality of the fresh-cut produce. The disruption of cellular membranes caused due to physical damage during minimal processing accelerates the loss of quality and shelf-life. Increase in respiration rate and ethylene are the two major reactions that are responsible for the rapid loss in quality and shelf-life. Gorny, Hess-Pierce, Cifuentes, and Kader (2002) compared the respiration and ethylene production of different cultivars of whole and sliced pear and showed that sliced samples had greater rate of respiration than whole fruit. Aguayo, Escalona, and Rtés (2004) studied metabolic behavior and quality changes of whole and fresh processed melon and found that wounding caused due to cutting resulted in an increase in carbon dioxide and ethylene production.

However, different edible coatings can be used to reduce the rate of respiration significantly. The reduction in respiration rate has been reported in the case of kiwifruit coated with aloe vera (Benítez, Achaerandio, Sepulcre, & Pujolà, 2013), in sweet cherry coated with aloe vera (Martínez-Romero et al., 2006), and in kiwifruit with alginate based coatings with grape fruit seed extracts (Mastromatteo, Mastromatteo, Conte, & Del Nobile, 2011). Considering thickness of coating is important, because if the thickness of coating is greater, it might restrict the transfer of gases to an extent that forces the cells to undertake anaerobic respiration. The production of ethanol and acetaldehyde as a result of anaerobic respiration may generate off-flavor and deteriorate the quality (Cisneros-Zevallos & Krochta, 2003).

4.5 Effect on nutrient content, product stability and overall quality and shelf-life of fresh-cut fruits and vegetables

The overall quality and shelf-life of fruits and vegetables is reduced by a number of factors, which may include water loss, texture deterioration, enzymatic browning, senescence processes, microbial growth, and so on. In the case of fresh-cut fruits, these events are further accelerated due to wounding of tissue by various unit operations during their preparation. Therefore, it is a major challenge for the fresh-cut fruit industry to maintain flavor, aroma, microbial, and overall quality of produce throughout the distribution chain. Fresh-cut processing wounds the fruit tissues, which may subsequently result in loss of nutrients and other possible detrimental changes during storage. Tissue damage during various processing steps like peeling, shredding, cutting, slicing, or dicing leads to degradation in fruit quality, in terms of both increased susceptibility of microbial attacks and disruption of tissue, and subsequent loss of nutritional components. Loss of essential components such as ascorbic acid, sugars, and beta-carotene during storage of fresh-cut fruits and vegetables is of concern as they are seen as the primary quality parameters. Ascorbic acid has nutritional importance as well as antioxidant properties.

Edible coatings have the potential to increase shelf-life and maintain freshness of fresh-cut produce by reducing the deleterious effect brought about by minimal processing. Edible coatings and antioxidants have been found to prevent loss of nutrient during cold storage of fresh-cut mangoes (Gonzalez-Aguilar et al., 2008). Edible coatings act as barriers against external environment and thereby increasing shelf-life. Shelf-life extension and quality is maintained due to reduction in gas exchange, loss of water, and retention of flavor and aroma. The data from a study carried out by Chien, Sheu, and Yang (2007) reveal that applying a chitosan coating effectively prolongs the quality attributes and extends the shelf-life of sliced mango fruit.

5 Commercialization of edible coatings for fresh-cut fruits and vegetables

The greatest hurdle to the commercial marketing of fresh-cut fruits is the limited shelf-life due to excessive tissue softening and cut surface browning. Researchers are continuously exploring the role of edible coatings to improve the quality and to extend the commercial shelf-life of different fresh-cut fruits and vegetables.

However, most of the studies on coating of foods (particularly fresh-cut products) are limited to laboratory scale only. Now, the need raised due to modernization is to make such studies viable on a commercial scale, with the aim to provide consumers with convenient but safe and wholesome food products. Edible films and coatings have long been investigated and used on a wide variety of foods and recently include fresh-cut produce to maintain optimal quality and to increase their shelf-life. Although only a limited number of coatings may be commercially available, the trend in their use and commercialization is on the rise.

6 Conclusion

Edible coating serves to complement or replace traditional preservation methods. The use of edible coating has shown promising results in reducing weight loss, microbial growth, and metabolism rate, delaying the ripening process, and extending the shelf-life of fresh and fresh-cut fruits and vegetables. Edible coatings can protect perishable foods including fresh-cut produce from deterioration. They help to maintain desirable quality characteristics of fresh-cut fruits and vegetables. Coatings act as a natural preservative and serve as carriers of various active agents. Edible coatings in combination with antimicrobial and antibrowning agents can effectively extended the shelf-life of fresh-cut fruits and vegetables. The development of novel techniques to improve the delivery properties of edible coatings is an important concern for research in the future. Most of the studies on use of edible coating on fresh-cut produce have been conducted at laboratory scale. However, efforts should be to commercialize the technology of edible coating as a protective method for fresh-cut produce. Further, more studies are essential to understand the process and consequences related to incorporation of active ingredients into edible coatings, and their mechanical, sensory, and functional properties.

References

Aguayo, Escalona, V. H., & Rtés, F. A. (2004). Metabolic behavior and quality changes of whole and fresh processed melon. *Journal of Food Science, 69*, 148–155.

Altunkaya, A., & Gökmen, V. (2009). Effect of various anti-browning agents on phenolic compounds profile of fresh lettuce (*L. sativa*). *Food Chemistry, 117*, 122–126.

Arnon-Rips, H., Porat, R., & Poverenov, E. (2019). Enhancement of agricultural produce quality and storability using citral-based edible coatings; the valuable

effect of nano-emulsification in a solid-state delivery on fresh-cut melons model. *Food Chemistry, 277*, 205–212.
Benítez, S., Achaerandio, I., Sepulcre, F., & Pujolà, M. (2013). Aloe vera based edible coatings improve the quality of minimally processed 'Hayward' kiwifruit. *Postharvest Biology and Technology, 81*, 29–36.
Bourtoom, T. (2008). Edible films and coatings: Characteristics and properties. *International Food Research Journal, 15*, 237–248.
Carvalho, R. L., Cabral, M. F., Germano, T. A., de Carvalho, W. M., Brasil, I. M., Gallão, M. I., et al. (2016). Chitosan coating with trans-cinnamaldehyde improves structural integrity and antioxidant metabolism of fresh-cut melon. *Postharvest Biology and Technology, 113*, 29–39.
Chien, P. J., Sheu, F., & Yang, F. H. (2007). Effects of edible chitosan coating on quality and shelf life of sliced mango fruit. *Journal of Food Engineering, 78*, 225–229.
Chiumarelli, M., & Hubinger, M. D. (2012). Stability, solubility, mechanical and barrier properties of cassava starch—Carnauba wax edible coatings to preserve fresh-cut apples. *Food Hydrocolloids, 28*, 59–67.
Cisneros-Zevallos, L., & Krochta, J. M. (2003). Dependence of coating thickness on viscosity of coating solution applied to fruits and vegetables by dipping method. *Journal of Food Science, 68*, 503–510.
Debeaufort, F., Quezada-Gallo, J. A., & Voilley, A. (1998). Edible films and coatings: Tomorrow's packagings: A review. *Critical Reviews in Food Science and Nutrition, 38*, 299–313.
Dhall, R. K. (2013). Advances in edible coatings for fresh fruits and vegetables: A review. *Critical Reviews in Food Science and Nutrition, 53*, 435–450.
Falguera, V., Quintero, J. P., Jiménez, A., Muñoz, J. A., & Ibarz, A. (2011). Edible films and coatings: Structures, active functions and trends in their use. *Trends in Food Science & Technology, 22*, 292–303.
Fan, X. J., Zhang, B., Yan, H., Feng, J. T., Ma, Z. Q., & Zhang, X. (2019). Effect of lotus leaf extract incorporated composite coating on the postharvest quality of fresh goji (*Lycium barbarum* L.) fruit. *Postharvest Biology and Technology, 148*, 132–140.
Ghavidel, R. A., Davoodi, M. G., Asl, A. F. A., Tanoori, T., & Sheykholeslami, Z. (2013). Effect of selected edible coatings to extend shelf-life of fresh-cut apples. *International Journal of Agriculture and Crop Sciences, 6*, 1171.
Ghidelli, C., Mateos, M., Rojas-Argudo, C., & Pérez-Gago, M. B. (2014). Extending the shelf life of fresh-cut eggplant with a soy protein–cysteine based edible coating and modified atmosphere packaging. *Postharvest Biology and Technology, 95*, 81–87.
Gonzalez-Aguilar, G. A., Celis, J., Sotelo-Mundo, R. R., De La Rosa, L. A., Rodrigo-Garcia, J., & Alvarez-Parrilla, E. (2008). Physiological and biochemical changes of different fresh-cut mango cultivars stored at 5°C. *International Journal of Food Science & Technology, 43*, 91–101.
González-Aguilar, G. A., Ruiz-Cruz, S., Cruz-Valenzuela, R., Rodrıguez-Félix, A., & Wang, C. Y. (2004). Physiological and quality changes of fresh-cut pineapple treated with antibrowning agents. *LWT-Food Science and Technology, 37*, 369–376.
Gorny, J. R., Hess-Pierce, B., Cifuentes, R. A., & Kader, A. A. (2002). Quality changes in fresh-cut pear slices as affected by controlled atmospheres and chemical preservatives. *Postharvest Biology and Technology, 24*, 271–278.
Guerreiro, A. C., Gago, C. M. L., Faleiro, M. L., Miguel, M. G. C., & Antunes, M. D. C. (2017). The effect of edible coatings on the nutritional quality of 'Bravo de Esmolfe' fresh-cut apple through shelf-life. *LWT-Food Science and Technology, 75*, 210–219.

Hashemi, S. M. B., Khaneghah, A. M., Ghahfarrokhi, M. G., & Eş, I. (2017). Basil-seed gum containing *Origanum vulgare* subsp. viride essential oil as edible coating for fresh cut apricots. *Postharvest Biology and Technology, 125*, 26–34.

Hassani, F., Garousi, F., & Javanmard, M. (2012). Edible coating based on whey protein concentrate-rice bran oil to maintain the physical and chemical properties of the kiwifruit (*Actinidia deliciosa*). *Trakia Journal of Sciences, 10*, 26–34.

Kechichian, V., Ditchfield, C., Veiga-Santos, P., & Tadini, C. C. (2010). Natural antimicrobial ingredients incorporated in biodegradable films based on cassava starch. *LWT-Food Science and Technology, 43*, 1088–1094.

Kester, J. J., & Fennema, O. R. (1986). Edible films and coatings: A review. *Food Technology, 60*, 47–59.

Khan, M. K. I., Cakmak, H., Tavman, Ş., Schutyser, M., & Schroën, K. (2014). Anti-browning and barrier properties of edible coatings prepared with electrospraying. *Innovative Food Science & Emerging Technologies, 25*, 9–13.

Kim, D. H., Kim, H. B., Chung, H. S., & Moon, K. D. (2014). Browning control of fresh-cut lettuce by phytoncide treatment. *Food Chemistry, 159*, 188–192.

Koh, P. C., Noranizan, M. A., Hanani, Z. A. N., Karim, R., & Rosli, S. Z. (2017). Application of edible coatings and repetitive pulsed light for shelf life extension of fresh-cut cantaloupe (*Cucumis melo* L. reticulatus cv. Glamour). *Postharvest Biology and Technology, 129*, 64–78.

Lawless, H. T., Rapacki, F., Horne, J., & Hayes, A. (2003). The taste of calcium and magnesium salts and anionic modifications. *Food Quality and Preference, 14*, 319–325.

Lin, D., & Zhao, Y. (2007). Innovations in the development and application of edible coatings for fresh and minimally processed fruits and vegetables. *Comprehensive Reviews in Food Science and Food Safety, 6*, 60–75.

Liu, H., Xie, F., Yu, L., Chen, L., & Li, L. (2009). Thermal processing of starch-based polymers. *Progress in Polymer Science, 34*, 1348–1368.

Maftoonazad, N., Ramaswamy, H. S., Moalemiyan, M., & Kushalappa, A. C. (2007). Effect of pectin-based edible emulsion coating on changes in quality of avocado exposed to *Lasiodiplodia theobromae* infection. *Carbohydrate Polymers, 68*, 341–349.

Martin-Diana, A. B., Rico, D., Frias, J., Henehan, G. T. M., Mulcahy, J., Barat, J. M., et al. (2006). Effect of calcium lactate and heat-shock on texture in fresh-cut lettuce during storage. *Journal of Food Engineering, 77*, 1069–1077.

Martínez-Romero, D., Alburquerque, N., Valverde, J. M., Guillén, F., Castillo, S., Valero, D., et al. (2006). Postharvest sweet cherry quality and safety maintenance by *Aloe vera* treatment: A new edible coating. *Postharvest Biology and Technology, 39*, 93–100.

Mastromatteo, M., Mastromatteo, M., Conte, A., & Del Nobile, M. A. (2011). Combined effect of active coating and MAP to prolong the shelf life of minimally processed kiwifruit (*Actinidia deliciosa* cv. Hayward). *Food Research International, 44*, 1224–1230.

McHugh, T. H., & Senesi, E. (2000). Apple wraps: A novel method to improve the quality and extend the shelf life of fresh-cut apples. *Journal of Food Science, 65*, 480–485.

Mohamed, A. Y. I., Aboul-Anean, H. E., & Amal, M. H. (2013). Utilization of edible coating in extending the shelf life of minimally processed prickly pear. *Journal of Applied Sciences Research, 9*(2), 1202–1208.

Nongtaodum, S., & Jangchud, A. (2009). Effects of edible chitosan coating on quality of fresh-cut mangoes (Falun) during storage. *Kasetsart Journal (Natural Science), 43*, 282–289.

Olivas, G. I., & Barbosa-Cánovas, G. V. (2005). Edible coatings for fresh-cut fruits. *Critical Reviews in Food Science and Nutrition, 45*, 657–670.

Ortiz-Duarte, G., Pérez-Cabrera, L. E., Artés-Hernández, F., & Martínez-Hernández, G. B. (2019). Ag-chitosan nanocomposites in edible coatings affect the quality of fresh-cut melon. *Postharvest Biology and Technology, 147,* 174–184.

Pagella, C., Spigno, G., & De Faveri, D. M. (2002). Characterization of starch based edible coatings. *Food and Bioproducts Processing, 80,* 193–198.

Perez-Gago, M. B., Serra, M., & del Rio, M. A. (2006). Color change of fresh-cut apples coated with whey protein concentrate-based edible coatings. *Postharvest Biology and Technology, 39,* 84–92.

Ponce, A. G., Roura, S. I., del Valle, C. E., & Moreira, M. R. (2008). Antimicrobial and antioxidant activities of edible coatings enriched with natural plant extracts: In vitro and in vivo studies. *Postharvest Biology and Technology, 49,* 294–300.

Pranoto, Y., Salokhe, V., & Rakshit, K. S. (2005). Physical and anti- bacterial properties of alginate-based edible film incorporated with garlic oil. *Food Research International, 38,* 267–272.

Qadri, O. S., Yousuf, B., & Srivastava, A. K. (2016). Fresh-cut produce: Advances inpreserving quality and ensuring safety. In M. W. Siddiqi, & A. Ali (Eds.), *Post-harvest management of horticultural crops: Practices for quality preservation* (pp. 265–290). Waretown, NJ: Apple Academic Press.

Radi, M., Firouzi, E., Akhavan, H., & Amiri, S. (2017). Effect of gelatin-based edible coatings incorporated with *Aloe vera* and black and green tea extracts on the shelf life of fresh-cut oranges. *Journal of Food Quality, 2017,* 1–10.

Rhim, J. W., Lee, J. H., & Ng, P. K. (2007). Mechanical and barrier properties of biodegradable soy protein isolate-based films coated with polylactic acid. *LWT-Food Science and Technology, 40,* 232–238.

Rojas-Graü, M. A., Soliva-Fortuny, R., & Martín-Belloso, O. (2009). Edible coatings to incorporate active ingredients to fresh-cut fruits: A review. *Trends in Food Science & Technology, 20,* 438–447.

Rojas-Grau, M. A., Tapia, M. S., & Martın-Belloso, O. (2008). Using polysaccharide-based ´ edible coatings to maintain quality of fresh-cut Fuji apples. *LWT- Food Science and Technology, 41,* 139–147.

Rojas-Graü, M. A., Tapia, M. S., Rodríguez, F. J., Carmona, A. J., & Martin-Belloso, O. (2007). Alginate and gellan-based edible coatings as carriers of antibrowning agents applied on fresh-cut Fuji apples. *Food Hydrocolloids, 21,* 118–127.

Romanazzi, G., Nigro, F., Ippolito, A., Divenere, D., & Salerno, M. (2002). Effects of pre-and postharvest chitosan treatments to control storage grey mold of table grapes. *Journal of Food Science, 67,* 1862–1867.

Sanchís, E., González, S., Ghidelli, C., Sheth, C. C., Mateos, M., Palou, L., et al. (2016). Browning inhibition and microbial control in fresh cut persimmon (*Diospyros kaki* Thunb. cv. Rojo Brillante) by apple pectin-based edible coatings. *Postharvest Biology and Technology, 112,* 186–193.

Shrestha, A. K., Arcot, J., & Paterson, J. L. (2003). Edible coating materials—Their properties and use in the fortification of rice with folic acid. *Food Research International, 36,* 921–928.

Sobral, P. D. A., Menegalli, F. C., Hubinger, M. D., & Roques, M. A. (2001). Mechanical, water vapor barrier and thermal properties of gelatin based edible films. *Food Hydrocolloids, 15,* 423–432.

Supapvanich, S., Prathaan, P., & Tepsorn, R. (2012). Browning inhibition in fresh-cut rose apple fruit cv. Taaptimjaan using konjac glucomannan coating incorporated with pineapple fruit extract. *Postharvest Biology and Technology, 73,* 46–49.

Suttirak, W., & Manurakchinakorn, S. (2011). Potential application of ascorbic acid, citric acid and oxalic acid for browning inhibition in fresh-cut fruits and vegetables. *Walailak Journal of Science and Technology, 7,* 5–14.

Tapia, M. S., Rodrıguez, F. J., Rojas-Grau, M. A., & Martın-Belloso, O. (2005). Formulation of alginate and gellan based edible coatings with antioxidants for fresh-cut apple and papaya. In *IFT annual meeting. Paper 36–43, New Orleans, USA.*
Toivonen, P. M., & Brummell, D. A. (2008). Biochemical bases of appearance and texture changes in fresh-cut fruit and vegetables. *Postharvest Biology and Technology, 48,* 1–14.
Tomadoni, B., Moreira, M. R., Pereda, M., & Ponce, A. G. (2018). Gellan-based coatings incorporated with natural antimicrobials in fresh-cut strawberries: Microbiological and sensory evaluation through refrigerated storage. *LWT-Food Science and Technology, 97,* 384–389.
Valencia-Chamorro, S. A., Pérez-Gago, M. B., del Río, M. A., & Palou, L. (2011). Antimicrobial edible films and coatings for fresh and minimally processed fruits and vegetables: A review. *Critical Reviews in Food Science and Nutrition, 51,* 872–900.
Velickova, E., Winkelhausen, E., Kuzmanova, S., Alves, V. D., & Moldão-Martins, M. (2013). Impact of chitosan-beeswax edible coatings on the quality of fresh strawberries (*Fragaria ananassa* cv Camarosa) under commercial storage conditions. *LWT- Food Science and Technology, 52,* 80–92.
Vilas-Boas, E. V. D. B., & Kader, A. A. (2006). Effect of atmospheric modification, 1-MCP and chemicals on quality of fresh-cut banana. *Postharvest Biology and Technology, 39,* 155–162.
Wu, S. (2019). Extending shelf-life of fresh-cut potato with cactus *Opuntia dillenii* polysaccharide-based edible coatings. *International Journal of Biological Macromolecules, 130,* 640–644.
Xiao, C., Zhu, L., Luo, W., Song, X., & Deng, Y. (2010). Combined action of pure oxygen pretreatment and chitosan coating incorporated with rosemary extracts on the quality of fresh-cut pears. *Food Chemistry, 121,* 1003–1009.
Yousuf, B., Qadri, O. S., & Srivastava, A. K. (2018). Recent developments in shelf-life extension of fresh-cut fruits and vegetables by application of different edible coatings: A review. *LWT- Food Science and Technology, 89,* 198–209.

Further reading

Iyidogan, N. F., & Bayindirli, A. (2004). Effect of L-cysteine, kojic acid and 4 hexylresorcinol combination on inhibition of enzymatic browning in Amasya apple juice. *Journal of Food Engineering, 62,* 299–304.

12

Active and intelligent packaging, safety, and quality controls

Bambang Kuswandi[*,†], **Jumina**[*,†]

*Chemo and Biosensors Group, Faculty of Pharmacy, University of Jember, Jember, Indonesia. †Department of Chemistry, Faculty of Mathematics and Natural Sciences, Gadjah Mada University, Yogyakarta, Indonesia

1 Introduction

Food packaging involves the design process, constructing, evaluating, and creating food packages as part of a food production system for distribution, storage, sale, and the end customer. It acts as a protection and barrier from the surrounding environment so that it preserves food in terms of quality and safety from the producer to the consumer's hand. Commonly, it also provides information about food composition, storage recommendations, labels, producer, and expire dates, tracking, etc. Hence, the food packaging main functions are containment, protection, communication, and convenience (Ghaani, Cozzolino, Castelli, & Farris, 2016; Kuswandi et al., 2011; Mangaraj, Goswami, Giri, & Joshy, 2012; Mangaraj, Goswami, Giri, & Tripathi, 2012; Yam, Takhistov, & Miltz, 2005). In addition, the packaging condition could be used to indicate and inform the food status, such as physiological (e.g., respiration), physical (e.g., desiccation), infestation (insects), microbial (e.g., spoilage bacteria), and chemical parameter (e.g., pH, oxidation).

Most food degradation is caused by two reasons, i.e., oxidation and microbial growth that occur on food, which in turn causes loss of freshness, and deterioration of food (Guillard, Issoupov, Redl, & Gontard, 2009; Kuswandi, Jayus, Abdullah, Heng, & Ahmad, 2012). Furthermore, enzymatic activity, nonoxidative reactions, the gain or loss of moisture, exposure to light, etc. may also be involved in food deterioration (Brody, Strupinsky, & Kline, 2001; Mangaraj & Goswami, 2009a, 2009b; Mangaraj, Sadawat, & Prasad, 2011). However, the food quality during packaging will be affected by transportation considerations, handling,

Fresh-cut Fruits and Vegetables. https://doi.org/10.1016/B978-0-12-816184-5.00012-4

and maintenance of storage conditions that lead to consumer satisfaction (Ghaani et al., 2016; Realini & Marcos, 2014).

Smart packaging is an alternative technology that offers to overcome many of these problems. It is a combination of both active and intelligent systems that could increase shelf-life, and communicates on the food quality throughout the distribution and supply chain to the end consumer. For instance, the combination of active function (ethanol emitters) and intelligent function (oxygen sensor) on the modified atmospheric packaging (MAP) of bread are useful packaging products to increase food shelf-life and the integrity detection of packaging (Hempel, O'Sullivan, Papkovsky, & Kerry, 2013). Here, the ethanol is used as a preservative to impede mycological growth, while the oxygen sensor is to detect the integrity of packaging. Smart packaging describes ways to provide enhanced functionality and additional features to packaging.

Smart packaging senses and informs related to internal or external conditions change. In a broader sense of its function, smart packaging can be classified into two categories; the first one is active packaging and the other is intelligent packaging, where active packaging systems are extending shelf-life and intelligent packaging systems communicate with the consumer. Smart packaging functions can be found in food packages, where they confirm integrity, authenticity, and actively, spoilage prevention. Furthermore, it enhances food attributes, such as taste or looks. It can also communicate additional food information to the user, and respond to food condition changes or package environment. More importantly, it can indicate seal integrity and act against counterfeiting.

Active food packaging is a food packaging that has an active function to enhance the food shelf-life or to reach some desirable properties (Biji, Ravishankar, Mohan, & Srinivasa Gopal, 2015). While intelligent food packaging is a packaging that monitors food conditions inside packaging and gives information regarding its quality and safety (Ghaani et al., 2016; Kuswandi et al., 2011). Thus, active food packaging focuses on the active function to enhance food quality, while intelligent packaging focuses on communication (Callaghan & Kerry, 2016; Floros, Dock, & Han, 1997; Ghaani et al., 2016).

Fig. 1 shows the smart packaging technologies, classification, and .functions in the improvement of food quality control and safety. However, many authors have regarded intelligent packaging itself as smart packaging (Biji et al., 2015; Kuswandi et al., 2011; Pacquit et al., 2006; Park, Kim, Lee, & Jang, 2015), since it has smart functions using sensors (biosensors, chemical sensors, and gas

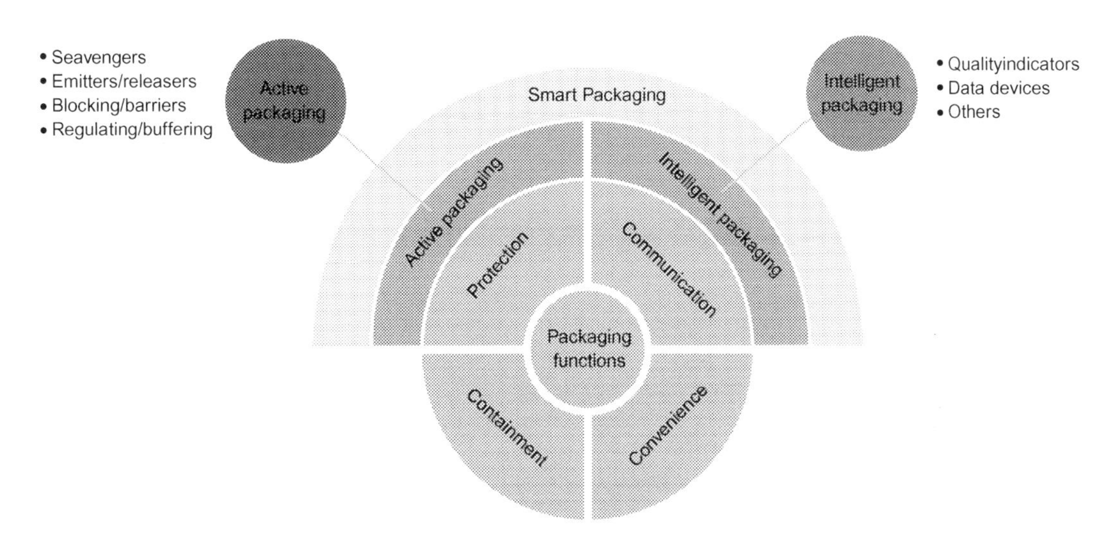

Fig. 1 Smart packaging, classification, and functions in the enhancement of food packaging functions.

sensor), or indicators (TTIs, freshness indicators, and integrity indicators), as well as RFID and barcodes (Kerry, Grady, & Hogan, 2006). Commercially, the smart packaging system is applied separately as the individual application, either as active packaging or intelligent packaging. This is due to active packaging acting to increase food shelf-life, since it is very difficult to detect the spoiled product simultaneously.

Commonly in intelligent packaging, the food quality and safety can be known by using sensor or indicator if there are any undesirable conditions during the product distribution and storage. However, the merge between active and intelligent systems in food packaging that have dual functions to create fully smart packaging is more beneficial for the consumer, where the real-time quality and the shelf-life extension of food could occur automatically and simultaneously on the packaging. Similarly, a recent study on consumer preferences regarding this issue showed that 36% recommended the smart packaging technology with nanotechnology, and 10% for the smart packaging alone in the case of cheese products (Callaghan & Kerry, 2016). This chapter focuses on the application of the active and intelligent system for food packaging. Their principle, design, construction, and applications are described including for fresh-cut fruits and vegetables.

2 Active packaging

In food packaging development, the active packaging term was first used by Labuza (1987), and is now commonly used in the areas of packaging studies. Active packaging is defined as packaging that performs some active functions that could enhance the food quality and safety, thus it is not only providing a barrier from the outside environment (Biji et al., 2015; Callaghan & Kerry, 2016; Hotchkiss, 1994; Labuza, 1987, 1990; Rooney, 1995a). In addition, active packaging can be described as a technology devoted to controlling the rate of respiration, the growth of microbial, delayed oxidation, and migration of moisture with the main purpose to enhance food quality, shelf-life, freshness, and safety (Ahvenainen, 1996).

Other definitions of active packaging have been proposed by other researchers (Brody et al., 2001; Kerry et al., 2006; Robertson, 2006; Rooney, 1995b; Yam et al., 2005); however, the active packaging basic concept is integrating specific substances in the food packaging in order to control or extend quality and enhance the shelf-life of food. The principles of active systems in food packaging are based on the physicochemical characteristics of the polymer material employed in packaging or on its surface of multilayer structures or associated with specific components (e.g., labels, pads, sachets, bottle caps) (Gontard, 2007; Gumiero, 2009). The active systems could be the compound used that is able to absorb oxygen, carbon dioxide, ethylene, flavors or odors, moisture, and compounds that are able to release or emit carbon dioxide, antioxidants, flavors, and antimicrobial agents (Biji et al., 2015; Realini & Marcos, 2014; Suppakul, Miltz, Sonneveld, & Bigger, 2003). Fig. 2 shows the working principles

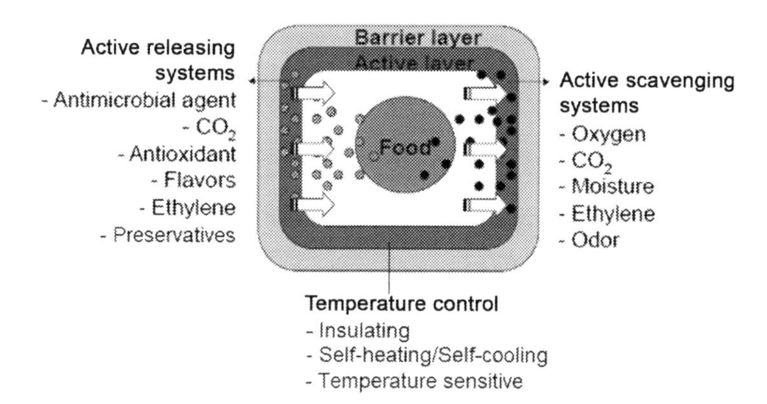

Fig. 2 The working principle of the active system in food packaging.

Table 1 Some applications of active food packaging including fruits and vegetables

Active system	Principles	Applications
Oxygen scavengers	1. Metal catalyst (e.g., platinum) 2. Metal/acid 3. Iron 4. Ascorbate/metallic salts 5. Enzyme	Cooked rice, biscuits, bread, cakes, snacks, pizza, pasta, cheese, cured meats, fish, dried foods, coffee, and beverages
Ethylene scavengers	1. $KMnO_4$ 2. Activated carbon, clays, zeolites	Fruits and vegetables and fresh cuts
Flavor/odor adsorbers	1. Ferrous salt, ascorbate 2. Activated carbon, clays, zeolites 3. Cellulose triacetate 4. Acetylated paper 5. Citric acid 6. Sodium bicarbonate	Fish, cereals, poultry, dairy, fruit, fruit juices, and fried snacks
Moisture absorbers	1. Activated clays, minerals 2. PVA blanket 3. Silica gel	Meats, poultry, fish, snacks, cereals, dried foods, sandwiches, fruit, and vegetables
Ethanol emitters	1. Alcohol spray 2. Encapsulated ethanol	Cakes, bread, biscuits, pizza crusts, fish, and bakery
Preservative releasers	1. Silver zeolite 2. Organic acids 3. Spice and herb extracts 4. BHA/BHT 5. Vitamin C and E 6. Chlorine dioxide/sulfur dioxide 7. Sorbates, benzoates, propionates	Meats, fish, cereals, bread, cheese, snacks, fruits, and vegetables
Temperature control packaging	1. Nonwoven plastics 2. Double-walled containers 3. Lime/water 4. Ammonium nitrate/water 5. Hydrofluorocarbon gas	Meats, fish, poultry, ready meals, and beverages

BHA, butylated hydroxyanisole; *BHT*, butylated hydroxytoluene; *PVA*, polyvinyl alcohol.

of the active system in food packaging. Thus, the packaging active functions create interaction with the internal environment inside food packaging that helps to enhance shelf-life and improve quality and safety of food. Various applications of active packaging systems are given in Table 1.

Active packaging can be further classified into four classes:
(i) active scavenging or absorber agent (nonmigratory);
(ii) active releasing or emitter agent;
(iii) blocking or barrier agent; and
(iv) regulating or buffering.

An example of nonmigratory active packaging is the moisture absorber, which acts without intentional migration. Here the packaging promotes a certain response from the packaged food, without its active substance migrating into the food inside the packaging (Dainelli, Gontard, Spyropoulos, van den Beuken, & Tobback, 2008). In the active releasing agent form, nonvolatile compounds or volatile agents emits or migrates at a controlled rate in the atmosphere inside the packaging (Dainelli et al., 2008). In the case where the relative humidity (RH) in the package achieves a certain level, a desiccant introduced in the food package starts to absorb moisture, so that its function is as a regulating agent. In addition, these four major types of active packaging can be classified based on their mode of action and function in the food packaging (Floros et al., 1997), as discussed in the following section.

2.1 Active scavenging and absorbents

The active scavenging systems are used to remove gases like CO_2, O_2, and ethylene from the food packaging. The use of absorbents, e.g., CO_2, O_2, and ethylene absorbent in modified atmospheric packaging (MAP) as the active packaging system for application in fresh-cut fruits and vegetables could enhance the food's shelf-life and maintaining the freshness in MAP (Kudachikar, Kulkarni, & Keshava Prakash, 2011; Thompson, 1998). Some applications of active scavengers and absorbents are described as follows.

2.1.1 Oxygen scavengers

In most cases, food spoilage is caused by oxidation or microorganism spoilage in the oxygen present inside food packaging (Cruz, Camilloto, & Pires, 2012). Therefore, The MAP can be used as a partial solution to this oxidation problem inside the food package due to the oxygen present. In this case, O_2 scavengers could be employed effectively to reduce the food quality loss due to oxidation of sensitive foods. The common oxygen scavenging agents are iron, ascorbic acid, enzymes, salts, or fatty acids, such as oleic acid and linolenic acid, and photo-sensitive dye, as well as yeast. These agents can be used individually or in a combination of two or more agents in order to increase their

Table 2 The principle and chemical reactions of different oxygen scavenging agents

Principles	Chemical reaction	References
Iron oxidation	$\frac{1}{2}O_2 + H_2O + 2e^- \rightarrow 2OH^-$ $Fe \rightarrow Fe^{+2} + 2e^-$ $Fe^{+2} + 2OH^- \rightarrow Fe(OH)_2$ $Fe(OH)_2 + \frac{1}{4}O_2 + \frac{1}{2}H_2O \rightarrow Fe(OH)_3$	Alvarez (2000), Frydrych, Foltynowicz, Kowalak, and Janiszewska (2007), Miltz and Perry (2005), Mills, Doyle, Peiro, and Durant (2006), Robertson (2006), and Smith, Hoshino, and Abe (1995)
Ascorbic acid oxidation	$AA + \frac{1}{2}O_2 \rightarrow DHAA + H_2O$ $AA + Cu^{+2} \rightarrow DHAA + 2Cu^{+} + 2H^{+}$ $2Cu^{+}2O_2 \rightarrow 2Cu^{+2} + 2O_2^{-}$ $2O_2^{-} + 2H^{+}Cu^{+2} \rightarrow O_2 + H_2O_2 + Cu^{+2}$ $H_2O_2 + Cu^{+2} + AA \rightarrow Cu^{+2} + DHAA + 2H_2O$	Cruz et al. (2012)
Photo sensitive dye oxidation	$Dye + Photon \rightarrow dye^*$ $dye^* + O_2 \rightarrow dye + O_2^*$ $O_2^* + acceptor \rightarrow oxide$	Rooney (1985) and Vermeiren, Devlieghere, Beest, de Kruijf, and Debevere (1999)
Enzymatic oxidation	$2\,G + 2O_2 + 2H_2O \rightarrow 2GA + 2H_2O_2$ $2H_2O_2 + Catalase \rightarrow 2H_2O + O_2$	Robertson (2006)

AA, ascorbic acid; *DHAA*, dehydro ascorbic acid; *G*, glucose; *GA*, gluconic acid.

effectiveness as oxygen scavengers (Cruz et al., 2012). The working principles and applications of the different oxygen scavenging agents for food packaging are given in Table 2, along with some example of commercially available active packaging, as depicted in Table 3, which can also be used for packaging of fresh-cut fruits and vegetables.

Currently, oxygen scavengers are widely employed in food packaging to prevent or slow down deterioration due to the food product oxidation or growth of microorganisms or tiny insects by absorbing the residual oxygen inside the food packaging (Brody et al., 2001). This oxygen scavenger can be employed individually or integrated with a MAP for more effective action of O_2 removal inside the packaging. Commercially, it is also common to use MAP in order to remove the atmospheric oxygen, and then employ a small amount of a low-cost oxygen scavenger to remove the rest of the oxygen inside MAP (Day, 1998; Guynot, Sanchis, Ramos, & Marín, 2003).

Table 3 Some example of commercial active packaging agents

O_2 scavengers (Manufacture)	CO_2 scavengers (Manufacture)	Ethylene scavengers (Manufacture)	Ethanol generators (Manufacture)
Ageless (Mitsubishi Gas Chemical)	Freshock, Ageless E (Mitsubishi gas chemical Co. Inc.)	Purafil (Purafil)	Ethicap (Antimold 102)
Keplon (Keplon)	Evert-fresh green bags (Evert-fresh Corp)	Air repair (Delta trak)	Negamold
Tamotsu (Oji Kako)	Evert-fresh United States (Ever-fresh type G)	Mrs Green extra life (Dennis green Ltd.)	Oitech
Freshilizer (Toppan)	Emco Fresh Tech. Ltd. (Oxyfresh)	Fain (Nippon container Co.)	ET pack
Vitalon (Toagosei Chemical Industry)	Lipmen (Lipmen)	Biofresh (Grofit Plastics)	Ageless SE
Secule (OxySorb) (Nippon Soda)		BO films (Odja Shoji Co.)	Fretek
Sansoless (Hakuyo)		Green bags (Evert fresh Co.)	

The O_2 scavengers have been studied to be effective in avoiding discoloration in tea, rancidity in foods, mold spoilage in bakery with high moisture content, as well coffee with oxidative flavor that changes its taste (Berenzon & Saguy, 1998; Gill & Mc Ginnis, 1995; Smith et al., 1995). They have also proved to be an effective agent in removing O_2 fully from a food package, resulting in increased seafood shelf-life (Goncalves, Mendes, & Nunes, 2004). Furthermore, it is reported that employing oxygen scavengers could reduce biogenetic amines formation in tuna fish (Ruiz-Capillas & Moral, 2005) and herring (Özogul, Taylor, Quantick, & Özogul, 2002a, 2002b). Similar results were also employed for yellowfin tuna (Guizani, Al-Busaidy, Al-Belushi, Mothershaw, & Rahman, 2005).

The sachets of oxygen scavenger are used in packaging for many foods, e.g., fresh and precooked pasta, meat products, catering, bakery products, cheese, coffee, nuts, and potato chips (Floros et al., 1997; Vermeiren et al., 1999). The sachets work more efficiently in the headspace of packaging that contains air, compared to vacuum packaging (Smith, Ooraikul, Koersen, Jacksont, & Lawrence, 1986). This is associated with the fact that some food has the capacity to absorb headspace gas that could make a slight vacuum inside the package so that the products became wrapped tightly with the film of packaging. Hence, this produces localized gaps between the food surface and the film, where the oxygen concentration could increase to a level high enough for mold growth.

In the early development stages, oxygen scavenging agent used techniques, such as labels or devices or sachets with self-adhesive (labeled "do not eat" to avoid accidental consumption) that were placed inside the food packages. The first polymer used as oxygen-absorber was as a polymer blend of polyethylene terephthalate (PET) with MXD6 polyamide catalyzed by cobalt oxidation. In this case, when it is applied with cobalt (200 ppm) as the stearate salt, this polymer blends cold blowing of bottle containers with a zero oxygen permeability for 1 year (Robertson, 2006). Furthermore, a film of oxygen scavenging has also been used that has iron in the structure, whereas others were used on organic compounds that need to be activated by ultra-violet light before absorbing oxygen (Cook, 2001; Damaj et al., 2009; Rooney, 2001; Tewari, Jayas, Jeremiah, & Holley, 2002). Here, the film oxygen scavenging capacity and its cost are considerably lower compared to the sachets of the iron-based oxygen scavenger (Day, 1998; Vermeiren et al., 1999). However, its potential toxicity as the added scavenger and unavailability of technical information related to their performance have been the main difficulties in applying these oxygen scavenging films in packaging (Damaj et al., 2009). In many cases, the oxygen absorbers used have improved the MAP technology effectiveness for food protection, particularly bakery and bread, from deterioration by aerobic microbes (Guynot et al., 2003; Salminen et al., 1996; Smith et al., 1986).

2.1.2 CO_2 scavengers

Commercially, CO_2 scavengers were employed for CO_2 removal as a result of respiration or fermented, and roasted foods, such as ground coffee, etc. (Labuza & Breene, 1989). The sachets of O_2 and CO_2 scavenger were employed in coffee to delay or slow-down oxidative flavor changes, to absorb the produced CO_2, since it would cause the package to burst (Smith, Ramaswamy, & Simpson, 1990). Commonly, the active compound of CO_2 scavenger is $Ca(OH)_2$, which reacts with CO_2 at high humidity to create $CaCO_3$ inside the food packaging (Cullen & Vaylen, 1994).

Usually, CO_2 scavengers are prevented from the pressure of the gas that produced inside packaging or expansion of volume in flexible packaging as a result of food fermentation or roasting (Charles, Sanchez, & Gontard, 2005; Lee, Shin, Lee, Kim, & Cheigh, 2001).

The O_2 and CO_2 absorbers' effect on the food quality has been studied for strawberry (Aday, Caner, & Rahval, 2011). Here, commercial CO_2 absorber, such as EMCO-A (mixed of 16% bentonite clay, 14% sodium chloride, 24% sodium carbonate, and 46% sodium carbonate peroxyhydrate) and EMCO-B (a mixture of 16% bentonite clay, 14% sodium chloride, 20% sodium carbonate, and 50% sodium carbonate peroxyhydrate). While commercial oxygen absorber, such as ATCO 210 (a mixture of reactive iron powder) where per sachet (2g) has a 210mL O_2 absorption capacity, and they have been applied for strawberry in this study. The results show that active packaging can reduce the rate of CO_2 accumulation and O_2 consumption. However, O_2 scavengers are more frequently used compared to CO_2 scavengers (Aday et al., 2011).

2.1.3 Ethylene scavengers

In fruits and vegetables, ethylene is a hormone for plant growth that pushes process of respiration to induce ripening, softening, and senescence, even though at very low concentration. It causes fruit and vegetable yellowing, lettuce russet spot, and has shelf-life reduction effect on many fruits and vegetables (Biji et al., 2015). Commonly, potassium permanganate ($KMnO_4$) ethylene is used as a scavenging agent that oxidizes ethylene to produce ethanol and acetate. Since $KMnO_4$ is a toxic compound, it cannot be directly integrated into packaging that has contact with the food directly. In order to use this toxic compound in the food packaging, $KMnO_4$ is added around 4%–6% to an inert substance that has a large surface area, e.g., silica gel, alumina, activated carbon, perlite, or vermiculite, and stored in a sachet, then it is added to packaging (Buonocore et al., 2003; Zagory, 1995).

Where ethylene is present in the packaging, it reacts with potassium permanganate and produces manganese dioxide, potassium hydroxide, and carbon dioxide. The quality changes of vegetables, i.e., diced onions, have been studied with and without a commercial gas absorbent using potassium permanganate and activated alumina (Howard, Yoo, Pike, & Miller, 1994). In this study, it has been found that the gas absorbent removed ethylene effectively and reduced the levels of sulfur volatiles and CO_2 in the diced onions packaging. The vegetables could be kept for 10days at 2°C using the gas absorbent based on potassium permanganate. In another study, it was found that the use ethylene absorbent, namely *Green keeper*, at 12 ± 1°C and 85%–90% RH for matured "Robusta" banana (75%–80%) packed in the MAP using LDPE film produced an improved shelf-life of up to 7weeks (Kudachikar et al., 2011).

Other ingredients that have been used as ethylene scavenging were fine porous material synthesized from zeolite, pumice, cristobalite (SiO_2), or clinoptilolite (hydrated NaKCaAl silicate) that were sintered together along with a small amount of metal oxide, then they were distributed in a plastic film (Zagory, 1995). Another type of ethylene scavenging was using activated carbon that could adsorb and then breakdown ethylene. For example, charcoal consisting PdCl as a metal catalyst was found effective in ethylene scavenging at 20°C and preventing ethylene accumulation. This agent has been used for reducing the softening rate in minimally processed fruits, such as bananas and kiwifruits, as well as reducing the loss of chlorophyll in spinach (Abe & Watada, 1991; Vermeiren et al., 1999). Furthermore, it has been proposed that a material consisting of a Pd-impregnated zeolite had the capacity of ethylene adsorption (4162 μL/g material) at 20°C and 100% RH, and it was found to be superior over $KMnO_4$-based scavenger agents when it was used in low concentrations, and in high RH conditions (Terry, Ilkenhans, Poulston, Rowsell, & Smith, 2007). Here, ethylene concentrations in packaging with banana, avocado, and strawberry were reduced below physiologically active level with the increasing amount of Pd-promoted material as ethylene scavenger.

2.2 Active emitter

2.2.1 CO_2 emitter

Generally, carbon dioxide at high concentrations is preferred for some perishable foods, such as fresh meat, poultry, fish, and nonclimacteric fruits to inhibit the growth of microorganisms on their surface. Carbon dioxide has a direct antimicrobial effect, it increases lag phase and time during the logarithmic microorganisms phase (Coma, 2008). Hence, commercial CO_2 emitters are widely available and are employed to enhance food's shelf-life. Furthermore, CO_2 is produced inside the packages by the food reaction with sodium carbonate and citric acid mixture inside the drip pad (Bjerkeng, Sivertsvik, Rosnes, & Bergslien, 1995). A pure CO_2 gas was most effective in controlling the microbial growth during MAP of Mozzarella cheese at 7°C compared to a gas mixture (50% N_2 and 50% CO_2), and nitrogen gas (100% N_2) (Alam & Goyal, 2011). A respiration rate decrease in carrots at 10% CO_2 has also been reported (Pal & Buescher, 1993). Furthermore, fruits and vegetables, e.g., ripening bananas, tomatoes, and pickling cucumbers showed a similar result with higher CO_2 concentrations inside the packages. Therefore, it is highly recommended to produce varying CO_2 concentrations to suit carbon dioxide tolerance of specific fruits and vegetables.

2.2.2 Ethanol emitter

Today, many ethanol-generating films or sachets that could be applied to food packaging have been patented and offered in the market worldwide (Suppakul et al., 2003). Some examples of marketed ethanol generators for food packaging are given in Table 3. In Japan, ethanol emitters are extensively employed in bakery products, due to the capability to retard or slow down the process of mold growth, which in turn enhances their shelf-life, such as pastries, cookies, cakes, bread, and pizza crust. *Ethicap* (an active ethanol emitter) has been used in different amounts for packaged prebaked buns, and it was found that yeasts and molds were absent in large amounts from the prebaked bun core at room condition during storage (Franke, Wijma, & Bouma, 2002). Here, the total mesophilic count was low at the beginning and increased within 1 week to an unacceptable level without ethanol emitter. In the presence of ethanol emitter during storage, the total mesophilic was found stabilized at a consumable level, and by adding *Ethicap*, the mold growth was delayed by 13 days.

The sachets composing 55% ethanol and 10% water, which were adsorbed onto 35% SiO_2 powder, then filled into a sachet of a paper-ethylene-vinyl acetate copolymer were used for antimicrobial effect on food packaging (Robertson, 2006). Here, in order to mask the alcohol odor, they contained vanilla or other desired flavors in traces. The sachet-contents work by moisture absorption from food, and then ethanol vapor is released. Hence, the food promotes the ethanol vaporization and the ethanol absorption from the package headspace. It has been studied to assess the quality and shelf-life of bread stored under MAP and under ambient conditions, where oxygen concentration was monitored continuously using an oxygen sensor (Hempel et al., 2013). Here, EE (ethanol emitters) or ES (Ethanol surface sprays) have been applied along with MAP treatments and used to establish their effect on the food shelf-life. It was found that the food packed in ES-MAP had a slow oxygen depletion, and depletion occurred completely after 14 days. In the case of EE-MAP, the oxygen depletion was found completely after 35 days. Thus, the use of ES and EE reduces microbiological counts in bread samples, proving the shelf-life of bread was extended in this case.

2.2.3 SO_2 emitter

Generally, SO_2 emitters work based on the metabisulfite hydrolysis mechanism and the reaction of calcium sulfite with moisture. Sulfur dioxide emitter has been used for grapes packaging to prevent mold development (Suppakul et al., 2003). The two different sheets of SO_2 release have been used for the white and the purple grapes packaging to investigate their effect on the decay and the grapes quality (Christie et al., 1997). After 4 days at 21°C of storage, it was shown that sulfite levels in the purple grape were lower than in the white grape, even though SO_2 was at a higher level. Here, a polymer has to be applied for SO_2 control release to obtain inactivation of fungi without any undesirable effects on the foods.

2.2.4 Antioxidant release

Commonly in food packaging, antioxidants are integrated into a film of packaging or coated on the packaging material surface to control oxidation of the fatty components and pigments (Vermeiren et al., 1999). In many cases, the coatings of the edible film using impregnation with antimicrobials or antioxidants could be employed for the direct treatment of meat surfaces (Kerry et al., 2006). Since there is a growing concern about food safety currently, the natural additives use in food is very interesting, such as the α-tocopherol use as an antioxidant alternative for protection of polymer during processing, which can be applied as an additive in active packaging for food (Siro et al., 2006).

In many cases, plastic films, such as polyolefins, are impregnated with antioxidants in order to make polymer stable and protect from oxidation (Robertson, 2006). Cellulose acetate films with different morphological features have been proposed to control the rate of L-ascorbic acid and L-tyrosine release as natural antioxidants (Gemili, Ahmet, & Altınkaya, 2010). Packaged lamb steaks have been studied under three conditions. One applied active film with rosemary, the second applied active film with oregano, and the third one used rosemary extract sprayed on the meat surface before high-oxygen atmosphere packaging (Camo, Beltrán, & Roncalés, 2008). Here, the samples were stored for 13 days under light at 1 ± 1°C. All treatments improved the lamb steaks' oxidative stability, where oregano active films were found more efficient over the rosemary. In this case, the shelf-life extended from 8 to 13 days related to fresh odor and color. Another active agent, composing a polypropylene film contained with rosemary extract as natural antioxidants, was found to improve the stability of both myoglobin and fresh meat against processes of oxidation compared with polypropylene only (Bentayeb, Rubio, Batlle, & Nerín, 2007; Nerin et al., 2006).

Butylated hydroxytoluene and butylated hydroxyanisole were also employed to protect fats from rancidity, as they have antioxidant capacities. Both of these compounds have been applied as an antioxidant in active packaging for food. This is due to the migration of both compounds from monolayer polyolefin films, having high mobility towards fat (Galindo-Arcega, 2004; Granda-Restrepo, Peralta, Troncoso-Rojas, & Soto-Valdez, 2009; Van Aardt, Duncan, Marcy, Long, & O'Keefe, 2007; Wessling, Nielsen, & Giacin, 2000; Wessling, Nielsen, Leufven, & Jagerstad, 1999). Furthermore, the compounds have been exposed to cereal and snack products packaging during storage that could extend their shelf-life (De Kruijf & van Beest, 2003; Robertson, 2006). However, the US Department of Health and Human Services (HSS) reported that Butylated hydroxyanisole is carcinogenic, so that it should be replaced with other safe compounds, e.g., vitamin E (Day, 2003).

The antioxidative activity could also be found in plant extracts and spices that consist of polyphenols, and can be used as antioxidant for oxidation of lipid in various foods (Lund, Hviid, & Skibsted, 2007; Sanchez-Escalante, Torrescano, Djenane, Beltran, & Roncales, 2003). Gallic, ascorbic acids, and lecithin have antioxidant qualities and could be applied as coatings to enhance the food quality and microbiological stability. For example, ascorbic acid; its antioxidant activity in meat and its products are related to oxidation of lipid, and it was found affected by its concentration, the transition metal ions presence, and the other antioxidants (Lund et al., 2007). Fresh beef steaks packaging has been applied with an antioxidant active agent consisting of oregano extract with different concentrations (0.5%, 1%, 2%, and 4%) (Camo, Lorés, Djenane, Beltrán, & Roncalés, 2011). Here, 1% oregano extract was shown to improve the beef steaks' shelf-life from 14 to 23 days, whereas a 4% oregano extract induced an unacceptable smell. The lipid oxidation retarding properties as active packaging with oregano extract has also been found earlier (Nerin et al., 2006).

2.3 Antimicrobial agent

Antimicrobial activity in active packaging can be classified as blocking or barrier functions that prevent microbial growth in food packaging. For example, the integration of active agents, such as antioxidants or antimicrobials, within the packaging material could give ability as antimicrobial active packaging (Camo et al., 2008). Various antimicrobial agent types and various packaging materials types for foods have been investigated to increase the antimicrobial efficacy of the active packaging system (Gontard, 2007). The parameters could be counted as food characteristics and antimicrobial compound, films nature or coatings, casting, processing, and residual activity of antimicrobial, storage temperature, packaging materials physical properties, mass transfer, etc. that could affect the design and construction of the antimicrobial package or film (Han, 2000). The active packaging that used oxygen-scavenger could also show the antimicrobial capabilities indirectly against aerobic microorganisms. In many cases, the antimicrobial packaging has been proven to stabilize food quality and improve meat shelf-life (Coma, 2008; Han, 2005; Lee, An, Lee, Park, & Lee, 2004). Antimicrobial packaging using gelatin-based coatings composing nisin, lysozyme, and EDTA for sausage and ham packaging could also be employed to control spoilage from pathogenic organisms, such as *Leuconostoc mesenteroides*, *L. monocytogenes*, *L. sake*, and *S. typhimurium*. In addition, Ag(I) ions impregnated in the network of silicate have also been used as an antimicrobial agent for food packaging.

Other materials, such as chitosan, a well-known compound derived from chitin, has antimicrobial capacity against bacteria, molds, and yeasts, as well as excellent properties of film-forming (Fernandez-Saiz, Ocio, & Lagaron, 2010; Vartiainen et al., 2004). Coating of chitosan has been found to reduce microbe contamination of oysters and shrimp (Chen, Liau, & Tsai, 1998; Simpson, Gagne, Ashie, & Noroozi, 1997), where it acts as blocking or barrier between the nutrients in food and microorganisms (Appendini & Hotchkiss, 2002). Another antibacterial agent that has been used in food packaging is chlorine dioxide (ClO_2). This gas is more effective in treating the rest of the food package contents compared to silver as an antimicrobial that acts on the food surface. However, this gas could change meat color and bleach the green of vegetables (LeGood & Clarke, 2006). Furthermore, the packaging films used in antimicrobial packaging that release volatile organic acids have the potential to reduce the slime-forming bacteria growth in fresh meat (Han, 2005).

Actually, there are many compounds with natural antimicrobial qualities, e.g., organic acids (sorbic acid, benzoic acid, propionic acid, lactic acid), bacteriocins (lactic and nisin), enzymes (peroxidase and lysozyme), seed extracts of grape and lemon, spice extracts (cinnamaldehyde, thymol, and p-cymene), metals (silver), and chelating agents (EDTA) that could be integrated into packaging material to block or inhibit bacteria growth in various foods (Devlieghere, Vermeiren, Bockstal, & Debevere, 2000; Falcone, Speranza, Del Nobile, Corbo, & Sinigaglia, 2005; Han,

2000; Mastromatteo, Mastromatteo, Conte, & Nobile, 2010). Other active compounds, such as allyl isothiocyanate contained in mustard oil, have been demonstrated as an inhibitor for the growth of mold on rye and wheat bread (Suhr & Nielsen, 2005).

Many studies have shown that food packaged in barrier plastic bags flushed with allyl isothiocyanate could enhance food's shelf-life (Suppakul et al., 2003). Labels impregnated with essential oil of cinnamon stuck onto the plastic packaging can act as an active packaging to increase late-maturing peaches' shelf-life (Montero-Prado, Rodriguez-Lafuente, & Nerin, 2011). Here, the infected fruit percentage in the active label was found to be reduced (13%) compared to the nonactive packaging (86%) at room temperature after 12 days of storage. In addition, significant differences were also obtained for weight loss and firmness during storage. The essential oils used as antimicrobials have been reviewed in the literature (Burt, 2004). Currently, a film composing allyl isothiocyanate extracted from brown mustard (*Brassica juncea*), mustard (*Brassica nigra*), and wasabi (*Eutrema wasabi* Maxim.) has been marketed in Japan (Dainelli et al., 2008; Suppakul et al., 2003). The film was entrapped in cyclodextrins to prevent thermal damage during the film preparation extrusion process. It has the ability to structurally change and release the antimicrobial agent into the surrounding atmosphere of the packaging (Lee, Sheridan, & Mills, 2005). Some example of antimicrobial agents that can be employed in food packaging is listed in Table 4.

Generally, the antimicrobial food packagings can be classed into four antimicrobial packaging types (Appendini & Hotchkiss, 2002):

(i) antimicrobial agents impregnated into a sachet and attached to the package;
(ii) antimicrobials adsorbed or coated onto surfaces of polymer;
(iii) antimicrobial agents integrated into packaging films directly; and.
(iv) packaging material having a matrix that consists of an antimicrobial agent.

The antimicrobial food packaging where the polymer surface is absorbed or coated with antimicrobial agents could inhibit the growth of microbial on the surface, thus improving food preservation (Appendini & Hotchkiss, 2002; Buonocore et al., 2003; Sebti, Coma, Deschamps, & Pichavant, 2001). The immobilization of antimicrobial agents on the packaging film was more effective as antimicrobial packaging compared with dip or spray due to continuous slow migration of the antimicrobial agent to the food surface from packaging material (Quintavalla & Vicini, 2002). Self-sterilizing packaging material could be developed with the use of

Table 4 Some example of antimicrobial agents that can be applied in food packaging

Type	Examples
Bacteriocins	Nisin
Thiosulfinates	Allicin
Spice extracts	Thymol, *p*-cymene
Proteins	Conalbumin
Organic acids	Propionic, benzoic, sorbic
Isothiocyanates	Allyl isothiocyanate
Chelating agents	EDTA
Parabens	Heptylparaben
Enzymes	Peroxidase, lysozyme
Fungicides	Benomyl
Antibiotics	Imazalil
Metals	Silver nanoparticle

antimicrobial polymers that reduce the peroxide treatment used in aseptic packaging and eliminate the chance of recontamination on processed food (Appendini & Hotchkiss, 2002). Antimicrobial additives thymol and carvacrol that are incorporated into packaging film at different concentrations could modify the barrier of oxygen and the PP film mechanical properties. For example, the addition of antimicrobial additives (8%) into PP film could increase the food quality and safety (Ramos, Jiménez, Peltzer, & Garrigós, 2012).

Safety is very important when edible films and coatings are used in food packaging. Therefore, the active ingredient selection is limited to edible compounds (Kerry et al., 2006). Edible films impregnated with lysozyme, chickpea albumin extract (CPAE), and disodium EDTA·$2H_2O$ are shown to have antimicrobial and antioxidant activities (Güçbilmez, Yemenicioflu, & Arslanoflu, 2007). Here, lysozyme and CPAE incorporation in food packaging could increase the film activity of free radical scavenging, while lysozyme and disodium EDTA.$2H_2O$, or CPAE and disodium EDTA.$2H_2O$, made the films more effective against *Escherichia coli* and *Bacillus subtilis*. Antimicrobial agents integration into food packaging films during manufacturing is more beneficial compared to the other methods. Since microbial contamination of food starts at the surface, the antimicrobial packaging film used is more efficient by slow migration to the surface of food and aiding to maintain sufficient antimicrobial concentration where they are needed (Coma, 2008). Sodium sulfite has been employed as an antimicrobial agent for fresh pork sausages (Martinez, Djenane, Cilla, Antonio Beltran, & Pedro, 2006). Acetaldehyde is also very helpful in reducing postharvest diseases due to its antifungal properties and improves fresh product sensory qualities (Almenar, Auras, Wharton, Rubino, & Harte, 2007). Here, acetaldehyde was found highly effective against *Alternaria alternata, Colletotrichum acutatum*, and *Botrytis cinerea* in the air.

2.4 Regulating functions

In the active packaging, regulation functions targeted to maintain food quality inside packaging during distribution, storage, and to the end consumer could apply to flavors and odors absorption or release, and absorption of moisture or humidity control. These regulating or buffering functions will be described as follows.

2.4.1 Release or absorption of flavors and odors

In some foods such as fruits and vegetables, aroma is a key factor in their acceptance, so controlling the aroma of headspace by delivering addition of fragrances or other sensitive ingredients are some solutions in active packaging approach to maintain food quality (LeGood & Clarke, 2006). In this context, packaging materials have a critical effect on the food quality, particularly related to vapor transfer from the packaging or migration, or followed by possible organic compounds enrichment in the food (Frank, Ulmer, Ruiz, Visani, & Weimar, 2001; Huber, Ruiz, & Chastellain, 2002; Strathmann, Pastorelli, & Simoneau, 2005). Scalping is often needed for removal of the aroma of some foods. Even though the food quality is damaged by flavor scalping, it can be employed selectively in absorbing undesired odors or flavors. (Hotchkiss, 1997; Van Willige, Linssen, Meinders, van der Stege, & Voragen, 2002).

Another approach in active packaging concept has been proposed for reducing grapefruit juice bitterness (Soares & Hotchkiss, 1998a, 1998b; Vermeiren et al., 1999). Commonly, the bitter taste is due to glycosidic flavanone naringin and the triterpenoid lactone limonin presence in grapefruit. Here, in order to reduce this bitterness in grapefruit, an active thin CA (cellulose acetate) layer was applied to the inside surface of the packaging that composed the fungal-derived enzyme naringinase (α-rhamnosidase and β-glucosidase) to be used in hydrolyzed naringin to nonbitter compounds, naringenin, and prunin. The CA films in contact with the grapefruit showed that immobilized naringinase (60%) could hydrolysis naringin in grapefruit juice at 7°C in 15 days and reduce the content on the CA film due to adsorption. Two other types of taints called amines and aldehydes can be also used in active food packaging (Rooney, 1995b). For off-odor application in active packaging, the interactions between citric acid (as an acidic compound) immobilized in polymers has been claimed to have amine-removing capabilities (Hoshino & Osanai, 1986; Vermeiren et al., 1999).

2.4.2 Humidity control/moisture absorption

Based on the food type, the level of humidity for maintaining its quality in a package would vary, depending on the moisture needed, either by absorption or releasing within the food package. Water can be accumulated in food packages due to temperature fluctuations in a package with high-moisture, drip loss of tissue fluid from fresh foods, and transpiration of horticultural products. In respiring food such as fruits and vegetables, the water should be controlled to reduce its activity, in order to eliminate the microorganisms growth on the food (Suppakul et al., 2003). Mostly, moisture absorbers work based on water adsorption by cellulose and its derivatives, zeolite, potassium chloride, sodium chloride, xylitol, and sorbitol (Azevedo, Cunhaa, Mahajan, & Fonseca, 2011; Dainelli et al., 2008).

In much food packaging, a silica gel sachet is employed inside the package as moisture absorbent for dry foods. Mostly, the polymers, such as polyacrylate and starch grafted copolymers, are used for absorbing water inside the package (Rooney, 1995a). These polymers are able to absorb liquid water 100–500 times their own weight (Hurme, Sipilainen-Malm, & Ahvenainen, 2002). Calcium chloride ($CaCl_2$) has been used to control RH in bell peppers packaging (Ben-Yehoshua, Shapiro, Chen, & Lurie, 1983). The desiccant (5 g) per fruit could maintain the RH between 80% and 88% inside the package that reduced weight loss and better maintenance of food quality. Furthermore, the addition of sorbitol to the MAP could help better maintenance of broccoli heads quality compared to control, since sorbitol has moisture absorbing capability (Deell, Toivonen, Cornut, Roger, & Vigneault, 2006). Absorbing pads consisting of polymer granules that have superabsorbent capacity sandwiched between two layers of a microporous or nonwoven polymer then sealed at the edges were employed to absorb water in flesh foods

(Robertson, 2006; Suppakul et al., 2003). A simplex lattice design for the desiccant mixture preparation (0.50 CaO, 0.26 calcium chloride, and 0.24 sorbitol mass fractions) has been optimized to be used as water absorbent (Azevedo et al., 2011). This desiccant mixture had high moisture absorbing capacity and was found effective to be incorporated in with packaging for highly perishable food, e.g., oyster mushroom.

3 Intelligent packaging

In food packaging, the food quality could change after it has been packed since the microenvironment inside the packaging could affect the food; therefore, there is a need for continuous monitoring of food quality during distribution, storage, and selling, or even to the end customer. The food quality testing method includes biological, physicochemical, and serological methods. However, these classical methods for monitoring and detection of food quality, contaminants, and safety require skilled persons and are laborious and time-consuming. Hence, there is a need for a simple, rapid, reliable, and sensitive methods for monitoring of food quality and safety (Kuswandi, 2018; Kuswandi et al., 2011; Thakur & Ragavan, 2013). Different applications of intelligent packaging systems are given in Table 5.

3.1 Sensors

A sensor can be defined as a device or tool used to quantify energy of matter as a signal for a chemical or physical property detection or measurement to which it responds (Kerry et al., 2006; Kress-Rogers, 1998). There are four main component of the sensor:

(i) receptor, or the sensing part of a sensor, which transforms the physicochemical signal into a form of energy;
(ii) transduction element, which transforms the physicochemical signal from the receptor into a useful analytical signal;
(iii) signal processing unit, where the transducer output is processed to get the desired output form; and
(iv) display unit, where quantifiable results are displayed in analog or digital form (Ghaani et al., 2016; Thakur & Ragavan, 2013).

Thus, the ideal sensor should have high sensitivity and selectivity, fully reversibility, fast response time, wide dynamic range, long-term stability, good reliability, and linearity (Ghaani et al., 2016; Hanrahan, Patil, & Wang, 2004). In these terms, the ideal

Table 5 Some examples of intelligent packaging for food, including fruits and vegetables

Intelligent packaging	Principle	Food applications
Sensors	1. Detection of biological and chemical compounds (enzymes, DNA, microbial) 2. Detection of chemical compounds (reagent dyes, pH indicators, gas, etc.)	Water, fish, poultry products, meat, milk, fruits, vegetables, beverages
Freshness indicators	1. Detection of biological compounds 2. Detection of chemical compounds 3. Ethylene gas detection 4. pH detection	Meat, fish, poultry, fruits, vegetables, milk, beverages
Integrity indicators	1. Oxygen indicators 2. CO_2 indicators	Fish, meat, poultry, fruits, vegetables
Time-temperature indicator	1. Diffusion 2. Enzymatic reactions 3. Microbial reactions 4. Photochemical reactions 5. Polymerization reactions 6. Chemical reactions	Fish, meat, poultry, perishable products, juices, fruits, vegetables
RFID	1. Radio frequency technology	For individual and bulk packed food products

sensor is lacking in application in food packaging, including its commercialization, due to several reasons, such as the size of the available sensors is difficult to integrate into the food packaging, the cost of sensors is relatively higher than the cost of food itself, and the commercial sensors are mainly for detecting a single component.

Sensors can be categorized based on the type of receptor, transducer, application, etc. Among all these classifications, this chapter focuses on classifications based on the type of receptor that can be classified into three type in respect to food packaging:

(i) chemical sensors;
(ii) biosensors, and
(iii) gas sensor.

See Fig. 3 (Adley, 2014; Kerry et al., 2006; Puligundla, Jung, & Ko, 2012; Thakur & Ragavan, 2013).

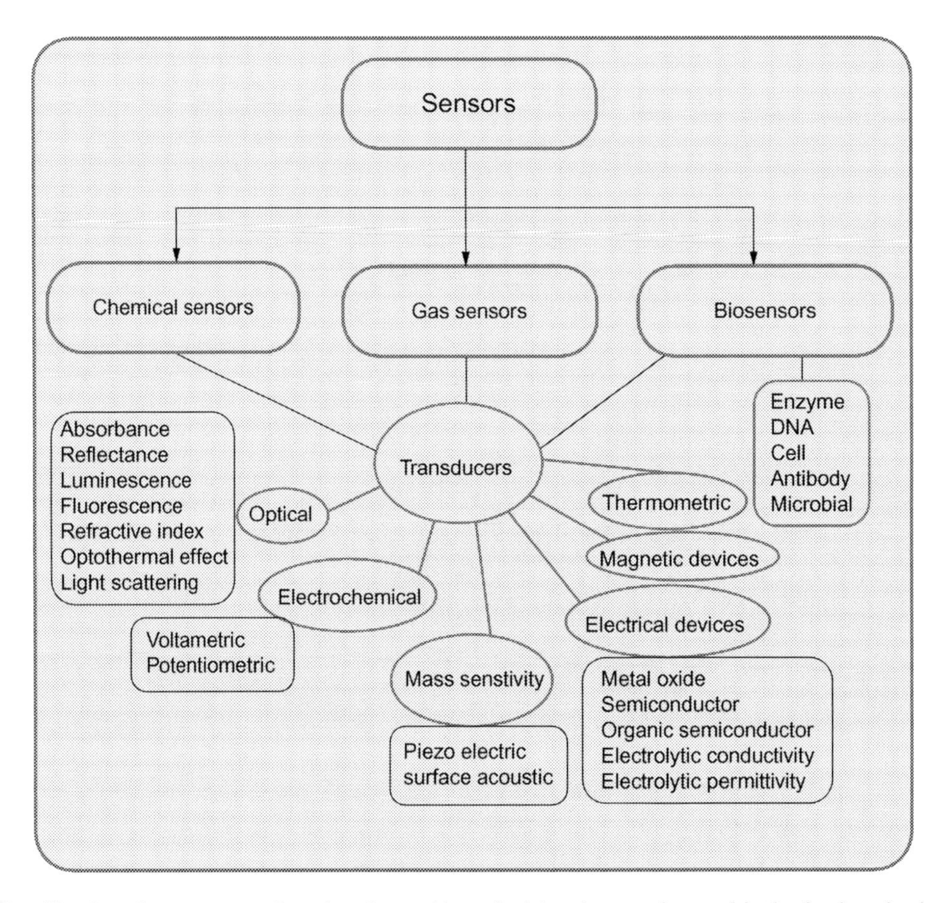

Fig. 3 Classification of sensors type based on its working principles that can be used in the food packaging.

3.1.1 Chemical sensors

Chemical sensors can be defined as an analytical device or tool that uses chemical reagent as a receptor or recognition element. The receptor has the capability to detect the activity, presence, concentration, or composition of specific chemical analytes or gases through surface adsorption that results in a change in surface properties (Vanderroost, Ragaert, Devlieghere, & Meulenaer, 2014). Then, the transducer converts this surface property change into a measurable signal proportional to the target analyte. In intelligent packaging, chemical sensors are generally employed to detect pH and volatile organic compounds (VOC) produces during distribution and storage, such as trimethylamine (TMA), dimethylamine (DMA), etc., and other gases (H_2, CO, NO_2, NH_3, CH_4, H_2S etc.) (Kuswandi et al., 2011).

In this case, the chemical sensors are used for direct detection of food quality inside the packaging. In this case, various different concepts for food quality inside packaging have been discussed in the literature. For example, chemical sensors have been developed for TVBN (total volatile basic nitrogen) (Pacquit et al., 2006; Pacquit et al., 2007), pH (Kuswandi et al., 2011; Kuswandi, Jayus, Abdullah, et al., 2012; Kuswandi, Jayus, Larasati, Abdullah, & Heng, 2012), CO_2 (Mattila, Tawast, & Ahvenainen, 1990), ammonia (Horan, 1998), amines (Wallach & Novikov, 1998), ethanol (Cameron & Talasila, 1995), and hydrogen sulfide (H_2S) (Smolander, Hurme, Ahvenainen, & Siika-Aho, 1998), by incorporating the chemical sensor into packaging that can be presented as a colorimetric sensor in the target analyte presence.

3.1.2 *Biosensors*

The biosensor can be defined as an analytical device or tool that detects, records, and transmits information regarding biochemical reactions (Kuswandi, Andres, & Narayanaswamy, 2001; Yam et al., 2005). In biosensors, the recognition system utilizes a biochemical mechanism interfacing with the detection system (Thakur & Ragavan, 2013). The receptor in a biosensor is from the biological components, e.g., enzymes, nucleic acids, antibodies, and cells, etc. (Thakur & Ragavan, 2013; Wang, 2006). This bioreceptor is an immobilized sensitive element (enzyme, antibodies, DNA, and nucleic acid) recognizing the target analyte (enzyme substrate or inhibitor, complementary DNA, antigen, etc.) Common immobilization methods are adsorption, microencapsulation, inclusion, cross-linking, and covalent bonding (Kuswandi et al., 2001).

In classical terms, commercial biosensors are mainly for clinical diagnostic applications and not yet for food applications, including food packaging (Kerry et al., 2006). Only in the case of enzymes or microbes used in TTI are they applied in intelligent food packaging (Kuswandi, 2018). However, several prototypes have been developed at lab scale. For instance, a barcode-based biosensor, namely Food Sentinel System, to detect pathogens in food has been developed by SIRA Technologies (Pasadena, California, United States) (Ayala & Park, 2000). Here, an antibody specific-pathogen is attached to a membrane forming the barcode part. When contaminating bacteria are present, these cause the dark bar formation in the barcode that makes the code unreadable to a barcode scanning reader. A diagnostic system has been developed by Toxin Alert (Ontario, California, United States), called Toxin Guard (Bodenhamer, Jackowski, & Davies, 2004). This system immobilized antibodies into plastic packaging films to detect pathogens. If a target pathogen encounters the antibodies, it displays a visual signal to warn the users. In addition, the system is aimed at gross contamination detection, as it is not sensitive enough for detection of pathogens at very low levels that may cause illness.

3.1.3 Gas sensors

The gas compositions inside the food packaging could change with time and/or temperature, as a result of biological or physiochemical reactions or leakages. Therefore, the gas concentrations inside the food packaging provide indirectly the food quality information. Gas sensors are used to detect the presence of gas analytes (O_2, CO_2) in food packaging. Optical and electrochemical, field effect transistors, organic conducting polymers, a metal oxide semiconductor, and piezoelectric crystal sensors are some of the gas sensors (Kerry et al., 2006; Kress-Rogers, 1998). By definition, the gas sensors can also be classified into chemical sensors, where the chemical reagent is used as the sensing system. However, in order to emphasize their applications in intelligent packaging, these gas sensors are described separately.

Optical oxygen sensors have been developed. Such systems are based on the absorbance changes or luminescence quenching caused by direct contact with the target analyte (Papkovsky, Smiddy, Papkovskaia, & Kerry, 2002). Furthermore, optochemical sensors are employed to detect the food quality by detecting the gas target, e.g., carbon dioxide, volatile amines, and hydrogen sulfide (Wolfbeis & List, 1995). The optical gas sensor methods are of three types, i.e., absorption based colorimetric sensor, fluorescence based on pH-sensitive indicator, and energy transfer approach based on phase fluorimetric detection (Mills, Qing Chang, & McMurray, 1992). In addition, pH-sensitive dyes can be used to develop gas sensors for the basic volatile amines detection in food with high content of protein (Kuswandi, Jayus, Abdullah, et al., 2012; Kuswandi, Jayus, Oktaviana, Abdullah, & Heng, 2014). In the case of the electrochemical oxygen sensors, however, the main disadvantages are oxygen consumption, CO_2 and hydrogen sulfide cross-sensitivity, and fouling (Kerry et al., 2006).

3.2 Indicators

Indicators can be defined as devices or tools that indicate the absence, presence, or concentration of an analyte, or the reaction degree between two or more analytes by a color change (Hogan & Kerry, 2008; Kuswandi, 2018). The difference from sensors is that they do not have a specific receptor or transducer, and they only provide information via visual changes related to the food conditions (Kuswandi, 2018; Mills, 2005). They provide information regarding any change occurring in a food packaging or the surrounding condition, such as time, temperature, pH, etc., usually by color change. There are different types of indicators including temperature indicators (TI), time-temperature indicators (TTI), freshness indicators, and integrity indicators, that can be used in intelligent packaging for fresh-cut fruits and vegetables. These indicators convey the information indirectly regarding the food quality inside the packaging to the users since they are mimicking or imitating the food quality change inside the packaging.

3.2.1 Temperature indicators

Temperature is one of the most important factors that influences the deterioration rate and microbial growth in food (Yam et al., 2005). A thermochromic ink dot is the most commonly used TI to indicate food is at the correct serving temperature following microwave heating or refrigeration. Thermochromic inks could change their color upon elevated temperature exposure. Mostly, these inks are used for beverage or microwavable food that lets consumers know whether the food is to be served hot or cold. These ink color patterns are either reversible or irreversible depending on temperature. Irreversible inks do not change once they have attained a specific color based on temperature and remain constant at a certain temperature on exposure. Reversible ink is capable of changing the color once achieved at a certain temperature and reversing back to its initial color as the temperature returns back (Vanderroost et al., 2014).

Usually, these thermochromic inks work based on leuco dyes, which are able to change their structure once exposed to a certain temperature, creating a color change. Thus, it can be applied as a reversible temperature indicator for fresh-cut fruits and vegetables storage, instead of working as a quality measurement of food (Robertson, 2012). However, their use in food packaging, particularly for fresh-cut fruits and vegetables, is rare.

3.2.2 Time-temperature indicators

Time-temperature indicators (TTIs) serve information regarding time where the food has been exposed to a temperature higher than the desired temperature during transportation and storage that is especially useful for temperature abuse alert for chilled or frozen foods (Pavelkova, 2013). There are two types of TTIs, i.e., partial history type and full history type (Kerry et al., 2006). Partial history indicator only provides information on whether food has been exposed to a higher temperature than its critical temperature. Full history indicators provide information about food packaging over time. TTIs work based on the change of chemical, electrochemical, enzymatic, mechanical, or microbiological principles, commonly presented as a color change or intensity color change.

Commonly, TTIs are constructed in the form of labels or tags placed on or attached to a shipping container or on each food packaging. In some cases, TTIs could also be employed as freshness indicators for online shelf-life estimation of perishable food (Kuswandi, 2018). TTIs can be divided into three categories, i.e., function; operating principle; temperature history (Fig. 4). Some of the recent works on TTIs for fruits and vegetable are given in Table 6. At present the commercially available TTIs work on diffusion, enzymatic, microbiological, photochemical, barcode, and polymer-based systems. The commercial TTI and their working principles that can be used for fresh-cut fruits and vegetables are listed in Table 7.

One example of diffusion-based TTI is 3M Monitor mark (3M Company). It contains a fatty acid ester that has a selected melting point mixed with a blue dye. When it is exposed to a higher temperature than its critical temperature, the free fatty acid ester melts and diffuses to make the indicator color change to blue.

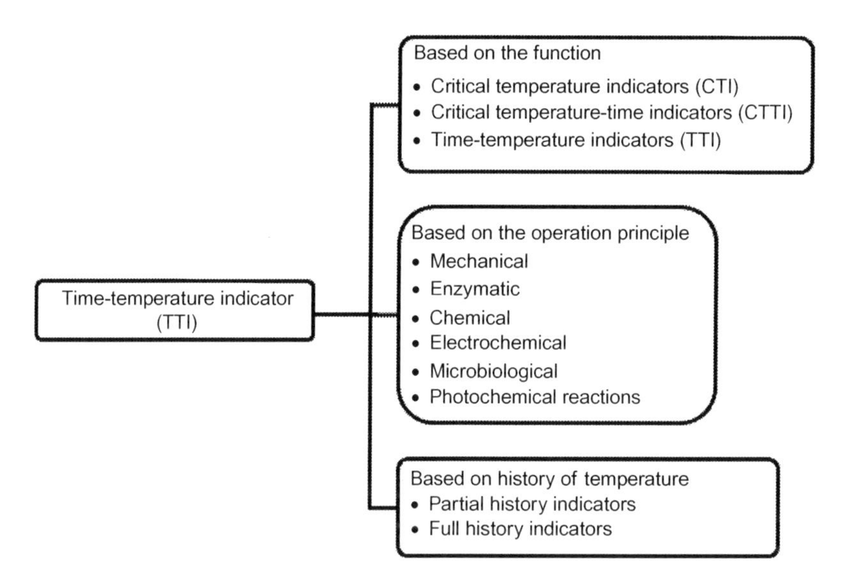

Fig. 4 Classification of time-temperature indicators (TTIs) based on the function, working principle, and temperature history.

Table 6 Some examples of TTIs that have been applied for fruits and vegetables

No.	Indicator component	Application	Principles and properties	Reference
1.	Lactic acid	Fruits, vegetables, and other oxygen sensitive foods	TTIs were characterized isothermally between 4°C and 45°C The TTIs considered as potential tools for food quality loss monitoring	Wanihsuksombat, Hongtrakul, and Suppakul (2010)
2.	Isopropyl palmitate (IPP) diffusion	Nonpasteurized angelica juice (NPA)	It is used for temperature above 13.5°C Diffusion was 9.7 and 7.2 mm at 15°C and 25°C, respectively. 7.0 mm diffusion of IPP was a point of threshold for the bacterial quality in juice	Kim et al. (2016)
3.	Dimethyl sulfoxide (DMSO) with RFID	Supply chain monitoring	The micro fluidic-critical temperature indicator is around 18–19°C for fresh-cut fruits It was applied with RFID The visible response was directly converted into RFID signal The response was fast (6 min) to the critical temperature	Lorite et al. (2017)
4.	Glucose 1-hydrate, powdered milk, yeast extract, peptone, and chlorophenol red	Perishable products	Gompertz equation based model was developed to simulate the color change of TTIs as the labels The maximum endpoint times of the tags were 6 days at 278 K and 12 h at 293 K	Zabala, Castan, and Martınez (2015)
5.	Glycerol tributyrate and aspergillus niger lipase	Fruits, vegetables, and some fish	The TTI involves reactions between glycerol tributyrate and aspergillus niger lipase Four TTIs containing substrates with different concentrations (0.55, 0.5, 0.45, 0.4 g/mL) have been developed	Wu et al. (2015)
6.	Urease and carbamide	Perishable fruits	A mathematical model was developed to describe the change of color with time and temperature	Wu et al. (2013)

Table 7 Some examples of commercial TTIs and their working principles

Trademark	Manufacture	Principles	TTIs	Online source
3M monitor mark	3M company	Diffusion	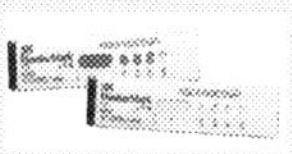	https://3m.com
Coolvu	Freshpoint	Dissolution process of a fine aluminum layer	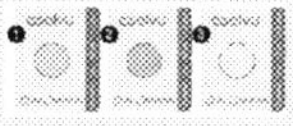	https://freshpointtti.com.keam.co.il/
Fresh-check	Temptime Corp.	Polymerization reaction		https://fresh-check.com
Keep-it	Keep-it technologies	Chemical reaction		https://www.keep-it.no
Timestrip	Timestrip Plc	Chemical reaction		https://timestrip.com
OnVu	Freshpoint and Ciba	Photochemical reaction		https://www.freshpointtti.com/links/default.aspx
Checkpoint	VITSAB international lab	Enzymatic	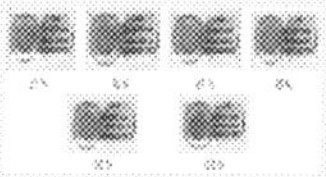	https://vitsab.com
TopCryo	TRACEO	Microbiological		https://frenchetor.cluster002.ovh.net/topcryo
FreshCode	Varcode ltd.	Barcode		https://www.varcode.com
Tempix	Tempix AB	Barcode		https://tempix.com

This is the irreversible color change that provides a permanent record of temperature abuse even when the temperature turns back to a normal temperature. The TTIs have been produced in a wide range from −15°C to 26°C. The range and response time of TTIs based on the type and ester concentration (Kuswandi et al., 2011). Another type of diffusion-based TTI is a 3M freeze watch used for food monitoring that is damaged when exposed to freezing temperature during transportation and storage. This TTI is commercially available at two temperature levels, 0°C and −4°C, and also works on the diffusion mechanism. Here, when the liquid in the ampule freezes, it causes the ampule to fracture and stains the indicator paper. A polymer-based time TTI has been developed by Temptime Corp. called Fresh-check for its commercial name. This TTI works based on a polymerization reaction, where diacetylene crystals polymerize via 1, 4 addition polymerization to the polymer with high color (Kerry et al., 2006; Realini & Marcos, 2014).

Keep-it technologies are producing a chemical reaction based commercial TTI called *Keep-it*. This TTI consists of an immobilized Fe^{3+}, and a mobile reagent ferrocyanide ($Fe(CN)_6^{4-}$) that are placed separately in different sealed compartments. It is activated by seal removal, whereby the mobile reagent reacts with the immobilized reagent, causing a visually colorimetric reaction. Commercial TTI, namely OnVu is a TTI produced by Freshpoint and Ciba that works based on photochemical reaction. This TTI contains benzo pyridines, which changes color over time at a temperature-dependent rate (Fuchs & Carrigan, 2010; Realini & Marcos, 2014). It activates when it is exposed to UV rays, causing it to turn dark blue. The TTI color gets faded with time when it is exposed to a temperature higher than the critical temperature.

VITSAB International lab has developed a TTI with the commercial name Checkpoint that works based on the enzymatic reaction. It consists of two separate compartments, one compartment consisting of substrate triglycerides and a pH indicator, and the other consisting of a lipolytic enzymes solution. These two compartments are separated by a seal. It is activated by breaking the seal between the two compartments. Then, there is a color change due to pH decrease caused by enzymatic hydrolysis of the substrate that produces acid, causing a pH drop that turns green to bright yellow (Galagan & Su, 2008; Koutsoumanis & Gougouli, 2015; Realini & Marcos, 2014).

TRACEO is a company that has developed a commercial TTI called Topcryo. This TTI works based on the microbiological reaction. It consists of 3 layers, where the first layer contains lactic acid bacteria (*C. malt aromatics*), a second layer contains a color indicator (acid fuchsin), and the third layer is the nutritive medium for microorganisms growth. When it is exposed to a temperature higher than the critical temperature, the layers are broken, resulting in enzymatic reaction by the bacteria that makes the color change from green to red (EFSA, 2013). TTIs that present as a barcode have been produced by Tempix AB, namely Tempix, and by Varcode Ltd., Namely FreshCode. The Tempix works in the range of −30°C to +30°C. If the handling temperature exceeds the recommended level, the barcode disappears and cannot be read by the reader.

3.2.3 Freshness indicators

Time-temperature indicators (TTIs) can also be employed as freshness indicators for food, including fruits and vegetables (Kuswandi, 2018). This is due to the fact that TTIs detect the temperature during transportation or storage that leads to spoilage of food. Most of these indicators are used to monitor the freshness of the food by changing their color (Kuswandi, 2018). The information on food quality is usually based on chemical changes or microbial growth within the food. Microbiological quality is indicated by the reaction of the indicator with microbial metabolites (Smolander, 2003). The different technologies used for assessing the food freshness, particularly for fruits and vegetables, including its fresh-cuts, are described below.

The major quality changes in fruits and vegetables during storage are sensory attributes (texture, flavor, color), nutritional value (vitamins, proteins, flavonoids, etc.) and development of pathogens. Any of these can give indications of the food quality inside the package to the consumer. Measurement of the nutritional value of the food is time-consuming and costly, so it is not advisable for detection of the quality of fruits and vegetables (Kuswandi et al., 2011; Kuswandi, 2018; Vanderroost et al., 2014). The shelf-life of fruits and vegetables depends upon the rate of respiration that can be expressed as the oxygen consumed or CO_2 evolved, whereas the result of the respiration process is the food color changes (Mangaraj et al., 2011). When the respiration rate increases, the production of ethylene increase as well. These properties can be employed as markers or indicators for the food freshness and quality. Some examples of freshness indicators developed based on freshness markers, such as CO_2, ethylene, and acid, etc. are given in Table 8. Ripe Sense is a commercial freshness indicator that indicates the fruit ripeness by a label that reacts with the aromatic compounds released during the ripening process during transport or storage. It changes its color according

Table 8 Some examples of freshness indicator that have been applied for fruits and vegetables

No	Marker	Food	Freshness indicator characteristics	Reference
1.	Ethylene	Fruits and vegetables	The indicator film was prepared on the filter paper by immobilized SiO_2 nanoparticles and the indicator solution that was synthesized from palladium sulfate and ammonium molybdate When exposed to C_2H_4 the color changed from white to dark blue	Kim and Shiratori (2006)
2.	Ethylene	Apple	Color ripeness indicator was developed based on the effect of ethylene reduction causing color changes Molybdenum (Mo) changed under the ethylene from white/light yellow to blue due to Mo (VI) to Mo (V) reduction	Lang and Hübert (2012)
3	Acetic acid	Guava	A low-cost indicator was developed using BPB(bromophenol blue)/cellulose membrane The pH was decreased as acetic acid developed during ripening process of guava that changed from blue to green	Kuswandi, Kinanti, Jayus, Abdullah, and Heng (2013)
4.	pH	Strawberry	A simple indicator was developed using MR (Methyl red)/cellulose membrane The pH was increased due to volatile acid reduced during ripening proses of strawberry that changed the indicator from yellow to purple	Kuswandi, Maryska, Jayus, Abdullah, and Heng (2013)
6.	pH	Grape	A simple indicator label based on chlorophenol red (CPR) immobilized onto the filter The pH increased as the volatile organic acids decreased; gradually that changed the indicator from white to beige then to yellow for overripe indication	Kuswandi and Murdyaningsih (2017)
5.	CO_2	Kimchi	CO_2 was detected for kimchi stored at 25°C compared to those stored at 15°C	Meng, Lee, Kang, and Ko (2015)

to the aromatic compounds concentration. It has been applied successfully to pears, kiwi fruit, mango, avocado, and melon (Kuswandi et al., 2011).

3.2.4 Integrity indicators

The damage in integrity of flexible plastic packaging is commonly caused by leaking seals (Hurme, 2003). Food package and seal integrity are tested using destructive test methods, such as biotest, dye penetration, an electrolytic and bubble tests. The disadvantages are that test methods are laborious and only suitable for food sampling, since it is not possible to check each food package individually (Kerry et al., 2006).

In this regard, the most common integrity indicators are time and gas indicators. These indicators provide information on integrity throughout the production and distribution chain (Biji et al., 2015). A commercially available integrity indicator is *Ageless Eye* indicator produced by Mitsubishi Gas Chemicals Company as an oxygen indicator. It indicates the presence or lack of oxygen inside the food package by changing its color. Due to the presence of oxygen in the package, it changes color from pink to blue, as well as turning pink if there is no oxygen in the package. The major disadvantage of this indicator is its reversibility. EMCO packaging (United Kingdom) has developed reversible and irreversible oxygen indicators to check the integrity of the packaging (EMCO, 2013). Some examples of integrity indicators that have been developed as a gas indicator are listed in Table 9.

3.3 Radio frequency identification

Radio frequency identification (RFID) tags are used for data carrying and transmission without human interference, which stores some number for identification where a reader can retrieve information regarding identification number from the database and can act accordingly (Todorvoic, Neag, & Lazarevic, 2014). It comprises three elements—tag, reader, and middleware. A tag consists of a microchip connected to an antenna. The microchip is used to store data. The reader sends radio signals and receives responses from the tag. The middleware, such as local network, web server, etc., connects the RFID hardware and their applications (Ghaani et al., 2016). RFID tags are mainly categorized as active, semipassive, and passive (Kuswandi, 2018), the differences are shown in Table 10. For many years, RFID technology has been used for identification and tracking purposes; it is the replacement for the barcode scanner for this particular operation, also

Table 9 Some examples of integrity indicators that have been developed as a gas indicator

Gas	Indicator composition	Characteristics	Reference
CO_2	Sodium salt of metacresol purple (MCP) in aqueous solution, hydroxyethyl cellulose (HEC) solution, sodium hydrogen carbonate, glycerol, tetrabutylammonium hydroxide (TBAH), ethyl cellulose, tributyl phosphate (TBP)	The water-based indicator that was developed from MCP faded slowly and uniformly over time, the indicator faded much more quickly and changed irreversibly to the acidified yellow form of the dye The response of the indicator was directly related to RH and roughly increased by 80% as the RH increased from 20% to 100% The indicator was largely humidity insensitive over the RH range of 20%–70%	Mills and Skinner (2010)
CO_2	Phenol red (PR), M-cresol, or purple (MCP), tetrabutylammonium hydroxide, ethylcellulose, ethanol,1-butanol solvent	Indicators were prepared by using inkjet printer on either plastic transparent film or cellulose paper Combined dyes (MCP+PR) gave a better sensitivity compared to individual dyes CO_2 concentration from 5% to 20%	Zhang and Lim (2016)
CO_2	Chitosan and 2-amino-2-methyl-1-propanol (AMP)	The chitosan suspension transparency decreased sharply up to approx. 40% (<pH6) 0.2% chitosan is best suited for CO_2 detection AMP (5%) added to enhance the visual contrast of the suspension transition from opaque to transparency (<90%)	Jung, Puligundla, and Ko (2012)
O_2	TiO_2, MB, triethanolamine (TEOA)	Irradiation of visible light did not bleach the indicator under aerobic or anaerobic High level of humidity did not affect the UV activation step The indicator was reversible, reusable, and could be applied for quantitative analysis	Lee et al. (2005)
O_2	Nanocrystalline SnO_2, glycerol, MB	Nanocrystalline SnO_2 based UV activated O_2 indicator The SnO_2 indicator was activated only by the UVB	Mills and Hazafy (2009)

Continued

Table 9 Some examples of integrity indicators that have been developed as a gas indicator—Cont'd

Gas	Indicator composition	Characteristics	Reference
O_2	Titanium dioxide (Degussa P25), MB, p-xylylviologen dibromide, p-butylviologen dibromide, thionine, 2,2′-Dicyano-1,1′-dimethyl-4,4′-dipyridinium dimesylate	The polyviologen based indicators need less UV radiation to develop a color change A clear color change could be seen for poly (*p*-xylylviologen dibromide) on fast oxidation Thionine showed greater stability to oxygen These indicators could be used to detect oxygen above 0.5% concentration	Roberts, Lines, Reddy, and Hay (2011)
O_2	Glycerol, TiO_2 nanoparticles, and methylene blue (MB)	The electrospun indicators were 4–5 times more sensitive to UV as compared to film indicators prepared by casting Sensitivity to UV radiation was enhanced by increasing the ethanol concentration Increasing the TiO_2 resulted in a decrease in color recovery rate	Mihindukulasuriya and Lim (2013)
O_2	P25 TiO_2, glycerol, thionine, and zein	Alginate (1.25%) introduction as the coating polymer avoids leaching The alginate film wavelength maximum was around 500 nm, whereas that of the zein film was at 640 nm The alginate-based indicator showed fast recovery behavior	Vu and Won (2013)

Table 10 Different types of active, semipassive, and passive tags RFID

Types	Active tag	Semipassive	Passive tag
Power	Internal battery	Chip battery operation	No power supply
Size	Large	Large	Small
Cost	20–100 $	10–50 $	10 $
Maintenance	Replacement	Replacement	No maintenance
Life	Based on battery life	Based on battery life	Up to 20 years
Memory	128 KB	128 KB	128 bytes
Range	Up to 100 m	Up to 30.5 m	Up to 6 m

it can operate at different frequencies depending upon reading range and the data transmission rate. The low frequency (LF) operates in the range of 125–143 KHz, and the reading range is 10 cm. While high frequency (HF) covers a frequency range of 3–30 MHz and operates in the range of 10–100 cm. The ultra-high frequency (UHF) operates in the range of 300 MHz–1 GHz and is operated over a range of 12 m. The microwave frequency (MW) covers a frequency of >1 GHz and is operated up to 30 m.

In some cases, the RFID tag is integrated with the TTI to know the temperature profile throughout the distribution chain (Want, 2004a, 2004b). To make the RFID system more intelligent, it should also provide information regarding the integrity, quality, and safety, as well as surrounding conditions during transportation or storage. This can be achieved by measuring the properties, e.g., temperature, relative humidity, gas concentration, volatile compounds, microbial content, and pressure (Kuswandi, 2018; Vanderroost et al., 2014). In fruit and vegetable applications, some of the RFID has been assisted with critical temperature indicators—for fresh-cut fruits, critical temperatures within 18–19°C (Lorite et al., 2017)—assisted with optical indicator for O_2 concentration (Martínez-Olmos et al., 2013), and traceability for kiwi fruit (Gautam et al., 2017). The RFID system and its operations are given in Fig. 5.

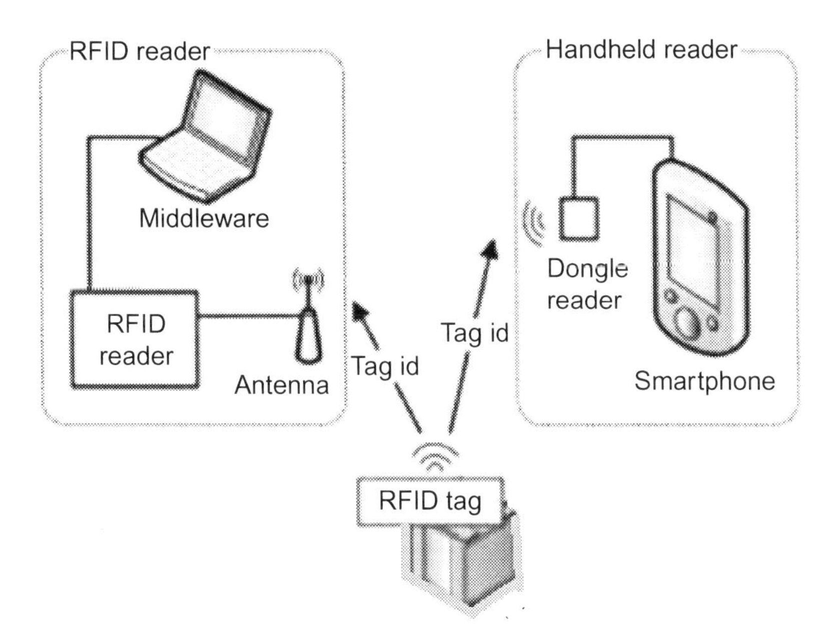

Fig. 5 The handheld reader and RFID to read the tag.

3.4 Barcodes and others

Usually, a barcode is arranged in parallel spaces and bars of systematic side lines that consist hidden encoded data. The signal is decoded and read by a scanner of the optical barcode, which interprets the signal to a system for further necessary action (Ghaani et al., 2016; Han, 2013). Thus, barcodes are the least expensive data carriers; they were introduced in 1970 for the universal barcode. Fig. 6 shows barcodes of different types employed in food packaging that are either 1-D barcodes or 2-D barcodes.

1-D barcodes hold information like identification number and the item number and require less storage space (Drobnik, 2015). Due to the increasing demand for storage of more data in less space, reduced space symbology (RSS) barcodes were developed. When the symbol would be too wide then it is arranged in stacks and is known as RSS expanded stacked barcode (Yam et al., 2005). 2-D barcodes store more data in less space as compared to 1-D barcodes. It is composed of dots, spaces, numeric, and alphanumeric arranged in a matrix form. PDF (Portable data file) 417 is a 2-D barcode that stores data up to 1.1 kB. The recent developments of 2-D barcodes (QR 2-D barcode) include specific

Fig. 6 Different types of barcodes (A) 1-D barcode, (B) PDF barcode, and (C) QR 2-D barcode.

characters along with numeric, alphanumeric, and byte/binary characters.

Besides the above tools, holograms are also useful devices that could be applied in food packaging to create intelligent systems. The hologram is an attractive and emerging tool to apply in food intelligent packaging. It could aid to protect the product brand name and to combat product counterfeiting. While improving the product brand image, they are also employed to avoid tampering of food products and packaging. With the help of the changing pattern of the hologram, the counterfeiters are unable to change the food product or the product label. If a counterfeit occurs by removing the hologram, the upper polyester film will be removed as well, leaving a sign regarding the product tempering (Pareek & Khunteta, 2014). They are commonly employed in the pharmaceutical industry for expensive drugs. However, their use in the food intelligent packaging is very rare. With current technological development in the food intelligent packaging system, it is possible to use holograms for food applications.

4 Conclusions and future trends

Various smart packaging forms, such as active and intelligent packaging, have been studied and marketed and are being integrated into food packaging to reach higher food quality and safety, including improving food shelf-life, consumer convenience, and supply chain efficiency. Active packaging helps to enhance food's shelf-life, while intelligent packaging continuously monitors the food quality throughout the supply chain and storage to the end customers. Applications of proper packaging technologies by the food industry, including fresh-cut fruits and vegetables, could be helpful to improve the shelf-life, quality, and safety of food as well as give information regarding food quality.

Studies on these active and intelligent packaging developments can create further improvement of the existing food packaging systems, including for fresh-cut fruits and vegetables, to offer consumer benefits and convenience in the near future. However, these studies generally focus on individual active or intelligent packaging, and very rarely on the combination of these smart packaging technologies into forms that have active and intelligent functions simultaneously. Therefore, there is a possible marriage between active and intelligent packaging to develop a full smart packaging system for potential applications for real-time quality information of the fruits and vegetables during storage and distribution.

In the commercial applications, there are gaps or barriers of these technologies due to the technical limitations and higher cost of the smart packaging. Therefore, a multidisciplinary approach is required to commercialize these smart packaging technologies with less cost, smaller devices, and reusable and durable sensors or indicators. The main challenge in active packaging is to produce active materials with improved physicochemical properties. For instance, the addition of nanoparticles as active compounds into the packaging materials would reduce the amount needed while improving its active properties, such as extending food shelf-life. At present, some of the researchers are working on a coupling of TTI to RFID for distribution chain monitoring system in order to inform and communicate food product, such as fresh-cut fruits quality in real-time (Lorite et al., 2017). Many further works are needed on integrating between active and intelligent packaging to produce truly smart packaging for fresh-cut fruits and vegetables that have active and intelligent functions to improve its quality and safety, and real-time monitoring as well as reducing food loss.

Acknowledgment

The author gratefully thanks the DRPM, Higher Education, Ministry of Research, Technology and Higher Education, the Republic of Indonesia for supporting this work via the Basic Research Project 2019 (Hibah Penelitian Dasar 2019).

References

Abe, K., & Watada, A. E. (1991). Ethylene absorbent to maintain quality of lightly processed fruits and vegetables. *Journal of Food Science, 56*, 1589–1592.

Aday, M. S., Caner, C., & Rahval, F. (2011). Effect of oxygen and carbon dioxide absorbers on strawberry quality. *Postharvest Biology and Technology, 62*, 179–187.

Adley, C. C. (2014). Past, present and future of sensors in food production. *Foods, 3*, 491–510.

Ahvenainen, R. (1996). New approaches in improving the shelf life of minimally processed fruit and vegetables. *Trends in Food Science and Technology, 71*, 179–187.

Alam, T., & Goyal, G. K. (2011). Effect of map on microbiological quality of mozzarella cheese stored in different packages at $7 \pm 1°C$. *Journal of Food Science and Technology, 48*(1), 120–123.

Almenar, E., Auras, R., Wharton, P., Rubino, M., & Harte, B. (2007). Release of acetaldehyde from β-cyclodextrins inhibits postharvest decay fungi in vitro. *Journal of Agricultural and Food Chemistry, 55*, 7205–7212.

Alvarez, M. F. (2000). Active food packaging: Review. *Food Science and Technology International, 6*, 97–103.

Appendini, P., & Hotchkiss, J. H. (2002). Review of antimicrobial food packaging. *Innovative Food Science & Emerging Technologies, 3*, 113–126.

Ayala, C. E., & Park, D. L. (2000). New bar code will help monitor food safety. *The Magazine of the Louisiana Agricultural Experiment Station, 43*, 8.

Azevedo, S., Cunhaa, L. M., Mahajan, P. V., & Fonseca, S. C. (2011). Application of simplex lattice design for development of moisture absorber for oyster mushrooms. *Procedia Food Science, 1*, 184–189.

Bentayeb, K., Rubio, C., Batlle, R., & Nerín, C. (2007). Direct determination of carnosic acid in a new active packaging based on natural extract of rosemary. *Analytical and Bioanalytical Chemistry, 389*(6), 1989–1996.

Ben-Yehoshua, S., Shapiro, B., Chen, Z. E., & Lurie, S. (1983). Mode of action of plastic film in extending life of lemon and bell pepper fruits by alleviation of water stress. *Plant Physiology, 73*(1), 87–93.

Berenzon, S., & Saguy, I. S. (1998). Oxygen absorbers for extension of crackers shelf-life. *LWT-Food Science and Technology, 31*, 1–5.

Biji, K. B., Ravishankar, C. N., Mohan, C. O., & Srinivasa Gopal, T. K. (2015). Smart packaging systems for food applications: A review. *Journal of Food Science and Technology, 52*(10), 6125–6135.

Bjerkeng, B., Sivertsvik, M., Rosnes, J. T., & Bergslien, H. (1995). Reducing package deformation and increasing filling degree in packages of cod fillets in CO_2 enriched atmospheres by adding sodium carbonate and citric acid to an exudate absorber. In P. Ackermann, M. Jagerstad, & T. Ohlsson (Eds.), *Foods and food packaging materials: Chemical interactions*. Cambridge: Royal Society of Chemistry.

Bodenhamer, W., Jackowski, G., & Davies, E. (2004). Toxin alert: Surface binding of an immunoglobulin to a flexible polymer using a water soluble varnish matrix. US patent 66992973.

Brody, A. L., Strupinsky, E. R., & Kline, L. R. (2001). *Active packaging for food applications* (pp. 131–196). Lancaster, PA: Technomic Publishing Company.

Buonocore, G. G., Del Nobile, M. A., Piergiovanni, L., Fava, P., Puglisi, M. L., & Nicolais, L. (2003). Naringinase immobilization in polymeric films intended for food packaging applications. *Journal of Food Science, 68*(6), 2046–2049.

Burt, S. (2004). Essential oils: Their antibacterial properties and potential applications in foods—A review. *International Journal of Food Microbiology, 94*, 223–253.

Callaghan, K. A. M. O., & Kerry, J. P. (2016). Consumer attitudes towards the application of smart packaging technologies to cheese products. *Food Packaging and Shelf Life, 9*, 1–9.

Cameron, A. C., & Talasila, T. (1995). Modified-atmosphere packaging of fresh fruits and vegetables. In *IFT annual meeting, book of abstracts* (p. 254).

Camo, J., Beltrán, A., & Roncalés, P. (2008). Extension of the display life of lamb with an antioxidant active packaging. *Meat Science, 80*, 1086–1091.

Camo, J., Lorés, A., Djenane, D., Beltrán, J. A., & Roncalés, P. (2011). Display life of beef packaged with an antioxidant active film as a function of the concentration of oregano extract. *Meat Science, 88*, 174–178.

Charles, F., Sanchez, J., & Gontard, N. (2005). Modeling of active modified atmosphere packaging of endives exposed to several postharvest temperatures. *Journal of Food Science, 70*(5), E443–E449.

Chen, C. S., Liau, W. Y., & Tsai, G. J. (1998). Antibacterial effects of N-sulfonated and N-sulfobenzoyl chitosan and application to oyster preservation. *Journal of Food Protection, 61*, 1124–1128.

Christie, G., Christov, V., Corrigan, P., Goubran, F., Holmes, R., Lyon, T., et al. (1997). Research into fungicide boxes for long term storage of grapes. In *World conference on packaging: Proceedings of the 13th international association of* *packaging research institutes, Victoria University of Technology, Melbourne, Vic., March 24-27* (pp. 459–469). Melbourne: Centre for Packaging, Transportation, and Storage.

Coma, V. (2008). Bioactive packaging technologies for extended shelf life of meat-based products. *Meat Science, 78*, 90–103.

Cook, P. (2001). Stealth scavenging systems: The scavenger in the polymer. In *International food technologists annual meeting book of abstracts*. New Orleans, LA: IFT.

Cruz, R. S., Camilloto, G. P., & Pires, A. C. S. (2012). Oxygen scavengers: An approach on food preservation. In A. A. Eissa (Ed.), *Structure and function of food engineering* (pp. 21–42). IntechOpen. https://doi.org/10.5772/48453.

Cullen, J. S., & Vaylen, N. E. (1994). *Carbon dioxide absorbent packet and process.* US patent 5322701.

Dainelli, D., Gontard, N., Spyropoulos, D., van den Beuken, E. Z., & Tobback, P. (2008). Active and intelligent food packaging: Legal aspects and safety concerns. *Trends in Food Science and Technology, 19*, S103–S112.

Damaj, Z., Naveau, A., Dupont, L., Hénon, E., Rogez, G., & Guillon, E. (2009). Co(II) (L-proline)$_2$ $(H_2O)_2$ solid complex: Characterization, magnetic properties, and DFT computations. Preliminary studies of its use as oxygen scavenger in packaging films. *Inorganic Chemistry Communications, 12*, 17–20.

Day, B. P. F. (1998). Active packaging of foods. *Campden and Chorleywood Food Research Association. New Technologies Bulletin*, 17–23.

Day, B. P. F. (2003). Active packaging. In R. Coles, D. McDowell, & M. Kirwan (Eds.), *Food packaging technology* (pp. 282–302). Boca Raton, FL: CRC Press [chapter 9].

De Kruijf, N., & van Beest, M. D. (2003). Active packaging. In D. R. Heldman (Ed.), *Encyclopedia of agriculture, food and biological engineering* (pp. 5–9). New York: Marcel Dekker.

Deell, J. R., Toivonen, P. M. A., Cornut, F., Roger, C., & Vigneault, C. (2006). Addition of sorbitol with $KMNO_4$ improves broccoli quality retention in modified atmosphere packages. *Journal of Food Quality, 29*(1), 65–75.

Devlieghere, F., Vermeiren, L., Bockstal, A., & Debevere, J. (2000). Study on antimicrobial activity of food packaging material containing potassium sorbate. *Acta Alimentaria, 29*, 137–146.

Drobnik, O. (2015). *Barcodes with IOS: Bringing together the digital and physical worlds.* New York: Manning [chapter 1].

EFSA CEF Panel (EFSA Panel on Food Contact Materials, Enzymes, Flavorings and Processing Aids). (2013). Scientific opinion on the safety evaluation of a time-temperature indicator system, based on Carnobacterium maltaromaticum and a color change indicator (acid fuchsin) for use in food contact materials. *EFSA Journal, 11*(7), 3307. 10 pp https://doi.org/10.2903/j.efsa.2013.3307.

EMCO Packaging. (2013). *Oxygen indicator labels. Oxygen indicating color change chemistry. http://www.emcopackaging.com/index.php/ products/oxygen-indicator-labels. Accessed 2 December 2016.*

Falcone, P., Speranza, B., Del Nobile, M. A., Corbo, M. R., & Sinigaglia, M. (2005). A study on the antimicrobial activity of thymol intended as natural preservative in food applications. *Journal of Food Protection, 68*, 1664–1670.

Fernandez-Saiz, P., Ocio, M. J., & Lagaron, J. M. (2010). Antibacterial chitosan-based blends with ethylene–vinyl alcohol copolymer. *Carbohydrate Polymers, 80*, 874–884.

Floros, J. D., Dock, L. L., & Han, J. H. (1997). Active packaging technologies and applications. *Food, Cosmetics and Drug Packaging, 20*(1), 10–17.

Frank, M., Ulmer, H., Ruiz, J., Visani, P., & Weimar, U. (2001). Complementary analytical measurements based upon gas chromatography/mass spectroscopy, sensor system and human sensory panel: A case study dealing with packaging materials. *Analytica Chimica Acta, 431*, 11–29.

Franke, I., Wijma, E., & Bouma, K. (2002). Shelf life extension of pre-baked buns by an active packaging ethanol emitter. *Food Additives and Contaminants, 19*(3), 314–322.

Frydrych, E., Foltynowicz, Z., Kowalak, S., & Janiszewska, E. (2007). Oxygen scavengers for packing system based on zeolite adsorbed organic compounds. *Studies in Surface Science and Catalysis, 170*, 1597–1604.

Fuchs, A., & Carrigan, A. V. (2010). *TTI indicator with balanced photochemical processes.* Patent number US 20100135353 A1.

Galagan, Y., & Su, W. F. (2008). Fadable ink for time temperature control of food freshness: Novel new time temperature indicator. *Food Research International, 41*, 653–657.

Galindo-Arcega, C. (2004). *Migraciondel BHT de pelıculas de PEBD y suefectoen la estabilidad del aceite de soya.* M.Sc. thesis Hermosillo: Centro de Investigacio'-nenAlimentacion y Desarrollo, A.C.

Gautam, R., Singh, A., Karthik, K., Pandey, S., Scrimgeour, F., & Tiwari, M. K. (2017). Traceability using RFID and its formulation for a kiwifruit supply chain. *Computers & Industrial Engineering, 103*, 46–58.

Gemili, S., Ahmet, Y., & Altınkaya, S. A. (2010). Development of antioxidant food packaging materials with controlled release properties. *Journal of Food Engineering, 96*, 325–332.

Ghaani, M., Cozzolino, C. A., Castelli, G., & Farris, S. (2016). An overview of the intelligent packaging technologies in the food sector. *Trends in Food Science & Technology, 51*, 1–11.

Gill, C. O., & Mc Ginnis, J. C. (1995). The effects of residual oxygen concentration and temperature on the degradation of the color of beef packaged under oxygen depleted atmospheres. *Meat Science, 39*, 387–394.

Goncalves, A., Mendes, R., & Nunes, M. (2004). Effect of oxygen absorber on the shelf life of gilthead seabream (*Spratusaurata*). *Journal of Aquatic Food Product Technology, 13*(3), 49–59.

Gontard, N. (2007). Antimicrobial paper based packaging. In *International antimicrobial in plastic and textile applications. Intertech PIRA conference, Prague, Czech Republic.*

Granda-Restrepo, D., Peralta, E., Troncoso-Rojas, R., & Soto-Valdez, H. (2009). Release of antioxidants from co-extruded active packaging developed for whole milk powder. *International Dairy Journal, 19*, 481–488.

Güçbilmez, C. M., Yemenicioflu, A., & Arslanoflu, A. (2007). Antimicrobial and antioxidant activity of edible zein films incorporated with lysozyme, albumin proteins and disodium EDTA. *Food Research International, 40*, 80–91.

Guillard, V., Issoupov, V., Redl, A., & Gontard, N. (2009). Food preservative content reduction by controlling sorbic acid release from a superficial coating. *Innovative Food Science & Emerging Technologies, 10*, 108–115.

Guizani, N., Al-Busaidy, M. A., Al-Belushi, I. M., Mothershaw, A., & Rahman, M. S. (2005). The effect of storage temperature on histamine production and the freshness of yellowfin tuna (*Thunnusalbacares*). *Food Research International, 38*, 215–222.

Gumiero, M. (2009). Innovative active packaging for food products: Research and developments. In: *14th workshop on the developments in the Italian Ph.D research on food science technology and biotechnology—University of Sassari, Oristano. 16–18 September 2009.*

Guynot, M. E., Sanchis, V., Ramos, A. J., & Marín, S. (2003). Mold-free shelf-life extension of bakery products by active packaging. *Journal of Food Science, 68* (8), 2547–2552.

Han, J. H. (2000). Antimicrobial food packaging. *Food Technology, 54*, 56–65.

Han, J. H. (2005). Antimicrobial packaging systems. In J. H. Han (Ed.), *Innovations in food packaging* (pp. 81–107). Amsterdam: Elsevier Academic Press.

Han, J. H. (2013). A review of food packaging technologies and innovations. In J. H. Han (Ed.), *Innovations in food packaging* (p. 3). Amsterdam: Elsevier Academic Press.

Hanrahan, G., Patil, D. G., & Wang, J. (2004). Electrochemical sensors for environmental monitoring: Design, development and applications. *Journal of Environmental Monitoring, 6*, 657–664.

Hempel, A. W., O'Sullivan, M. G., Papkovsky, D. B., & Kerry, J. P. (2013). Use of smart packaging technologies for monitoring and extending the shelf-life quality of modified atmosphere packaged (MAP) bread: Application of intelligent oxygen sensors and active ethanol emitters. *European Food Research and Technology, 237*, 117–124.

Hogan, S., & Kerry, J. (2008). Smart packaging of meat and poultry products. In J. Kerry (Ed.), *Smart packaging technologies for fast moving consumer goods* (pp. 33–59). West Sussex, England: John Wiley & Sons Ltd.

Horan, T. J. (1998). *Method for determining bacterial contamination in food package.* US Patent 5,753,285.

Hoshino, A., & Osanai, T. (1986). *Packaging films for deodorization.* Japanese patent 86209612.

Hotchkiss, J. S. (1994). Recent research in MAP and active packaging systems. In *Abstract of the 27th annual convention.* North Sydney: Australian Institute of Food Science and Technology.

Hotchkiss, J. H. (1997). Food-packaging interactions influencing quality and safety. *Food Additives and Contaminants, 14*(6–7), 601–607.

Howard, L. R., Yoo, K. S., Pike, L. M., & Miller, G. H. (1994). Quality changes in diced onions stored in film packages. *Journal of Food Science, 59*, 110–112, 117.

Huber, M., Ruiz, J., & Chastellain, F. (2002). Off-flavor release from packaging materials and its prevention: A foods company's approach. *Food Additives and Contaminants, 19*, 221–228.

Hurme, E. (2003). Detecting leaks in modified atmosphere packaging. In R. Ahvenainen (Ed.), *Novel food packaging techniques* (pp. 276–286). Cambridge: Woodhead Publishing Ltd.

Hurme, E., Sipilainen-Malm, T., & Ahvenainen, R. (2002). Active and intelligent packaging. In T. Ohlsson, & N. Bengtsson (Eds.), *Minimal processing technologies in the food industry* (pp. 87–115). Boca Raton, FL: CRC Press [chapter 5].

Jung, J., Puligundla, P., & Ko, S. (2012). Proof-of-concept study of chitosan based carbon dioxide indicator for food packaging applications. *Food Chemistry, 135*(4), 2170–2174.

Kerry, J. P., Grady, M. N. O., & Hogan, S. A. (2006). Past, current and potential utilisation of active and intelligent packaging systems for meat and muscle-based products: A review. *Meat Science, 74*, 113–130.

Kim, J. U., Ghafoor, K., Ahn, J., Shin, S., Lee, S. H., Shahbaz, H. M., et al. (2016). Kinetic modeling and characterization of a diffusion-based time-temperature indicator (TTI) for monitoring microbial quality of non-pasteurized angelica juice. *LWT-Food Science and Technology, 67*, 143–150.

Kim, J. H., & Shiratori, S. (2006). Fabrication of color changeable film to detect ethylene gas. *Japanese Journal of Applied Physics, 45*(5A), 4274–4278.

Koutsoumanis, K. P., & Gougouli, M. (2015). Use of time temperature integrators in food safety management. *Trends in Food Science & Technology, 43*, 236–244.

Kress-Rogers, E. (1998). Chemo sensors, biosensors and immune sensors. In E. Kress-Rodgers (Ed.), *Instrumentation and sensors for the food industry* (pp. 581–669). Cambridge: Woodhead Publishing Ltd.

Kudachikar, V. B., Kulkarni, S. G., & Keshava Prakash, M. N. (2011). Effect of modified atmosphere packaging on quality and shelf life of 'Robusta' banana (*Musa sp.*) stored at low temperature. *Journal of Food Science and Technology, 48*(3), 319–324.

Kuswandi, B. (2018). Freshness sensors for food packaging. In *Reference module in food science*, Elsevier. https://doi.org/10.1016/B978-0-08-100596-5.21876-3.

Kuswandi, B., Andres, R., & Narayanaswamy, R. (2001). Optical fibre biosensors based on immobilised enzymes. *The Analyst, 126*, 1469–1491. https://doi.org/10.1039/b008311i.

Kuswandi, B., Jayus, R. A., Abdullah, A., Heng, L. Y., & Ahmad, M. (2012). A novel colorimetric food package label for fish spoilage based on polyaniline film. *Food Control, 25*(1), 184–189.

Kuswandi, B., Jayus, Larasati, T. S., Abdullah, A., & Heng, L. Y. (2012). Real-time monitoring of shrimp spoilage using on-package sticker sensor based on natural dye of curcumin. *Food Analytical Methods, 5*(4), 881–889.

Kuswandi, B., Jayus, Oktaviana, R., Abdullah, A., & Heng, L. Y. (2014). A novel on-package sticker sensor based on methyl red for real-time monitoring of broiler chicken cut freshness. *Packaging Technology and Science, 27*(1), 69–81.

Kuswandi, B., Kinanti, D. P., Jayus, Abdullah, A., & Heng, L. Y. (2013). Simple and low-cost freshness indicator for strawberries packaging. *Acta Manilana, 61*, 147–159.

Kuswandi, B., Maryska, C., Jayus, Abdullah, A., & Heng, L. Y. (2013). Real time on-package freshness indicator for guavas packaging. *Journal of Food Measurement and Characterization*, https://doi.org/10.1007/s11694-013-9136-5.

Kuswandi, B., & Murdyaningsih, E. A. (2017). Simple on package indicator label for monitoring of grape ripening process using colorimetric pH sensor. *Journal of Food Measurement and Characterization, 11*(4), 2180–2194.

Kuswandi, B., Wicaksono, Y., Abdullah, A., Jayus, Heng, L. Y., & Ahmad, M. (2011). Smart packaging: Sensors for monitoring of food quality and safety. *Sensing and Instrumentation for Food Quality and Safety, 5*, 137–146.

Labuza, T. P. (1987). Oxygen scavenger sachets. *Food Research, 32*, 276–277.

Labuza, T. P. (1990). Active food packaging technologies. *Food Science and Technology Today, 4*(1), 53–56.

Labuza, T. P., & Breene, W. M. (1989). Applications of active packaging for improvement of shelf-life and nutritional quality of fresh and extended shelf-life foods. *Journal of Food Processing & Preservation, 13*, 1–69.

Lang, C., & Hübert, T. (2012). A color ripeness indicator for apples. *Food and Bioprocess Technology, 5*, 3244–3249.

Lee, C. H., An, D. S., Lee, S. C., Park, H. G., & Lee, D. S. (2004). A coating for use as an antimicrobial and antioxidative packaging material incorporating nisin and α-tocopherol. *Journal of Food Engineering, 62*, 323–329.

Lee, S. K., Sheridan, M., & Mills, A. (2005). Novel UV-activated colorimetric oxygen indicator. *Chemistry of Materials, 17*(10), 2744–2751.

Lee, D. S., Shin, D. H., Lee, D. U., Kim, J. C., & Cheigh, H. S. (2001). The use of physical carbon dioxide absorbents to control pressure buildup and volume expansion of kimchi packages. *Journal of Food Engineering, 48*, 183–188.

LeGood, P., & Clarke, A. C. (2006). *Smart and active packaging to reduce food waste*. Rotherham: National Metals Technology Centre (NAMTEC). 7 December 2006.

Lorite, G. S., Selkala, T., Sipola, T., Palenzuela, J., Jubete, E., Vinuales, A., et al. (2017). Novel, smart and RFID assisted critical temperature indicator for supply chain monitoring. *Journal of Food Engineering, 193*, 20–28.

Lund, M. N., Hviid, M. S., & Skibsted, L. H. (2007). The combined effect of antioxidants and modified atmosphere packaging on protein and lipid oxidation in beef patties during chill storage. *Meat Science, 76*, 226–233.

Mangaraj, S., & Goswami, T. K. (2009a). Modified atmosphere packaging—An ideal food preservation technique. *Journal of Food Science and Technology, 46*(5), 399–410.

Mangaraj, S., & Goswami, T. K. (2009b). Modified atmosphere packaging of fruits and vegetables for extending shelf-life—A review. *Fresh produce, 3*(1), 1–31.

Mangaraj, S., Goswami, T. K., Giri, S. K., & Joshy, C. G. (2012). Design and development of a modified atmosphere packaging system for guava (*cv. Baruipur*). *Journal of Food Science and Technology, 51*(11), 2925–2946.

Mangaraj, S., Goswami, T. K., Giri, S. K., & Tripathi, M. K. (2012). Perm selective MA packaging of litchi (*cv. Shahi*) for preserving quality and extension of shelf-life. *Postharvest Biology and Technology, 71*, 1–12.

Mangaraj, S., Sadawat, I. J., & Prasad, S. (2011). Assessment of quality of pears stored under laminated modified atmosphere packages. *International Journal of Food Properties, 14*, 1–14.

Martinez, L., Djenane, D., Cilla, I., Antonio Beltran, J., & Pedro, R. (2006). Effect of varying oxygen concentrations on the shelf-life of fresh pork sausages packaged in modified atmosphere. *Food Chemistry, 94*, 219–225.

Martínez-Olmos, A., Fernandez-Salmeron, J., Lopez-Ruiz, N., Rivadeneyra Torres, A., Capitan-Vallvey, L. F., & Palma, A. J. (2013). Screen printed flexible radiofrequency identification tag for oxygen monitoring. *Analytical Chemistry, 85*(22), 11098–11105.

Mastromatteo, M., Mastromatteo, M., Conte, A., & Nobile, M. A. D. (2010). Advances in controlled release devices for food packaging applications. *Trends in Food Science & Technology, 21*, 591–598.

Mattila, T., Tawast, J., & Ahvenainen, R. (1990). New possibilities for quality control of aseptic packages: Microbiological spoilage and seal defect detection using head-space indicators. *Lebensmittel-Wissenschaft & Technologie, 23*, 246–251.

Meng, X., Lee, K., Kang, T. Y., & Ko, S. (2015). An irreversible ripeness indicator to monitor the CO2 concentration in the headspace of packaged kimchi during storage. *Food Science and Biotechnology, 24*(1), 91–97.

Mihindukulasuriya, S. D., & Lim, L. T. (2013). Oxygen detection using UV-activated electrospun poly(ethylene oxide) fibers encapsulated with TiO2 nanoparticles. *Journal of Materials Science, 48*, 5489–5498.

Mills, A. (2005). Oxygen indicators and intelligent inks for packaging food. *Chemical Society Reviews, 34*, 1003. https://doi.org/10.1039/b503997p.

Mills, A., Doyle, G., Peiro, A. M., & Durant, J. (2006). Demonstration of a novel, flexible, photocatalytic oxygen-scavenging polymer film. *Journal of Photochemistry and Photobiology A: Chemistry, 177*, 328–331.

Mills, A., & Hazafy, D. (2009). Nano crystalline SnO_2-based, UVB-activated, colorimetric oxygen indicator. *Sensors and Actuators B: Chemical, 136*(2), 344–349.

Mills, A., Qing Chang, Q., & McMurray, N. (1992). Equilibrium studies on colorimetric plastic film sensors for carbon dioxide. *Analytical Chemistry, 64*, 1383–1389.

Mills, A., & Skinner, G. A. (2010). Water-based colorimetric optical indicators for the detection of carbondioxide. *Analyst, 135*, 1912–1917.

Miltz, J., & Perry, M. (2005). Evaluation of the performance of iron-based oxygen scavengers, with comments on their optimal applications. *Packaging Technology and Science, 18*(1), 21–27.

Montero-Prado, P., Rodriguez-Lafuente, A., & Nerin, C. (2011). Active label-based packaging to extend the shelf-life of "Calanda" peach fruit: Changes in fruit quality and enzymatic activity. *Postharvest Biology and Technology, 60*, 211–219.

Nerin, C., Tovar, L., Djenane, D., Camo, J., Salafranca, J., & Beltrán, J. A. (2006). Studies on the stabilization of beef meat by new active packaging containing natural antioxidants. *Journal of Agricultural and Food Chemistry, 54*, 7840–7846.

Özogul, F., Taylor, K. D. A., Quantick, P., & Özogul, Y. (2002a). Changes in biogenic amines in herring stored under modified atmosphere and vacuum pack. *Journal of Food Science, 67*(7), 2497–2501.
Özogul, F., Taylor, K. D. A., Quantick, P., & Özogul, Y. (2002b). Biogenic amines formation in Atlantic herring (*Clupea harengus*) stored under modified atmosphere packaging using a rapid HPLC method. *International Journal of Food Science and Technology, 37*, 515–522.
Pacquit, A., Frisby, J., Diamond, D., Lau, K. T., Farrell, A., Quilty, B., et al. (2007). Development of a smart packaging for the monitoring of fish spoilage. *Food Chemistry, 102*(2), 466–470.
Pacquit, A., Lau, K. T., McLaughlin, H., Frisby, J., Quilty, B., & Diamond, D. (2006). Development of a volatile amine sensor for the monitoring of fish spoilage. *Talanta, 69*(2), 515–520.
Pal, R. K., & Buescher, R. W. (1993). Respiration and ethylene evolution of certain fruits and vegetables in response to carbon dioxide in controlled atmosphere storage. *Journal of Food Science and Technology, 30*, 29–32.
Papkovsky, D. B., Smiddy, M. A., Papkovskaia, N. Y., & Kerry, J. P. (2002). Nondestructive measurement of oxygen in modified atmosphere packaged hams using a phase-fluorimetric sensor system. *Journal of Food Science, 67*, 3164–3169.
Pareek, V., & Khunteta, A. (2014). Pharmaceutical packaging : Current trends and future. *International Journal of Pharmacy and Pharmaceutical Sciences, 6*(6), 480–485.
Park, Y. W., Kim, S. M., Lee, J. Y., & Jang, W. (2015). Application of biosensors in smart packaging. *Molecular & Cellular Toxicology, 11*, 277–285.
Pavelkova, A. (2013). Time temperature indicators as devices intelligent packaging. *Acta Universitatis Agriculturae et Silviculturae Mendelianae Brunensis, 56*, 245–252.
Puligundla, P., Jung, J., & Ko, S. (2012). Carbon dioxide sensors for intelligent packaging applications. *Food Control, 25*, 328–333.
Quintavalla, S., & Vicini, L. (2002). Antimicrobial food packaging in meat industry. *Meat Science, 62*, 373–380.
Ramos, M., Jiménez, A., Peltzer, M., & Garrigós, M. C. (2012). Characterization and antimicrobial activity studies of polypropylene films with carvacrol and thymol for active packaging. *Journal of Food Engineering, 109*, 513–519.
Realini, C. E., & Marcos, B. (2014). Active and intelligent packaging systems for a modern society. *Meat Science, 98*, 404–419.
Roberts, L., Lines, R., Reddy, S., & Hay, J. (2011). Investigation of polyviologens as oxygen indicators in food packaging. *Sensors and Actuators B: Chemical, 152*(1), 63–67.
Robertson, G. L. (2006). Active and intelligent packaging. In *Food packaging: Principles and practices* (2nd ed., pp. 285–312). Boca Raton, FL: CRC Press [chapter 14].
Robertson, G. L. (2012). *Food packaging: Principles and practice.* Boca Raton, Florida, USA: Taylor & Francis.
Rooney, M. L. (1985). Oxygen scavenging from air in package headspaces by singlet oxygen reactions in polymer media. *Journal of Food Science, 47*, 291–294. 298.
Rooney, M. L. (1995a). Active packaging in polymer films. In M. L. Rooney (Ed.), *Active food packaging* (pp. 1–37). London: Blackie Academic and Professional.
Rooney, M. L. (1995b). Active packaging: Science and application. In J. Welti-Chanes, G. V. Barbosa-Canovas, & J. M. Aguilera (Eds.), *Engineering and food for the 21st century.* Boca Raton, FL: CRC Press [chapter 32].
Rooney, M. L. (2001). Overview of oxygen scavenging technologies. In *International food technologists annual meeting book of abstracts.* New Orleans, LA: IFT.
Ruiz-Capillas, C., & Moral, A. (2005). Sensory and biochemical aspects of quality of whole big-eye tuna (*Thunnus obesus*) during bulk storage in controlled atmospheres. *Food Chemistry, 89*, 347–354.

Salminen, A., Latva-Kala, K., Randel, K., Hurme, E., Linko, P., & Ahvenainen, R. (1996). The effect of ethanol and oxygen absorption on the shelf-life of packaged slice rye bread. *Packaging Technology and Science, 9*, 29–42.

Sanchez-Escalante, A., Torrescano, G., Djenane, D., Beltran, J. A., & Roncales, P. (2003). Combined effect of modified atmosphere packaging and addition of lycopene rich tomato pulp, oregano and ascorbic acid and their mixtures on the stability of beef patties. *Food Science and Technology International, 9*, 77–84.

Sebti, I., Coma, V., Deschamps, A., & Pichavant, H. F. (2001). Antimicrobial edible packaging based on cellulosic ethers, fatty acids and nisin incorporation to inhibit *Listeria monocytogenes* and *Staphylococcus aureus*. *Journal of Food Protection, 64*, 470–475.

Simpson, B. K., Gagne, N., Ashie, I. N. A., & Noroozi, E. (1997). Utilization of chitosan for preservation of raw shrimp. *Food Biotechnology, 11*, 25–44.

Siro, I., Fenyvesi, B., Szente, L., Meulenaer, B., Devlieghere, F., & Orgova, J. (2006). Release of alpha-tocopherol from antioxidative low-density polyethylene film into fatty food stimulant: Influence of complexation in beta-cyclodextrin. *Food Additives and Contaminants, 23*, 845–853.

Smith, J. P., Hoshino, J., & Abe, Y. (1995). Interactive packaging involving sachet technology. In M. L. Rooney (Ed.), *Active food packaging* (pp. 143–173). London: Champman and Hall [chapter 6].

Smith, J. P., Ooraikul, B., Koersen, W. J., Jacksont, E. D., & Lawrence, R. A. (1986). Novel approach to oxygen control in modified atmosphere packaging of bakery products. *Food Microbiology, 3*, 315–320.

Smith, J. P., Ramaswamy, H. S., & Simpson, B. K. (1990). Developments in food packaging technology. Part II-storage aspects. *Trends in Food Science and Technology, 1*, 111–118.

Smolander, M. (2003). The use of freshness indicators in packaging. In R. Ahvenainen (Ed.), *Novel food packaging techniques* (pp. 127–143). Cambridge: Wood head Publishing Ltd.

Smolander, M., Hurme, E., Ahvenainen, R., & Siika-Aho, M. (1998). Indicators for modified atmosphere packages. In *24th IAPRI symposium*, New York, USA.

Soares, N. F. F., & Hotchkiss, J. H. (1998a). Bitterness reduction in grapefruit juice through active packaging. *Packaging Technology and Science, 11*, 9–18.

Soares, N. F. F., & Hotchkiss, J. H. (1998b). Naringinase immobilization in packaging films for reducing naringin concentration in grapefruit juice. *Journal of Food Science, 63*, 61–65.

Strathmann, S., Pastorelli, S., & Simoneau, C. (2005). Investigation of the interaction of active packaging material with food aroma compounds. *Sensors and Actuators B, 106*, 83–87.

Suhr, K. I., & Nielsen, P. V. (2005). Inhibition of fungal growth on wheat and rye bread by modified atmosphere packaging and active packaging using volatile mustard essential oil. *Journal of Food Science, 70*(1), M37–M44.

Suppakul, P., Miltz, J., Sonneveld, K., & Bigger, S. W. (2003). Active packaging technologies with an emphasis on antimicrobial packaging and its applications. *Journal of Food Science, 68*(2), 408–420.

Terry, L. A., Ilkenhans, T., Poulston, S., Rowsell, L., & Smith, A. W. J. (2007). Development of new palladium-promoted ethylene scavenger. *Postharvest Biology and Technology, 45*, 214–220.

Tewari, G., Jayas, D. S., Jeremiah, L. E., & Holley, R. A. (2002). Absorption kinetics of oxygen scavengers. *International Journal of Food Science and Technology, 37*(2), 209–217.

Thakur, M. S., & Ragavan, K. V. (2013). Biosensors in food processing. *Journal of Food Science and Technology, 50*(4), 625–641.

Thompson, A. K. (1998). *Controlled atmosphere storage of fruits and vegetables*. New York: CAB International.

Todorvoic, V., Neag, M., & Lazarevic, M. (2014). On the usage of RFID tags for tracking and monitoring shipped perishable goods. *Procedia Engineering, 69,* 1345–1349.

Van Aardt, M., Duncan, S., Marcy, J., Long, T., & O'Keefe, S. (2007). Release of antioxidants from poly(lactide-co-glycolide) film into dry milk products and food simulating liquids. *International Journal of Food Science and Technology, 42,* 1327–1337.

Van Willige, R. W. G., Linssen, J. P. H., Meinders, M. B. J., van der Stege, H. J., & Voragen, A. G. J. (2002). Influence of flavor absorption on oxygen permeation through LDPE. *Food Additives and Contaminants, 19*(3), 303–313.

Vanderroost, M., Ragaert, P., Devlieghere, F., & Meulenaer, B. D. (2014). Intelligent food packaging: The next generation. *Trends in Food Science & Technology, 39,* 47–62.

Vartiainen, J., Motion, R., Kulonen, K., Ratto, M., Skytta, E., & Advenainen, R. (2004). Chitosan-coated paper: Effects of nisin and different acids on the antimicrobial activity. *Journal of Applied Polymer Science, 94,* 986–993.

Vermeiren, L., Devlieghere, F., Beest, M., de Kruijf, N., & Debevere, J. (1999). Developments in the active packaging of foods: Review. *Trends in Food Science and Technology, 10*(3), 77–86.

Vu, C. H. T., & Won, K. (2013). Novel water-resistant UV-activated oxygen indicator for intelligent packaging. *Food Chemistry, 140*(1–2), 52–56.

Wallach, D. F. H., & Novikov, A. (1998). *Methods and devices for detecting spoilage in food products.* WIPO Patent WO/1998/020337.

Wang, J. (2006). *Analytical electrochemistry.* New York: John Wiley & Sons Ltd [chapter 6].

Wanihsuksombat, C., Hongtrakul, V., & Suppakul, P. (2010). Development and characterization of a prototype of a lactic acid based time temperature indicators for monitoring food product quality. *Journal of Food Engineering, 100,* 427–434.

Want, R. (2004a). Enabling ubiquitous sensing with RFID. *Computer, 37*(4), 84–86.

Want, R. (2004b). *The magic of RFID.* Vol. 2. Association of Computing Machinery, ACM Queue. Available from *http://www.acmqueue.com.*

Wessling, C., Nielsen, T., & Giacin, J. (2000). Antioxidant ability of BHT and α-tocopherol impregnated LDPE film in packaging of oatmeal. *Journal of the Science of Food and Agriculture, 81,* 194–201.

Wessling, C., Nielsen, T., Leufven, A., & Jagerstad, M. (1999). Retention of a-tocopherol in low-density polyethylene (LDPE) and polypropylene (PP) in contact with foodstuffs and food-simulating liquids. *Journal of the Science of Food and Agriculture, 79,* 1635–1641.

Wolfbeis, O. S., & List, H. (1995). *Method for quality control of packaged organic substances and packaging material for use with this method. US patent 5407829.*

Wu, D., Hou, S., Chen, J., Sun, Y., Ye, X., Liu, D., et al. (2015). Development and characterization of an enzymatic time-temperature indicator (TTI) based on *Aspergillus niger* lipase. *LWT-Food Science and Technology, 60,* 1100–1104.

Wu, D., Wang, Y., Chen, J., Ye, X., Wu, Q., Liu, D., et al. (2013). Preliminary study on time temperature indicator (TTI) system based on urease. *Food Control, 34,* 230–234.

Yam, K. L., Takhistov, P. T., & Miltz, J. (2005). Intelligent packaging: Concepts and applications. *Journal of Food Science, 70*(1), R1–R10.

Zabala, S., Castan, J., & Martınez, C. (2015). Development of a time temperature indicator (TTI) label by rotary printing technologies. *Food Control, 50,* 57–64.

Zagory, D. (1995). Ethylene-removing packaging. In M. L. Rooney (Ed.), *Active food packaging* (pp. 38–54). London: Chapman and Hall [chapter 2].

Zhang, Y., & Lim, L. T. (2016). Inkjet-printed CO_2 colorimetric indicators. *Talanta, 161,* 105–113.

Further reading

Abad, E., Palacio, F., Nuin, M., Zárate, A. G., Juarros, A., Gómez, J. M., et al. (2009). RFID smart tag for traceability and cold chain monitoring of foods: Demonstration in an intercontinental fresh fish logistic chain. *Journal of Food Engineering, 93*, 394–399.

Akbar, A., & Anal, A. K. (2014). Zinc oxide nanoparticles loaded active packaging, a challenge study against *Salmonella typhimurium* and *Staphylococcus aureus* in ready-to-eat poultry meat. *Food Control, 38*, 88–95.

Bahadlr, E. B., & Sezginturk, M. K. (2015). Applications of commercial biosensors in clinical, food, environmental, and biothreat/biowarfare analyses. *Analytical Biochemistry, 478*, 107–112.

Baiano, A., Marchitelli, V., Tamagnone, P., & Del Nobile, M. A. (2004). Use of active packaging for increasing ascorbic acid retention in food beverages. *Journal of Food Science, 69*(9), E502–E508.

Baleizao, C., Nagl, S., Schaferling, M., Berberan-Santos, M. N., & Wolfbeis, O. S. (2008). Dual fluorescence sensor for trace oxygen and temperature with unmatched range and sensitivity. *Analytical Chemistry, 80*, 6449–6457.

Barbiroli, A., Bonomi, F., Capretti, G., Iametti, S., Manzoni, M., Piergiovanni, L., et al. (2012). Antimicrobial activity of lysozyme and lactoferrin incorporated in cellulose-based food packaging. *Food Control, 26*, 387–392.

Basch, C., Jagus, R., & Flores, S. (2013). Physical and antimicrobial properties of tapioca starch-HPMC edible films incorporated with nisin and/or potassium sorbate. *Food and Bioprocess Technology, 6*(9), 2419–2428.

Borchert, N. B., Kerry, J. P., & Papkovsky, D. B. (2013). A CO_2 sensor based on Pt-porphyrin dye and FRET scheme for food packaging applications. *Sensors and Actuators B: Chemical, 176*, 157–165.

Brizio, A. P. D. R., & Prentice, C. (2014). Use of smart photochromic indicator for dynamic monitoring of the shelf life of chilled chicken based products. *Meat Science, 96*, 1219–1226.

Brody, A. L., Bugusu, B., Han, J. H., Sand, C. K., & McHugh, T. H. (2008). Innovative food packaging solutions. *Journal of Food Science, 73*, 8.

Brody, A. L., Strupinsky, E. R., & Kline, L. R. (2006). *Active packaging for food application*: (pp. 26–45). Boca Raton, FL: CRC Press [chapter 2].

Bültzingslöwen, C. V., McEvoy, A. K., McDonagh, C., MacCraith, B. D., Klimant, I., & Krause, C. (2002). Solegel based optical carbon dioxide sensor employing dual luminophore referencing for application in food packaging technology. *Analyst, 127*, 1478–1483.

Byrne, L., Lau, K. T., & Diamond, D. (2002). Monitoring of headspace total volatile basic nitrogen from selected fish species using reflectance spectroscopic measurements of pH sensitive films. *The Analyst, 127*, 1338–1341.

Byun, Y., Darby, D., Cooksey, K., Dawson, P., & Whiteside, S. (2011). Development of oxygen scavenging system containing a natural free radical scavenger and a transition metal. *Food Chemistry, 124*, 615–619.

Cerisuelo, J. P., Bermúdez, J. M., Aucejo, S., Catalá, R., Gavara, R., & Hernãndez-Muñoz, P. (2013). Describing and modeling the release of an antimicrobial agent from an active PP/EVOH/PP package for salmon. *Journal of Food Engineering, 116*(2), 352–361.

Charles, F., Sanchez, J., & Gontard, N. (2003). Active modified atmosphere packaging of fresh fruits and vegetables: Modeling with tomatoes and oxygen absorber. *Journal of Food Science, 68*, 1736–1742.

Chu, C. S., & Lo, Y. L. (2008). Fiber-optic carbon dioxide sensor based on fluorinated xerogels doped with HPTS. *Sensors and Actuators B, 129*, 120–125.

Clark, L. C., Jr., & Lyons, C. (1962). Electrode system for continuous monitoring in carbiovascular surgery. *Annals of the New York Academy of Sciences, 102*, 29–45.

Concha-Meyer, A., Schobitz, R., Brito, C., & Fuentes, R. (2011). Lactic acid bacteria in an alginate film inhibit *Listeria monocytogenes* growth on smoked salmon. *Food Control, 22*(3/4), 485–489.

Crowley, K. (2005). A gas-phase colorimetric sensor for the detection of amine spoilage products in packaged fish. In *Proceedings of proceedings of IEEE sensors 2005 conference* (pp. 754–757). Irvine, CA: The Printing House, Inc.

Dave, D., & Ghaly, A. E. (2011). Meat spoilage mechanisms and preservation techniques: A critical review. *American Journal of Agricultural and Biological Sciences, 6*(4), 486–510.

Demarco, D. R., & Lim, D. V. (2002). Detection of *Escherichia coli* O157:H7 in 10 and 25-gram ground beef samples with an evanescent-wave biosensor with silica and polystyrene waveguides. *Journal of Food Protection, 65*, 596–602.

Devi, R., Yadav, S., Nehra, R., Yadav, S., & Pundir, C. S. (2013). Electrochemical biosensor based on gold coated iron nanoparticles/chitosan composite bound xanthine oxidase for detection of xanthine in fish meat. *Journal of Food Engineering, 115*(2), 207–214.

Dias, M. V., Soares, N. F. F., Borges, S. V., de Sousa, M. M., Nunes, C. A., de Oliveira, I. R. N., et al. (2013). Use of allyl isothiocyanate and carbon nanotubes in an antimicrobial film to package shredded, cooked chicken meat. *Food Chemistry, 14* (3), 3160–3166.

Emiroglu, Z. K., Yemis, G. P., Coskun, B. K., & Candogan, K. (2010). Antimicrobial activity of soy edible films incorporated with thyme and oregano essential oils on fresh ground beef patties. *Meat Science, 86*(2), 283–288.

Ercolini, D., Ferrocino, I., La Storia, A., Mauriello, G., Gigli, S., Masi, P., et al. (2010). Development of spoilage microbiota in beef stored in nisin activated packaging. *Food Microbiology, 27*(1), 137–143.

Erdem, E., Zeng, H., Shi, J., & Wells, D. L. (2007). Application of RFID and sensing technology for improving frozen food quality management. In T. Blecker, & G. Q. Huang (Eds.), *RFID in operations and supply chain management: Research and applications* (pp. 221–245). Erich Schmidt Verlag.

Farouk, M. M., Boles, J. A., Mikkelsen, V. L., McBeth, E. E., De Manser, G. C., & Swan, J. E. (1997). Effect of adjuncts on the color stability of bologna and fresh beef sausages. *Journal of Muscle Foods, 8*, 383–394.

Fernández, A., Picouet, P., & Lloret, E. (2010). Reduction of the spoilage-related microflora in absorbent pads by silver nanotechnology during modified atmosphere packaging of beef meat. *Journal of Food Protection, 73*(12), 2263–2269.

Ferrocino, I., La Storia, A., Torrieri, E., Musso, S. S., Mauriello, G., Villani, F., et al. (2013). Antimicrobial packaging to retard the growth of spoilage bacteria and to reduce the release of volatile metabolites in meat stored under vacuum at 1°C. *Journal of Food Protection, 76*(1), 52–58.

Fitzgerald, M., Papkovsky, D. B., Smiddy, M., Kerry, J. P., O'Sullivan, C. K., & Buckley, D. J. (2001). Non-destructive monitoring of oxygen profiles in packaged foods using phase fluorimetric oxygen sensor. *Journal of Food Science, 66*, 105–110.

Galotto, M. J., Valenzuela, X., Rodriguez, F., Bruna, J., & Guarda, A. (2012). Evaluation of the effectiveness of a new antimicrobial active packaging for fresh Atlantic salmon (*Salmo salar* L.) shelf life. *Packaging Technology and Science, 25*(6), 363–372.

Geng, T., Uknalis, J., Tu, S. I., & Bhunia, A. K. (2006). Fiber optic biosensor employing Alexa- Fluor conjugated antibody for detection of *Escherichia coli* O157:H7 from ground beef in four hours. *Sensors, 6*, 796–807.

Gomez, A. L., Cartagena, F. C., Muro, J. U., Boluda-Aguilar, M., Hernandez, M. E. H., Serrano, M. A. L., et al. (2015). Radiofrequency identification and surface acoustic wave technologies for developing the food intelligent packaging concept. *Food Engineering Reviews*, *7*, 11–32.

Hereu, A., Bover-Cid, S., Garriga, M., & Aymerich, T. (2012). High hydrostatic pressure and biopreservation of dry-cured ham to meet the food safety objectives for *Listeria monocytogenes*. *International Journal of Food Microbiology*, *154* (3), 107–112.

Higueras, L., López-Carballo, G., Hernández-Muñoz, P., Gavara, R., & Rollini, M. (2013). Development of a novel antimicrobial film based on chitosan with LAE (ethyl-Nα- dodecanoyl-L-arginate) and its application to fresh chicken. *International Journal of Food Microbiology*, *165*(3), 339–345.

Hong, H. S., Kim, J. W., Jung, S. J., & Park, C. O. (2005). Thick film planar CO_2 sensors based on Na b-alumina solid electrolyte. *Journal of Electroceramics*, *15*(2), 151–157.

Hsu, Y. C., Chen, A. P., & Wang, C. H. (2008). A RFID-enabled trace ability system for the supply chain of live fish. In *Proceedings of the IEEE international conference on automation and logistics* (pp. 81–86).

Huber, C., Nguyen, T. A., Krause, C., Humele, H., & Stangelmayer, A. (2006). Oxygen ingress measurement into PET bottles using optical-chemical sensor technology. *Monatsschrift fur Brauwissenschaft*, *59*, 5–15.

Jawaheer, S., White, S. F., Rughooputh, S. D. D. V., & Cullen, D. C. (2003). Development of a common biosensor format for an enzyme based biosensor array to monitor fruit quality. *Biosensors & Bioelectronics*, *18*, 1429–1437.

Jouki, M., Yazdi, F. T., Mortazavi, S. A., Koocheki, A., & Khazaei, N. (2014). Effect of quince seed mucilage edible films incorporated with oregano or thyme essential oil on shelf life extension of refrigerated rainbow trout fillets. *International Journal of Food Microbiology*, *174*, 88–97.

Kaniou, I., Samouris, G., Mouratidou, T., Eleftheriadou, A., & Zantopolous, N. (2001). Determination of biogenic amines in fresh and unpacked and vacuum-packed beef during storage at 4°C. *Food Chemistry*, *74*, 515–519.

Kim, M. J., Jung, S. W., Park, H. R., & Lee, S. J. (2012). Selection of an optimum pH-indicator for developing lactic acid bacteria-based time-temperature integrators (TTI). *Journal of Food Engineering*, *113*, 471–478.

Kim, M. J., Shin, H. W., & Lee, S. J. (2016). A novel self-powered time-temperature integrator (TTI) using modified biofuel cell for food quality monitoring. *Food Control*, *70*, 167–173.

Lee, D. S. (2005). Packaging containing natural antimicrobial or antioxidative agents. In J. H. Han (Ed.), *Innovations in food packaging* (pp. 92–93). Amsterdam: Elsevier Academic Press.

Mangaraj, S., & Goswami, T. K. (2011). Measurement and modelling of respiration rates of guava (cv. Baruipur) for modified atmosphere packaging. *International Journal of Food Properties*, *14*(3), 609–628.

Mangaraj, S., Goswami, T. K., & Mahajan, P. V. (2009). Application of plastic films in modified atmosphere packaging of fruits and vegetables—A review. *Food Engineering Reviews*, *1*, 133–158.

Marcos, B., Aymerich, T., Garriga, M., & Arnau, J. (2013). Active packaging containing nisin and high pressure processing as post-processing listericidal treatments for convenience fermented sausages. *Food Control*, *30*(1), 325–330.

Mexis, S. F., Badeka, A. V., Riganakos, K. A., & Kontominas, M. G. (2010). Effect of active and modified atmosphere packaging on quality retention of dark chocolate with hazelnuts. *Innovative Food Science & Emerging Technologies*, *11*, 177–186.

Mexis, S. F., Chouliara, E., & Kontominas, M. G. (2009). Combined effect of an O_2 absorber and oregano essential oil on shelf-life extension of Greek cod roe paste (tarama salad) stored at 4°C. *Innovative Food Science and Emerging Technologies, 10,* 572–579.

Mizutani, F., Sato, Y., Hirata, Y., & Yabuki, S. (1998). High-throughput flow-injection analysis of glucose and glutamate in food and biological samples by using enzyme/polyion complex-bilayer membrane-based electrodes as the detectors. *Biosensors & Bioelectronics, 13,* 809–815.

Mohan, C. O., Ravishankar, C. N., Srinivasa Gopal, T. K., Ashok Kumar, K., & Lalitha, K. V. (2009). Biogenic amines formation in seer fish (*Scomberomorus commerson*) steaks packed with O_2 scavenger during chilled storage. *Food Research International, 42,* 411–416.

Nopwinyuwong, A., Trevanich, S., & Suppakul, P. (2010). Development of a novel colorimetric indicator label for monitoring freshness of intermediate-moisture dessert spoilage. *Talanta, 81*(3), 1126–1132.

Okuma, H., Okazaki, W., Usami, R., & Horikoshi, K. (2000). Development of the enzyme reactor system with an amperometric detection and application to estimation of the incipient stage of spoilage of chicken. *Analytica Chimica Acta, 411,* 37–43.

Pospiskova, K., Safarik, I., Sebela, M., & Kuncova, G. (2013). Magnetic particles–based biosensor for biogenic amines using an optical oxygen sensor as a transducer. *Microchimica Acta, 180,* 311–318.

Prakash, O., Talat, M., Hasan, S. H., & Pandey, R. K. (2008). Enzymatic detection of mercuric ions in ground-water from vegetable wastes by immobilizing pumpkin (*Cucumis melo*) urease in calcium alginate beads. *Bioresource Technology, 99* (10), 4524–4528.

Rahman, S. F. (1999). Post harvest handling of foods of animal origin. In S. F. Rahman (Ed.), *Handbook of food preservation* (pp. 47–54). New York: Marcel Dekker. ISBN: 0-8247-0209-3.

Randell, K., Ahvenainen, R., Latva-Kala, K., Hurme, E., Mattila-Sandholm, T., & Hyvönen, L. (1995). Modified atmosphere-packed marinated chicken breast and rainbow trout quality as affected by package leakage. *Journal of Food Science, 60,* 667–672.

Rodriguez, A., Nerin, C., & Batlle, R. (2008). New cinnamon-based active paper packaging against *Rhizopusstolonifer* food spoilage. *Journal of Agricultural and Food Chemistry, 56,* 6364–6369.

Rokka, M., Eerola, S., Smolander, M., Alakomi, H. L., & Ahvenainen, R. (2004). Monitoring of the quality of modified atmosphere packaged broiler chicken cuts in different temperature conditions B. Biogenic amines as quality-indicating metabolites. *Food Control, 15,* 601–607.

Shu, H. C., Kanson, H. H., & Mattiason, B. (1993). D-lactic acid in pork as a freshness indicator monitored by immobilized D-lactate dehydrogenase using sequential injection analysis. *Analytica Chimica Acta, 283,* 727–737.

Smith, J. P. (1988). Shelf life extension of a fruit filled bakery product using ethanol vapor. In *CAP '88 international conference on controlled/modified atmosphere/vacuum packaging, 1988 conference proceedings.* Princeton, NJ: Schotland Business Research, Inc.

Smolander, M., Hurme, E., Latva-Kala, K., Luoma, T., Alakomi, H. L., & Ahvenainen, R. (2002). Myoglobin-based indicators for the evaluation of freshness of unmarinated broiler cuts. *Innovative Food Science and Emerging Technologies, 3,* 279–288.

Soares, N. F. F., Rutishauser, D. M., Melo, N., Cruz, R. S., & Andrade, N. J. (2002). Inhibitation of microbial growth in bread through active packaging. *Packaging Technology and Science, 15,* 129–132.

Sua, L., Jia, W., Hou, C., & Lei, Y. (2011). Microbial biosensors: A review. *Biosensors and Bioelectronics*, *26*, 1788–1799.

Taoukis, P. S., & Labuza, T. P. (2003). Time-temperature indicators (TTIs). In R. Ahvenainen (Ed.), *Novel food packaging techniques* (pp. 103–126). Cambridge: Woodhead Publishing Limited.

Tsironi, T., Giannoglou, M., Platakou, E., & Taoukis, P. (2016). Evaluation of time temperature integrators for shelf-life monitoring of frozen seafood under real cold chain conditions. *Food Packaging and Shelf Life*, *10*, 46–53.

Valero, D., Valverde, J. M., Martínez-Romero, D., Guillén, F., Castillo, S., & Serrano, M. (2006). The combination of modified atmosphere packaging with eugenol or thymol to maintain quality, safety and functional properties of table grapes. *Postharvest Biology and Technology*, *41*, 317–327.

Verma, N., Kumar, S., & Kaur, H. (2010). Fiber optic biosensor for the detection of Cd in milk. *Journal of Biosensors & Bioelectronics*, *1*, 102. https://doi.org/10.4172/2155-6210.1000102.

Wang, L. F., & Rhim, J. W. (2016). Grapefruit seed extract incorporated antimicrobial LDPE and PLA films: Effect of type of polymer matrix. *LWT- Food Science and Technology*, *74*, 338–345.

13

Microwave and ohmic heating of fresh cut fruits and vegetable products

Aamir Hussain Dar*, Rafiya Shams†, Qurat ul Eain Hyder Rizvi‡, Ishrat Majid§

**Department of Food Technology, Islamic University of Science and Technology, Awantipora, India. †Sher-e-Kashmir University of Agricultural Sciences and Technology, Jammu, India. ‡Eternal University, Baru Sahib, India. §Lovely Professional University, Phagwara, India*

1 Introduction

Emerging techniques have developed as an alternative to traditional thermal processing, which was mostly done for the preservation and processing of food materials. Conventional heating methods are commonly used for heating of food products. Aseptic processing can result in various quality losses to the food components due to slow convection or conduction of heat (Zell, Lyng, Cronin, & Morgan, 2009). In this context, several emerging techniques, namely inductive heating, microwave heating, ohmic heating, and other innovative techniques, have been developed as alternatives to traditional heat treatment. As suggested by Drouzas and Schubert (1996), the thermal destruction of essential nutrients and flavor components is the major drawback of conventional heating operations. Thus, methods of preservation should be introduced that help to retain the health benefits obtained from fruits and vegetables. Vegetables and fruits are essential sources of dietary nutrients, like minerals, fiber, and vitamins. They also serve as important sources of natural antioxidants, which are essential for human health (Antunes, Dandlen, Cavaco, & Miguel, 2010). Hence, care needs to be taken to conserve antioxidant compounds, which may be lost due to conventional heating.

The important antioxidants present in vegetables include carotenoids, phenolic compounds, mainly flavonoids, and vitamins E and C (Podsędek, 2007). The domestic cooking as well

Fresh-cut Fruits and Vegetables. https://doi.org/10.1016/B978-0-12-816184-5.00013-6

as industrial processing of certain vegetables is likely to change the functional properties (e.g., composition, antioxidant activity, antioxidant content, and bioavailability of antioxidants). Generally, the antioxidant compounds present in thermally treated vegetables are not very active as compared to their corresponding raw material, due to detrimental effects of heat processing (Nayak, Liu, & Tang, 2015). Antioxidant content may be affected by cooking due to the creation or destruction of redox-active metabolites and due to antioxidant release (Wachtel-Galor, Wong, & Benzie, 2008). The thermal processing effect on antioxidant capacity—loss of nutrients or availability—must be examined in order to choose suitable food processing techniques. Drying, as one of the oldest preservation methods, covers a major field in vegetable and fruit processing (Ahmed et al., 2013). This method affects the physical morphology of plant tissues and causes disruption of cellular structure resulting in certain macroscopic changes, like case hardening, textural changes, and shrinking, among others, there by altering the mass transfer process also (Martín, Albors, Martínez-Navarrete, Chiralt, & Fito, 2000).

To retain nutrient content, as well as to make sure that processed vegetables and fruits retain their fresh qualities, certain advanced methods must be used like microwave heating as well as ohmic heating. In these methods, the pretreatment before microwave fruit drying possess numerous advantages both from the technological and consumer point of view, like in the enhanced reconstitution of the dehydrated products (Contreras, Martín-Esparza, & Martínez-Navarrete, 2012). Osmotic pretreatment or vacuum impregnation improves microwaves while generating internal heat in food materials. (Contreras, Martín, Martínez-Navarrete, & Chiralt, 2005; Orsat, Changrue, & Raghavan, 2006). To achieve the requisite temperatures in not more than a minute at relatively low strength of electric field ($E < 100\,V/cm$); vegetables as well as fruits possess adequate conductivity (Sarang, Sastry, & Knipe, 2008; Wang & Sastry, 1997a, 1997b). As per recent research, plant materials are mostly used and mostly suitable for microwave and ohmic heat treatment (Leadley, 2008).

2 Microwave heating

High-frequency radiation microwave ranges between the wavelengths of 1 m–0.1 mm in the air. Such radiations are represented in millions of cycles per second or megahertz (MHz), where frequency ranges from 300 MHz to 300 GHz. One part of the electromagnetic radiation spectrum represents microwave radiation. Further entities of the spectrum consist of gamma rays

(10^{-11} to 10^{-16} m), X-rays (10^{-8} to 10^{-11} m), UV (10^{-7} m), visible (10^{-6} m), and long and short radio waves (7 $\sim$ 10″4 to 10$\sim$8 m). Microwaves are classified as nonionizing radiations as they have longer wavelengths than ionizing radiations and do not have enough energy to ionize other components. Nonionizing waves are characterized by frequencies ranging between 10^{1} to 10^{10} c/s and wavelengths between 10^{-7} to 10^{7} m. Industrial microwaves operate at both frequencies of 2450 and 915 MHz and domestic microwaves are generally operated at a frequency of 2450 MHz (Datta & Anantheswaran, 2000). Over recent decades, heating by microwaves has been used in the food processing field due to its numerous applications like pasteurization, thawing, sterilization, tempering, baking, drying of food products, etc. (Gupta & Wong, 2007; Metaxas & Meredith, 1983). Heating by microwave has various advantages in processing of food because of its tendency to achieve more uniform heating, high heating rates, safe handling, considerable drop in cooking time, ease of operation, and low maintenance (Salazar-Gonzalez, San Martin-Gonzalez, Lopez-Malo, & Sosa-Morales, 2012; Zhang, Tang, Mujumdar, & Wang, 2006). In addition, heating by microwave causes less alteration in nutritional behavior as well as flavor of food (Vadivambal & Jayas, 2010).

2.1 Mechanism

Microwave heating is produced by the capability of the product to absorb microwave energy and transform it to heat. Heating by microwave generally occurs by two principles, namely dipole rotation and ionic polarization. The main cause of dielectric heating is the presence of water or moisture, due to its polar nature. Goldblith described the principle of microwave heating caused by polarized dipolar molecules to arrange itself with the frequently varying alternating electrical field. The molecular oscillation in the microwave field occurs around their axes in response to the reversal of electric field in 915 or 2450 million times/s. Such oscillations produce intermolecular friction resulting in the volumetric heating of the material. These polar molecules in water or food exhibiting intermolecular friction can be defined as "lossy" (obtained from the dielectric loss tangent, defined as the angular change of the polar material due to the change in direction of electric field). The "lossiness" changes due to nature of the material, radiation frequency, and temperature. Due to increase in lossiness of a component, microwave energy absorption is also increased and, hence, the heat production. The most known lossy material is water, which gets heated quickly in a microwave field.

Microwave heating also occurs in the presence of a high-frequency oscillating electric field responsible for generation of heat, because of oscillatory ions migration in the food (Datta & Davidson, 2000). Many factors are responsible for heat distribution and microwave heating, among which the depth of penetration and dielectric properties are most important. In microwave processing, heat is produced all over the product uniformly, causing rapid heat rates, as compared to traditional thermal processes in which heat is mostly moved from the surface to the interior of the material (Gowen, Abu-Ghannam, Frias, & Oliveira, 2006).

2.2 Block diagram of microwave system

Only part of the electrical energy is converted into microwave energy by microwave system because heat energy is directly passed to the molecules inside food due to which microwave ovens are efficient and quick. When the door is opened, microwave generation stops due to two safety interlock switches present in the microwave.

Just like light bounces back and forth off a mirror, the microwaves bounce in the same way on reflective metal walls of the food compartment. When the microwaves approach the food itself, they do not bounce off, but penetrate inside the food, as do radio waves. The vibration inside the food material occurs more quickly as the microwaves travel through it. The food becomes hotter as the vibration inside the food increases. Thus, the microwave energy is passed on to the food molecules, which in turn heats the food up. Heating rate depends on moisture content, shape, and weight of food. This system consists of an outlet to remove internal vapor produced inside the microwave system, which helps to shrink and dry fruits.

2.3 Magnetron

High-frequency microwave oscillations are produced by an electronic tube known as a magnetron. Such oscillations help in generation of the microwaves. No external devices other than the power supply are required in self-oscillating magnetron. Due to the absence of a grid, the magnetron is referred to as a diode. At the center of the magnetron tube, the cathode and filament are placed, which in turn are supported by the filament leads. The magnetron anode is formed from a solid copper cylindrical block. The filament and cathode structure are kept fixed in position with the help of large and rigid filament leads. The cathode is made up of a high-emission material and is indirectly heated. The resonant cavities consist of 8–20 cylindrical holes around its circumference. The internal structure is divided into as many segments as the cavities that are present, as the narrow slot runs from each cavity into the central portion of the tube. Each cavity behaves like a parallel resonant circuit. Microwave heating is a complicated process depending on various material parameters such as dielectric properties, density, sample composition, volume, dimension, shape, and specific heat capacity.

2.4 Dielectric properties of foods

The dielectric properties help us to understand the capability of a product to convert microwave energy into heat. The dielectric properties of a material measure the potential for charge movement within the material in response to an external electric field. Dielectric properties predict the interaction of nonmagnetic components like foods possessing electromagnetic radiation. The actual part of dielectric property, referred to as dielectric constant (permittivity, ε'), depicts the quantity of energy stored by a particular product in a particular electric field relative to empty space, and the imaginary part of dielectric property, referred as loss factor (ε'') or the dielectric loss indicates how much energy is converted into heat.

$$\varepsilon^* = \varepsilon' - j\varepsilon'' \quad (1)$$

where $j = \sqrt{-1}$.

Another factor showing the dielectric properties is loss tangent:

$$(\tan \delta = \varepsilon''/\varepsilon') \quad (2)$$

This is the ratio of dielectric loss to dielectric constant and is referred to as the capability of a medium to change electromagnetic energy into heat at specific frequency and temperature (Buffler, 1993). The dielectric properties are usually affected by density, electromagnetic waves frequency, salt content, water content, solute percentage, operating temperature, and the state (gas, solid, or liquid) of the material under observation (Yaghmaee & Durance, 2001). Dielectric properties are essential tools in depicting the behavior of a product exposed to microwaves.

Depending on the absorption of microwave, materials are divided into three parts:

(i) transparent or low dielectric loss materials, in which a small amount of microwaves are passed through the material;
(ii) conductors or opaque, where material reflect the microwaves; and
(iii) high dielectric loss materials or absorbers, where material strongly absorbs the microwave.

The power penetration depth (Dp) is well-defined as the distance at which the power density drops to a value of 1/e from its value at the surface and is given by the following formula (Metaxas & Meredith, 1983):

$$Dp = \frac{c}{\sqrt{2}\pi f \left[\kappa' \left\{ \sqrt{1 + \kappa''/\kappa'^2} - 1 \right\} \right]^{1/2}} \quad (3)$$

where ω is the angular frequency, c is the velocity of light given as ($c = \varepsilon_0 \mu_0$), and 0 μ is the free space permeability ($\mu_0 = 4\pi \times 10^{-7}$ H/m).

Eq. (3) is mostly used for nonmagnetic food materials ($\mu_r = 1$). The power which changes with the square of electric field is expressed as follows:

$$q = \frac{1}{2} \omega \varepsilon_0 \kappa'' |E|^2 \quad (4)$$

where E is the intensity of electric field.

In addition to the dielectric properties and depth of penetration, the other factors that influence microwave processing of foods include frequency of microwave, design of microwave oven (size of oven and geometry), position of material inside the microwave oven, composition, load, density, size and shape of food product, and moisture content, (Icier & Baysal, 2004a). Generally, the quantity of moisture content of a food product has a significant role in defining the dielectric properties of the product, as microwaves are better absorbed by water.

2.5 Factors affecting dielectric properties of food materials

Food material does not absorb microwaves because of its atomic or electronic polarization, although they may be absorbed due to its ionic or dipole polarization. At frequencies below 1 GHz, ionic losses are important whereas dipole polarization is predominant at frequencies above 1 GHz (Ryynanen, 1995). With increase in frequency, the dielectric constant of pure water decreases. In addition, for moist foods, the dielectric loss is raised, with increase in frequency. The dielectric properties of material are influenced by various factors, like electromagnetic heating conditions (frequency, temperature), material nature (composition like salt content, moisture content, and other constituents, as well as physical structure), and the maturity or age of the food product associated to storage period (Sosa-Morales, Valerio-Junco, Lopez-Malo, & Garcia, 2010).

2.6 Influence of composition

Chemical composition is an important factor in the determination of dielectric properties of food. The influence of salt (or ash) and water content largely depends on the method where they are restricted or bound in their movement by the other constituents of food. The organic compounds of food material are dielectrically inert (e053 and e00501) and, equated with water or aqueous ionic fluids, are considered energy transparent (Mudgett, 1985). With salt addition, the dielectric loss is increased at a certain frequency. In electromagnetic field presence, salt behaves act as conductor, therefore the dielectric loss factor is increased and permittivity is decreased, as studied by Icier and Baysal (2004a).

Depending on the state of water (free or bound), the dielectric property changes. The polar molecules of water in bound state orient less freely than that of free water, due to the presence of electric field. As the temperature of the melting zone rises, the dielectric properties of frozen materials with high water content also increases. As thawed and frozen foods are heated, it rapidly heats up the warm part and at the same time some ice still remains in the food component and thus results in nonuniformity of materials (Ryynanen, 1995). The dielectric properties of the foods are also changed with density, structure, and particle size of the sample. Apparent density of a particulate or granular material is also responsible for the change in dielectric properties (Icier & Baysal, 2004a). The dielectric characteristics of food components like vegetables and fruits depend largely on their moisture content. Microwave energy is largely absorbed by water present in the food and thus, the greater the amount of water, the more efficient the heating. Water is an excellent example of a polar dielectric in its pure form. As the dielectric constant of free water is greater (78 at ambient temperature and 245 GHz), the dielectric constant is mainly affected by the amount of free water present in the food material. Although dielectric loss and dielectric constant values are usually low for oils and fat, the dielectric loss increases with rise in temperature (Icier & Baysal, 2004a).

2.7 Influence of temperature

The vital process responsible for the frequency dependence of the dielectric properties is the polarization rising from the orientation with the forced electric field of materials having permanent dipole moments. The dielectric constant depending on temperature is somewhat difficult, and with change in temperature, it is increased or decreased depending on the material. The relaxation time is decreased with increase in temperature, and at higher frequencies, the loss factor peak is shifted. Thus, with rise in temperature, the dielectric constant is also increased in a dispersion region, while as the loss factor decreases or increases, depending on frequency of operation, is lower or higher than the relaxation frequency. Above and below the dispersion region, as the temperature rises, the dielectric constant is decreased.

Loss factors and dielectric constants of fresh vegetables and fruits have been discovered by a number of researchers (Nelson, Forbus Jr, & Lawrence, 1994; Tran, Stuchly, & Kraszewski, 1984). The dielectric constant reveals the probable monotonic decline in value in the frequency range of 2–20 GHz. In all vegetables and fruits, the loss factor is decreased with increase in frequency from 2 GHz, approaches at a region between 1 and 3 GHz, and then rises as frequency reaches 20 GHz. At lower frequencies, such behavior is dominated by bound water relaxation, by ionic conductivity, and by free water relaxation close to the highest frequency range. Some common examples of the frequency dependence detected for different vegetables and fruits includes dielectric constant of apple, avocado, banana, mango, orange, papaya carrot, cucumber, kiwifruit, lemon, lime, and onion as 57, 47, 64, 64, 73, 69, 59, 71, 70, 73, 72, and 61, respectively, at a frequency of 915 MHz; 54, 45, 60, 61, 69, 67, 56, 69, 66, 71, 70, and 64, respectively, at 2.45 MHz; exhibits the expected tissue density of 0.76, 0.99, 0.94, 0.96, 0.92, 0.96, 0.99, 0.88, 0.97, 0.97 g cm, respectively; and moisture content (on wet basis) of 88%, 71%, 78%, 86%, 87%, 88%, 87%, 97%, 87%, 91%, 90%, and 92%, respectively.

2.8 Measurement of dielectric properties

Various techniques are used to measure the dielectric properties such as coaxial probe, cavity resonant methods, free space, time domain spectroscopy, lumped circuit, parallel plate, and transmission line (Icier & Baysal, 2004b; Sosa-Morales et al., 2010). The major properties of these measurement methods have been investigated (Icier & Baysal, 2004b; Sosa-Morales et al., 2010). For example, the dielectric properties of semisolids and liquids in the frequency range of 200 MHz–20 GHz, even >100 GHz, are measured using open-ended coaxial probe. The benefits of this method are that:

(1) a sample preparation is not required;
(2) it does not cause destruction to certain materials; and
(3) it is easy to use.

Besides these benefits, this method has some limitations, such as:

(1) more surface requirement for solids and large samples;
(2) less accuracy; and
(3) less loss resolution (Icier & Baysal, 2004b; Sosa-Morales et al., 2010).

On the other hand, for solids and liquids, the transmission line technique is efficient, having frequency <100 MHz. It is sensitive but more precise than that of coaxial probe system, but has more limited accuracy than resonators, and preparation of sample is time consuming and complex.

Another commonly used method, the cavity resonator, is suggested for calculating the dielectric properties of solid products, having frequency 1 MHz–100 GHz, and its simplicity in sample formulation makes it suitable for a wide range of temperatures. The limitations of this method include complex analysis, and frequency data of broadband are not required. In addition to these methods, the most expensive method termed as time domain spectroscopy may also be used for accurate and rapid measurements of the dielectric properties of smaller sized and homogeneous foods within 10 MHz–10 GHz frequency range.

2.9 Effects of microwave heating on chemical composition of fruits and vegetables

In microwave heating, various alterations related to chemical constituents of food components are mostly associated to bioactive components, antioxidant activity, and antinutritional constituents mainly: hemagglutinin activity, saponins, tannins, phytic acid, and trypsin inhibitor. Many researchers have studied the maintenance of bioactive constituents and antioxidant activity of vegetables by microwave processing. Akdaş and Bakkalbaşı (2016) determined that microwave heating in absence of water is appropriate for cooking of kale for the maintenance of total carotenoids, total chlorophylls, and ascorbic acid at levels of 99.8%, 4.7%, and 89.4%, respectively. Tian et al. (2016) reported that losses of total chlorogenic acid (20.01%), total anthocyanin (14.01%), and phenolics (negligible) due to microwave heating in purple-fleshed potatoes in absence of water were less than other heating methods. Xu et al. (2014) determined that microwave heating does not cause much loss of ascorbic acid of red cabbage and causes slight loss in phenolic content (by adding water and sample in the ratio of 10:300), while significant losses are caused after boiling and stir-frying.

Microwave heat treatment is responsible for retaining more bioactive constituents of vegetables than those processed by other cooking techniques due to their shorter heat treatment time, and without soaking pretreatment (Tian et al., 2016; Xu et al., 2014). Therefore, it is necessary to use a small amount of water in microwave heat treatment for the retention of bioactive components and antioxidant activity. However, there is significant loss of nutrients when food material is treated with large amounts of water during microwave heat treatment. Dolinsky et al. (2015) observed that microwave cooking of certain vegetables using water may result in sufficient decrease in polyphenolic content (hydrolysable polyphenol and soluble polyphenol) of green beans (decrease of 22.9%), tomato (decrease of 21.9%), and kale (decrease of 23.4%), whereas steaming with small osmotic exchanges is efficient for retaining their polyphenolic contents to a large extent. De Lima et al. (2017) observed that cassava treated with steam causes retention of more antioxidant content (retention of 308.6%) and phenolic constituents (retention of 236.1%) than by microwave heat treatment (retention of 273.4% of antioxidant activity and 164.4% of phenolic).

Microwave cooking increases rate of retention of bioactive components of vegetables to wider range as water may be responsible for rupture and softening of the lignocellulosic structure, resulting in the release of soluble bioactive components from the food matrix, while losses may increase due to thermal liability and leaching. Various research is also been done on the alterations of antinutritional factors, such as phytic acid, haemagglutinin activity, tannins, trypsin inhibitor, and saponins due to microwave heat treatment. Microwave cooking causes a rise in vitro protein digestibility of foods while lowering the antinutritional factors efficiently.

2.10 Effects of microwave heating on sensory attributes of fruits and vegetables

The changes related to sensory attributes of vegetables and fruits during microwave heating is seen in textural and color properties. During thermal processing, changes in food color are influenced by enzymatic browning, nonenzymatic browning, depletion of pigments, and oxidation of ascorbic acid (Ling, Tang, Kong, Mitcham, & Wang, 2014). Munoz, Achaerandio, Yang, and Pujola (2017) observed that after microwave heat treatment, a sufficient reduction in the shear force for potato tubers occurs, as cohesive forces between cells is weakened by microwave heating. Many studies on the changes in color of vegetables with microwave heating have been described. Pellegrini et al. (2010) stated that microwave heating has proven to be one of the most suitable cooking techniques for color retention of both frozen brassica vegetables (cauliflower, broccoli, and Brussels sprouts) and fresh fruits as compared to steaming and boiling. The b values for three different cooking methods are 18.59 for microwave,17.63 for steaming and 15.44 for boiling (Akdaş & Bakkalbaşı, 2016; Armesto, Gomez-Limia, Carballo, & Martínez, 2016).

Furthermore, vegetables treated with microwave heating achieve the highest scores for color. Microwave cooking gives better color intensity, L, for fresh kale when treated at 20 min (40.26–34.55) and 30 min (40.26–33.26) studied by Armesto et al. (2016). Related results has also been shown by dos Reis et al. (2015), where the L value of fresh broccoli (from 46.1 to 39.9) and fresh cauliflower (from 56.7 to 49.4) has been significantly reduced due to microwave cooking. On the other hand, Zhong, Dolan, and Almenar (2015) observed that the L value of the broccoli (frozen) increases due to microwave heating (from ~21 to ~24) and Xu et al. (2014) found that microwave heating slightly increases the L value of red cabbage (26.57–28.99). Such variation in the results might be due to the longer cooking times used in microwave heating by Armesto et al. (2016).

2.11 Microwave pasteurization and sterilization

Sterilization and pasteurization are done to inactivate or destroy microbes to increase the storage life and food safety (Nott & Hall, 1999). The food product is maintained at a definite time and temperature to ensure that pathogenic microbes are successfully destroyed. Pasteurization is a phenomenon that helps to destroy the pathogenic microbes such as such as vegetative bacteria by thermal processing. Pasteurization causes the deactivation of enzymes responsible for cloud loss in various juices. Microwave pasteurization causes minimum effects on the flavor, nutritive value, or color of food products (Pereira & Vincente, 2010).

Microwave sterilization effects on the quality characteristics of vegetables and fruit mostly involves bioactive substances, color, texture, and activity of enzyme. Piasek et al. (2011) have shown the effect of microwave sterilization on the decrease of total anthocyanins in aronia ranging from 39.7% to 59.1%, which is less (66.1%–99.8%) than that of other heat treatment processing done at 100°C. On the other hand, Marszałek, Mitek, and Skąpska (2015) have shown that conventional heat treatment is more destructive as compared to that of microwave heat treatment for strawberry puree, as loss in total anthocyanin content (19.2%), total polyphenolic content (5.7%), and total vitamin C content (3.4%) in strawberry puree using microwave processing at 90°C (at atmospheric pressure) for 10 s has been obtained. This was less than that (60.2%, 14.0%, and 61.7%, respectively) of conventional heating for 15 min at 90°C. The reason behind this is that the microwave time (10 s for strawberry puree, 7 s for aronia) exposure was much lower than that of thermal treatment time (15 min for strawberry puree, 1–5 h for aronia) (Marszałek et al., 2015; Piasek et al., 2011).

Microwave sterilization treatment usually does not cause changes with the short exposure time for bioactive components of food material. However, Lu, Turley, Dong, and Wu (2011) observed that microwave heat treatment causes lesser loss of lycopene and ascorbic acid content of grape tomato 13.52% and 6.83% and respectively. Therefore, antioxidant activity and bioactive substances of food material are not affected by microwave sterilization. Inactivation of enzymatic activity of the food material is mainly caused by microwave sterilization. Umudee, Chongcheawchamnan, Kiatweerasakul, and Tongurai (2013) found that in oil palm fruits, the disruption of enzymatic lipolysis reaction is caused by microwave sterilization. When the temperatures used are up to 45°C and 60°C, followed by microwave radiation, the lipase activity in wheat germ is reduced by 60% and 100%, respectively, as observed by Chen et al. (2016). Microwave heating at 120°C decreases polyphenol oxidase activity and peroxidase activity in strawberry puree by 100% and 98%, respectively (Marszałek et al., 2015).

For color and textural attributes of food material, microwave sterilization usually causes negligible effects. Lu et al. (2011) showed that firmness of grape tomato is not changed much due to low treatment (ranging from 2.43 to 2.82 N). Microwave sterilization does not cause significant changes in the textural properties; examples include coriander, grape tomato, and jalapeño pepper, in which texture is not affected when microwave sterilization is done for less than 1 min (De La Vega-Miranda, Santiesteban-Lopez, Lopez-Malo, & SosaMorales, 2012; Lu et al., 2011). However, the texture properties are slightly affected by lengthy microwave sterilization processes. Conversely, for color change, De La Vega-Miranda et al. (2012) showed that microwave sterilization causes significant changes in color of coriander with the lightness (L) reducing from 25.98 to 19.61 and from 30.38 to 24.35 for jalapeño pepper. In addition, Lu et al. (2011) showed that microwave heating does not causes much change in color attributes (L, a and b) of grape tomato. The variation between these results might be due to red vegetables used by Lu et al. (2011) and the green vegetable used by De La Vega-Miranda et al. (2012). Lycopene, mainly responsible for the red vegetables, is not much affected by microwave heat treatment.

2.12 Microwave blanching

Blanching is defined as the enzyme (heat resistant) deactivation phenomena, which helps in retaining color, reduction in initial microbial growth, cleansing the product, product preheating prior to processing, and gas exhausting from plant tissue (Shaheen, El-Massry, El-Ghorab, & Anjum, 2012), and helps to release carotenoids, thus enhancing their bioavailability and extractability (Arroqui, Rumsey, Lopez, & Virseda, 2002). Blanching is done by immersing the food product in hot water, boiling solutions, or steam, comprising of salts or acids. Microwave blanching retains maximum chlorophyll contents, ascorbic acid, and color, as compared to that of steam and water blanching. Microwave blanching of herbs like rosemary and marjoram has been done by immersing the herbs in a small amount of water and exposure to microwave processing (Singh, Raghavan, & Abraham, 1996). Microwave blanched samples retains better quality attributes as compared to microwave unblanched dried samples (Singh et al., 1996). Likewise, Salmonella typhimurium has been reduced by 4–5 log cycles of microbial load by microwave treatment using water in fresh coriander foliage and jalapeno peppers (De La Vega-Miranda et al., 2012).

Seed coats or testa are also removed by blanching, which helps to decrease the moisture content and enzymic activity, which in turn slightly interferes with further treatment into specific materials. Dust, foreign material, and discolored or damaged seeds are removed by blanching. Latorre, Bonelli, Rojas, and Gerschenson (2012) showed that polyphenoloxidase and peroxidase enzymes in red beet are deactivated by microwave heat treatments without or with water blanching and water soaking. It has been demonstrated that tissue color change and elastic properties are decreased by all heat treatments. It has been suggested that microwave blanching of red beets without water soaking (at 100, 150, and 200 W for 5 min) causes weight loss and shrinkage, whereas microwave blanching along with water soaking (at power level of 250, 350, 450 W, or powers more than 900 W) prevents shrinkage in red beets. It has been observed that hydration characteristics and cell wall integrity is superior in red beet microwave blanched (along with water soaking) samples. Hence, this method can be used as an alternative to water blanching.

2.13 Microwave thawing

Over long periods, freezing has been used as an efficient method of food preservation. Prior to use, frozen products are thawed by various techniques. Conventional thawing techniques have certain limitations, like more thawing time and more space requirement, which cause biological and chemical degradation of food material. To reduce chemical degradation, microbial growth, dehydration, and excessive drip loss, thawing time must be as low as possible. Another thawing technique is microwave thawing, requiring smaller thawing time, but with a limitation of runaway or uneven heating. Runaway heating can be understood by taking an example that certain meat parts may remain frozen while other parts may cook by microwave thawing. Hence, it is essential to optimize the heat produced by the microwaves. Due to greater power absorption and uneven power distribution in liquid regions, nonuniformity in heating occurs. In the thawing of biological components, uneven temperatures inside a microwave is influenced by the dielectric characteristics of the biological components, shape, size, frequency, and the magnitude of the microwaves (Taher & Farid, 2001). Using lower power levels or power cycling in a continuous manner are efficient means to reduce uneven heating (Chamchong & Datta, 1999).

In commercial microwave ovens, uniform heating occurs when temperature is equilibrated/maintained inside the material due to the fraction of time when the microwaves are off. Oszmianski, Wojdylo, and Kolniak (2009) observed the impact of prefreezing, like addition of pectin, L-ascorbic acid, sucrose, and different thawing and freezing conditions on the polyphenolic percent of three strawberry varieties. It has been suggested that microwave thawing has various significant effects on many polyphenolic components, like β-catechin, proanthocyanins, anthocyanins, and ellagic acid. Large numbers of dissolved components present in food causes gradual thawing and freezing over a wide temperature range and results in a heterogeneous region of coexisting liquid and solid phases referred as the mushy zone (Basak & Ayappa, 2002); whereas in pure material, a phase change occurs at a single temperature.

2.14 Microwave drying

In drying, the aim is to eliminate moisture from food without changing its chemical and physical composition. In addition, it is essential to protect the food material and increase its storage stability, which is mainly attained by the dehydration. Drying of foods is achieved by various drying processes like convection drying, smoking, solar (open air) drying, spray drying, freeze drying, fluidized-bed drying, puffing, etc. (Cohen & Yang, 1995). Microwave drying has benefits of improving the food product quality and rapid drying rates. It is appropriate for foods with high water content, like cabbage, carrot, and mushroom, due to high dielectric properties of water, which readily absorbs the microwave energy (Prakash, Jha, & Datta, 2003). The heat produced by microwave heating produces vapor pressure inside the material, which further delivers the moisture to the surface of material (Turner & Jolly, 1991).

Microwave drying does not cause case hardening due to moisture delivering effect. Thus, enhanced drying rates and improved quality of product without increased surface temperature are obtained. Moreover, due to high initial capital investment requirement, implementation of microwave drying at the large scale is quite slow. Microwave drying has certain limitations like nonuniform heating of the foods as the moisture content reduces, and undesirable textural changes as high rapid mass transfer occurs (Zhang et al., 2006). However, microwave drying is significant during the falling rate period, in which the diffusion is rate-limiting, causing structure shrinkage and decreased surface water content. But in microwave drying, as volumetric heating occurs, an internal pressure gradient is established due to the vapors produced inside, which forces the water outside. Hence, microwave drying prevents material shrinkage. Microwave energy along with other drying techniques enhances the quality of food products and the drying efficiency as compared to microwave drying only or by other traditional methods only (Zhang et al., 2006). Microwave drying causes changes in the product quality including structural properties like porosity, density, and specific volume, textural properties, sensory properties like flavor, taste and odor, optical properties like appearance and color, rehydration characteristics like rehydration capacity and rehydration rate, and nutritional properties like proteins and vitamins.

Soysal (2004) performed microwave drying of parsley, where drying occurred mostly in the constant rate period and the falling rate period. Drying time decreased with increasing microwave output power level. Soysal (2004) studied microwave drying of parsley leaves and observed that color parameters were not affected by microwave power level. In other research, microwave drying of carrot slices occurred during falling rate period and not in constant rate period. It has been observed that rapid transfer of mass occurs from the center to the surface at high microwave power level, due to more heat generation. It has also been observed that rehydration ratio and β-carotene content decreases due to increase in slice thickness. Internal pressure is increased inside the samples due to high volumetric heating resulting in bubbling and boiling of sample. Thus, rehydration ratio and β-carotene content is observed to be decreased (Wang & Xi, 2005).

In order to decrease water content of spinach leaves up to 99%, microwave drying was carried out by Ozkan, Akbudak, and Akbudak (2007). The energy consumption remained constant above 350 W microwave power level. Microwave drying has been observed to be more significant at processing conditions of 750 W and 350 s. Under such conditions, optimum drying properties of spinach (including drying time, utilization of energy, ascorbic acid level and color criteria) has been achieved (Ozkan et al., 2007). Lombrana, Rodriguez, and Ruiz (2010) examined sliced mushroom drying using a single mode microwave at frequency of 2.45 GHz. During microwave drying, the moisture content present at the edges of samples heated more slowly than the moisture content present in the center of the sample, causing temperature profile inversion. The results revealed that at optimum microwave heating (120 W) and low pressure, the quality of mushroom is better and the rate of drying is high. At reduced microwave power level (60 W), better quality of mushroom has been achieved at the cost of slow drying rate, whereas at atmospheric pressure or at high microwave power level (240 W), ineffective drying has been observed resulting in the entrapment of moisture and the formation of large voids inside the sample. Hence, drying at low pressure with moderate microwave power is suggested for mushroom drying.

Much research has been done on the alteration of food materials in nutritional properties during microwave drying. Nawirska-Olszanska, Stępień, and Biesiada (2017) reported that at power level of 100 W, the t bioactive compounds (chlorophyll aþb, carotenoids), polyphenols, and antioxidative characteristics of microwave dried pumpkin slices were more than that using the power level of 250 W. Wojdyło, Figiel, Lech, Nowicka, and Oszmianski (2013) showed that during vacuum-microwave drying of sour cherries, the amount of antioxidant activity, phenolic compounds, and color (which reveals anthocyanins content) is less when drying at elevated temperature than that of reduced temperatures. Maximum antioxidant capacity, total phenolics, especially anthocyanins, and minimum change in color in sour cherries could be achieved by decreasing vacuum-microwave drying power level from 480 to 120 W (Wojdyło et al., 2013).

In microwave drying, elevated temperatures may damage food nutrients to a large extent—mostly heat sensitive constituents.

Rehydration and textural properties are influenced by microwave drying. Nawirska-Olszanska et al. (2017) observed that microwave drying of pumpkin slices decreases compression work with increase in microwave power level, probably because of partial dissolution of the middle lamellae of the material. However, Sarimeseli (2011) reported that the rehydrating capacity of microwave dried leaves of coriander decreases with increase in microwave power level (180–900 W), which might be due to cellular disintegration of the leaves. Therefore, the tissue integrity of the food is maintained by microwave drying at low power level. Microwave dried pomegranate (*Punicagranatum L.*) arils and eggplant dried by microwave-far infrared combination drying results in more porous structure than that of processed by hot-air drying with rapid rate of drying, that greatly damages the tissue integrity (Aydogdu, Sumnu, & Sahin, 2015; Horuz & Maskan, 2015).

Considering better rehydration capacity, shrinkage, time of drying, and color, the optimal condition of microwave drying for pomegranate (*Punicagranatum L.*) arils has been found at 350 W (Horuz & Maskan, 2015). However, the reduced degree of shrinkage is correlated with a low bulk density. Thus, pomegranate (*Punicagranatum L.*) arils and eggplant slices treated by microwave drying could attain higher rehydration ratio and lower shrinkage and bulk densities than hot-air drying processes (Aydogdu et al., 2015; Horuz & Maskan, 2015). Microwave dried edamame decreases compression force as compared to that of hot-air drying (Lv, Zhang, Bhandari, Yang, & Wang, 2017). Moreover, Zielinska, Sadowski, and Błaszczak (2015) reported the same observations about the gumminess, chewiness, and hardness of blueberries (*Vacciniumcorymbosum L.*), which is numerous times less by microwave-vacuum drying than by hot-air drying.

2.14.1 *Microwave drying of tomato slice*

Tomato slices were prepared by microwave drying. At the initial point of drying, change in sample weight was recorded at 2 min time intervals. The drying process was stopped when the 10% of water content was recorded. The final amount of water present in every sample was recorded in order to determine the amount of water after every weighing interval. Among various quality parameters of dried tomato slices, the color as the chief source predicts the level of effects on respective conditions and drying methods. Drying using 1.13 W/g coupled with 50°C hot air ventilation at the temperatures of 40°C and 50°C in absence of microwave heating was shown to be better for retaining the quality color parameters of tomato slices.

2.14.2 Microwave drying of strawberries

Strawberries have a very pleasant bittersweet taste and delicate sweet-smelling flavor. Moreover, strawberries are easily damaged by spoilage because of high water content—around 90% (wet basis). Strawberries are especially used in syrups, preserved foods, biscuits, dairy products, jams and cookies, or can be consumed directly as dehydrated or fresh fruit. Chile exports presently 16% of its total production to the United States of dried fruits, 14% to Mexico and 10% to Europe (Central Bank, Chile 2014). Hence, the dehydration of types of fruits have clear economic incentive. The operation of vacuum-microwave drying system is carried out at time temperature combination of 2.5 h at a temperature range of 40–60°C. A suitable position of magnetron with an optimal microwave power level enhances the drying uniformity of fruit.

2.14.3 Microwave drying of lemon slices

At room temperature, lemon samples are placed for 4 h to reach thermal equilibrium with the environment. For each observation, lemons slices with 4 mm thickness are placed perpendicular to the fruit axis and around 200 g of the samples in an air-ventilated oven (lemon slices having peel) as a monolayer on a tray, uniformly. Drying of lemon slices is done in an air-ventilated oven dryer at temperatures of 75°C, 60°C, and 50°C. Temperature of drying has a greater effect on drying time where the time-temperature combination is maintained at 1100, 2120, and 4350 min and 75°C, 60°C, and 50°C, respectively. There was no constant rate period and it has been observed that the whole drying treatment occurs in the falling rate period.

2.14.4 Microwave drying of grapes

The drying of grapes is referred as long process time. Many pretreatments are applied to grapes such as grapes dipped in potassium carbonate solution plus ethyl oleate. At 60°C, microwave drying of grapes provides the best results. Temperature sensors are used to maintain the temperature of the microwave. The temperature sensors are responsible to control the microwave power in order to ensure that grapes should be within the set temperature value. The two stages of the drying process have been approved. In the first stage, the samples were introduced in the microwave oven until drying occurs mainly in the constant rate period, with water removal of approximately 55%. From the internal system, moisture is successfully removed with the help of exhaust fans. The compound contents and nutrient element of microwave-vacuum dried grapes has been compared with that of sun-dried fruits and fresh fruits, where sun dried grapes were prepared by traditional methods. Further, microwave heating has proven to decrease the drying rate of various agricultural materials significantly. Drying parameters, when controlled properly, such as microwave power and inlet air temperature, specific energy time, fresh fruit sugar, temperature, and fresh fruit moisture content have been analyzed using multiple linear regression analysis, to improve the content of the dried grapes, and surface plots.

3 Ohmic heating

3.1 Mechanism

Ohmic heating is a thermal method where heat is produced internally by passing an alternating electrical current (AC) within a body like a food system, which behaves as an electrical resistance. It is also termed as direct resistance heating or resistance heating. The ohmic heating principles are often simple. During resistance heating, AC voltage is applied to the product placed between two electrodes. The rate of heating is directly proportional to the square of the electric field strength, nature of food being heated, and the electrical conductivity. The strength of electric field is maintained by the applied voltage or by maintaining the electrode gap, while the electrical conductivity of food materials differs widely, but could be maintained by adding electrolytes. An ample amount of heat is produced to sterilize and pasteurize foods (Ranesh, 1999). Usually, pasteurization includes heating high acidic (pH <4.5) foods to 90–95°C for 30–90 s to deactivate enzymes and spoilage causing microorganisms (molds, yeasts, vegetative bacteria). Low acidic (pH >4.5) foods can harbor *Clostridium botulinum* growth, and, on the basis of pH and other characteristics of the food material, needs heating to 121°C for a minimum time period of 3 min (lethality $Fo=3$ min) to attain sterility (12D colony reduction).

Resistance heating is basically a high-temperature short-time process, which heats about 80% solids material from ambient temperature to 129°C in about 90 s (Zuber, 1997), to reduce the amount of high temperature overprocessing. Additionally, ohmic heating heats the particulates more rapidly than the carrier liquid, known as the heating inversion (Kim et al., 1996), which is impossible by conductive heating processes. Power generation is directly proportional to the square of the applied electric field (E, V/cm) and the electrical conductivity of the material depicts the major parameters of the ohmic heating regarding the variables inherent of the product, process, and equipment.

3.2 Electrical conductivity

Ohmic heating is often an appropriate thermal method for processing of particulates in liquid foods, as the particulates are heated at faster or similar rates than the liquid foods (Ruan, Chen, Chang, Kim, & Taub, 1999; Zitoun, 1997). Moreover, numerous important factors affect the heating of particulate and liquids mixtures. For commercial resistance heating, the main parameters are temperature, rate of flow, holding time, and rate of heating of the process (Palaniappan & Sastry, 1991a, 1991b). The factors affecting the heating of the food include the shape (spheres, cubes, rods, discs, twists, rectangles), size (2.54 cm^3), orientation, density (20%–80%), capacity, and electrical and thermal conductivity of the particle, electrolytes present, specific heat capacity of the carrier medium. The major processing variables are the electric field intensity, which differs according to the electrical conductivity and the voltage applied, depending on ionic dissociation, temperature, texture, viscosity, solid content, nonconductive components present, like sugar, gases, and fat, and the cell structure (Sarang et al., 2008).

The electrical conductivity shows the capability of material to conduct electricity through a unit of area per unit time having potential gradient (Goullieux & Pain, 2005) and with SI unit Siemens per meter (S/m), and is given as follows (Zell, Lyng, Morgan, & Cronin, 2009):

$$\sigma = \frac{L}{A} \times \frac{1}{V} \tag{5}$$

where L is the distance between the electrodes (m), I is the alternating electric current (A), A is the cross sectional area perpendicular to the electric current passage (m^2), and V is the voltage applied (V).

Electrical conductivity is the main factor in the resistance heating system, that helps to determine the finest parameters to be used and the intensity of process (Goullieux & Pain, 2005; Kaur & Singh, 2016). Its value is not constant and depends on the temperature of material, increasing in a proportional manner with the former (Sakr & Liu, 2014). Electrical conductivities between 0.01 and 10 S/m at 25°C are mostly used for ohmic heating (Mercali, Schwartz, Marczak, Tessaro, & Sastry, 2014). The electrical conductivity and its dependency on temperature are the main parameters in resistance heating to calculate the heating rate of the product. Usually, materials with higher conductivity have a greater rate of heating, with differences in heating rates in different materials possibly due to differences in specific heat (Palaniappan & Sastry, 1991a, 1991b). For materials with more than one phase, like mixtures of particulates and liquid, the respective electrical conductivity of all the phases needs to be kept in mind. The solid particulates have lower electrical conductivities that the carrier liquid. However, the heating pattern is a complex function of the relative electrical conductivities of the particulates as well as liquids. The liquid is more rapidly heated than the particulate during ohmic heating if a single particulate has lower electrical conductivity than the carrying liquid. Additionally, when the particulate density in the mixture is raised, the rate of heating for the particulates also rises, and may even exceed that for the liquid (Sastry & Palaniappan, 1992a, 1992b). With the rise in temperature, the electrical conductivity of liquids or particulates rises linearly (Wang & Sastry, 1997a, 1997b). Variation in the electrical resistance between the two phases is the main cause of complex heating characteristics of the system.

The three electrical conductivities for food materials are defined as follows:

- $s > 0.05\,S/m$: good conductivity foods: eggs, fruit juices, condiments, milk desserts, wine, hydrocolloids, etc.
- $0.005 < s < 0.05\,S/m$: low conductivity foods, which require higher electrical field strength: margarine, marmalade, etc.
- $s < 0.005\,S/m$: poor conductivity foods, which requires very high electrical field strength and are mostly difficult to process by ohmic heating: foam, liquor, fat, etc.

The electric conductivity of food materials mostly rises with increase in temperature, frequency, voltage gradient, and water content.

The strength of the electric field can be influenced by input voltage and the distance between electrodes. The electrical conductivity of food materials varies from 0.1 to 3 S/m, except oils and syrups, which are considered to be good electrical insulators and are not affected by moderate ohmic heating conditions (Stirling, 1987). Various parameters such as composition of food (e.g., acid, water content, and salt) and temperature can influence the electrical conductivity. Moreover, varying the formulation of a product to optimize the electrical conductivity has a diminishing effect on organoleptic characteristics of the product, and thus it is necessary to adjust the rate of heating by maintaining the electric field strength, i.e., by adjusting input voltage or electrode gap.

Additionally, the thermal conductivity of food materials is influenced by the rise in temperature in the same way as that of a resistor, and food products are more electroconductive at elevated temperatures (Ramaswamy, Marcotte, Sastry, & Abdelrahim, 2014). Electrical conductivity, depending on the temperature, causes thermal runaway and needs to be considered when design an ohmic heating system (Gally et al., 2017; Palaniappan & Sastry, 1991a, 1991b). In double phase food materials, as particle concentration increases, values of electrical conductivity decrease uniformly during ohmic heating.

3.3 Ohmic heater design

Resistance heating systems consist of two or more electrodes to produce current upon the fluid. When designing ohmic heating equipment, the electrodes play a critical role. Various designs of ohmic heaters are available; depending on the electrodes positions and locations, the design can be either a static system in a container vessel or with continuous flow through them (Icier & Ilicali, 2005). Generally, two broad designs of ohmic heating system are available (Simpson, 1994):

- electrode arrangement; and
- electrode design.

3.3.1 Electrode arrangement

In an ohmic heating system, the material (multiphase mixture, liquid, or solid) is the conductive medium. Solid electrodes are generally used. Such electrodes are in close contact with the material, into which electric power supply is introduced. The electrodes are usually parted by a plate spacer or tube that is insulated electrically. It involves three generic configurations:

1. Batch configuration (discontinuous, without flow): where the electrodes are parallel, plane and coaxial.
2. Transverse configuration: where the material travels perpendicularly to the electric field and parallel to the electrodes and the electrodes are mostly coaxial or plane and slightly spaced.
3. Collinear configuration: where the material fluid travels from one electrode to another, parallel to the electric field and with wider electrodes spacing.

Each of the configurations has its benefits and limitations as per their specific technical considerations and applications. Electrodes are generally organized in one of three different configurations where the operation is optimized.

Batch configuration

For a stationary medium in a batch system, which is electrically and thermally insulated and shows a linear dissimilarity between electrical conductivity and temperature, the temperature time course is expressed by the formula as follows:

$$T(t) - T_{\text{in}} = \frac{1}{m}\left[\exp(\alpha t) - 1\right] \text{with}\, \alpha = E^2 \frac{m\sigma_{\text{in}}}{\rho c} \tag{6}$$

where, T_{in} is the starting temperature.

The heating rate is then:

$$\frac{\partial T}{\partial t} = E^2 \frac{m\sigma_{\text{in}}}{\rho c} \exp(\alpha t) \tag{7}$$

The static or batch heater has various applications in:

(i) optimization of formulation on electrical parameters;
(ii) model validations and observations of solid/liquid samples or composite mixture behavior during resistance heating; and
(iii) as an HTST simulator.

When liquid and particle conductivity varies, orientation and shape of particle to the electric field in the liquid generates an underheated or an overheated liquid and shadow effects for more particles (Davies, Kemp, & Fryer, 1999; de Alwis, Halden, & Fryer, 1989). This batch heater permits for essential characteristics to be measured, such as process homogeneity, heating time, and electrical conductivity of the product. This in turn offers a tool to observe the heating effects on the end product quality or to test the optimal initial product composition. This system is easier and efficient to determine the optimum conditions for a continuous resistance heating system in order to control the processes like heating, holding, and cooling (Goullieux, Pain, & Baudez, 1997). For product formulation and electrical conductivity estimation in laboratories, the major benefits are ease of operation, less product quantity, and the capability to treat large number of products.

Sterilization by resistance heating within a multilayered laminate pouch for extended period space missions has been reported by Sastry, Heskitt, Jun, and Somavat (2013). Batch ohmic cells placed on a conveyor belt are used to manufacture breadcrumbs. Ohmic heating is used for 60% of the yearly production of breadcrumbs (panko) in various countries like Japan. This process conserves cost of production by decreasing the energy required for heating, and eases the operation (Parker, 2001). Parker's device was constructed to keep bread dough within a container between two electrodes. The dough is baked by ohmic heating without forming a crust, and the chamber constrains the dough, resulting in acceptable structure of grain. Some Japanese manufacturers of tofu are concerned with the properties of batch resistance heating and have begun to observe its potential to make tofu for domestic and industrial purposes (Noguchi, 2004). For various applications at small scale, such as in laboratories, the major benefits are ease of application, lower product quantity, and capability to treat large items.

Transverse ohmic heating

In transverse ohmic heating, the current flux and electric field applied are at right angles to the flow of mass and the electrical field strength is constant. These are mainly used for fluids with low conductivities (<5 S/m) and also have an advantage where there are huge solids particles with minimum shear force because of unrestricted flow channel (Simpson, 1994). The uniformity of electric field is maintained in this geometry enhancing uniform heating. The design functions at standard voltages (as 240 or 415 V). If the electric field strength is kept constant $E(z) = E_c$, the temperature rise is determined by the following:

$$T(z) - T_{in} = \frac{1}{m}(\exp(\alpha'' z) - 1) \text{ with } \alpha'' = E_c^2 \frac{m\sigma_{in}}{\rho Cp} \frac{T}{L} \quad (8)$$

Stirling (1987) presents the use of 50 kW transverse-mode ohmic heater with a standard plate-type regenerative cooler. In this design, the electrodes are kept parallel, with minimum space between plastic insulating spacers and channeled within the fluid flow. The current increases as the product contact surface is greater and the voltage between the electrodes is less (<96 V) (Polny & Thaddeus, 1996). Thus, electrode erosion, overheating, and boiling has been observed. These heaters thus have limited applications to fluids with no particles, like beverages and milk. A heater for the resistive heating of a fluid with different plate configurations is presented by Ayadi et al. (2005) who further reported the fouling effect by milk products. Wild-Indag established an INDAG High Power Heating System with two ceramic sections in transversal configuration.

Collinear ohmic heating

These heaters are a good option for large conductivity, presenting more spacing between electrodes. The electrodes are placed in the fluid stream providing a completely unrestricted flow channel. For various purposes, the design needs more voltage than the parallel plate. In addition, the low current distribution and areas of large current density present at the leading edges of the electrodes can create localized boiling and arcing. In a collinear resistance heating system, the current flux and electric field are parallel to the flow. In case of isotropic material flowing through thermally isolated pipes or between plates (adiabatic boundary conditions) of uniform residence time and length L, if the free convection and heat axial diffusion (higher Peclet number) are ignored, the solution of Eq. (5) can be calculated by integration and variable separation. If the current density is kept constant $J(z) = Jc$, the temperature rise is given as:

$$T(z) - T_{\text{in}} = \frac{1}{m}\left(\sqrt{1+2}\,\alpha' z - 1\right) \text{ with } \alpha' = \frac{\int_c^2}{\sigma_{\text{in}}} \frac{m}{\rho C} \frac{T}{L} \quad (9)$$

By comparing Eqs. (8), (9), it can be predicted that the similar heat electrical production provides high temperature elevation when both electric field strength and current density are kept constant.

A flowing cross sectional area can be variable or constant. Heating of fluid is generally done by jet heater. Each jet is an electrical resistance, which gets heated by passing an electric current through it. Usually, the largely spaced electrodes are arranged in electrode housings (cantilever tube) or are annular on each side of the spacer tube. The product contact surface is less and the voltage is more (up to 4500 V). This structure is mostly used for inclined, vertical, or horizontal tubes (Reitler & Rudolph, 1986). APV Baker (England) gave a commercial process mostly for high-temperature, short-term applications (Biss, Combes, & Skudder, 1989; Skudder & Biss, 1987). Recently Emmepiemme (Italy) developed annular electrodes facilitating integral flow of current adapted to large-particle food mixtures and ohmic pump for displacement and continuous heating of foods.

For coagulation of liquids, Ghnimi, Malaspina, and Zaid (2009) patented a jet applicator design including organic fluids, milk, and liquid egg. The flow of fluid down through the electrode, produces a jet of fluid, and falls on the electrode. Reznik (1997, 2000) patented a spacer tube with a straight, convergent, and divergent section adapted to liquids with no particulates (like juices) and a conical shaped electrode to prevent coagulation on the wall. The conduit course of comparatively small cross-section area generates a flash treatment in a turbulent flow regime with high heating rate (550°C/s). Thus, temperature increases rapidly and then remains constant during heating. Bibun (Japan) developed ohmic heaters for processing of surimi and salt-solubilized washed fish mince continuously, where the thin surimi paste is passed on a conveyor and the conveyor roll is fed using an electric current. The band between roll and paste remains moistened and is continuously cleaned.

3.3.2 Elcctrode design

During ohmic heating, the key factor that needs to be considered is choosing an appropriate electrode (Assiry, Gaily, Alsamee, & Sarifudin, 2010; Sarkis, Mercali, Tessaro, & Marczak, 2013). Earlier designs were based on the use of various conductive electrode materials like stainless steel, graphite, aluminum, titanium, and platinized-titanium electrodes are mostly selected depending on correction resistance and price, which influences the effectiveness of the ohmic heater. In certain cases, where the product quality is not important, low carbon electrodes are mostly used; for superior quality materials, metals like stainless steel are used, at the same time the frequency of the power supply needs to be enhanced significantly to avoid apparent metal dissolution and corrosion (Stancl & Zitny, 2010).

3.4 Effects of ohmic heating on chemical composition of fruits and vegetables

Like other processing methods, food material undergoes various changes during ohmic heating, which influences its taste, color, and consistency (Yildiz, Icier, & Baysal, 2010). During ohmic heating, as food materials are in direct contact with the electrodes through an electrical conductive medium, under specific operating conditions undesirable ions may get into the food and may result in electrode corrosion (Herting, Wallinder, & Leygraf, 2008). The electrical conductivity is an essential parameter in resistance heating, as it is a prerequisite for the heat generation (Shirsat, Lyng, Brunton, & Mckenna, 2004). As the conductivity depends on the temperature, it varies with the heating process. Cell structures caused by heat release ions, leading to significant variation in the food conductivity and thus affect the ohmic heating process (Darvishi, Khostaghaza, & Najafi, 2013).

Ohmic heating of vegetable and fruit tissue has been revealed to enhance hot-air rate of drying, shift desorption isotherms, and enhances extraction yield of juice as compared to conventionally heated, microwave heated and untreated samples. The impact of resistance heating on tissue of sweet potato would increase the rate of vacuum drying as compared to nontreated samples.

Ohmically heat-treated sweet potato cubes followed by freeze drying results in faster drying rate with maximum decrease in drying rate of 24%. Further, there is considerable reduction in vacuum drying time when minimally treated with ohmic heating, which may have a vital role in product quality and economy. During blanching of vegetables using ohmic heat treatment, it is quite possible to use larger pieces of vegetables than with traditional heating, where the limiting factor is thermal conductivity. Larger pieces have varying surface to volume ratio as compared to smaller pieces, leading to the reduced damage of soluble components (Mizrahi, 1996). In low-frequency conditions, the quantity of freely soluble substances increases in white radish, due to increased cell decomposition, which may explain the quick initial heating at low frequencies (Imai, Uemura, Ishida, Yoshizaki, & Noguchi, 1995).

In peach pieces, at a low frequency, cell membrane lysis occurs due to texture degradation and increased electrical conductivity. Higher frequencies decreases these effects, but the time increases until the optimum final temperature is attained (Shynkaryk, Ji, Alvarez, & Sastry, 2010). Cell lysis has also been detected in apples and potatoes and increases at elevated temperatures with an optimum electric field and an electric field strength less than 100 V/cm (Lebovka, Praporscic, Ghnimi, & Vorobiev, 2005). Comparable effects have been detected during mushroom blanching (Sensoy & Sastry, 2004). Quinces juice, a pectin-rich fruit, obtained by conventional or ohmic heating does not show significant difference in its flow characteristics Bozkurt and Icier extracted yield from sugar beet cuts along with subsequent high-voltage pulse treatment. It is possible to enhance the extraction of juice significantly by thermal decomposition of the tissue matrix along with cell membrane electropermeabilization (Praporscic, Ghnimi, & Vorobiev, 2005).

The nutritional stability of food products is affected by processing conditions. It has been suggested that the flavoring compounds octanal, pinene, myrcene, and limonene decompose slightly during ohmic heat treatment as compared to traditional heating (Leizerson & Shimoni, 2005b). In case of acerola puree, no difference has been observed between the two methods in the kinetics of anthocyanins content decomposition (Mercali, Jaeschke, Tessaro, & Marczak, 2013), whereas the ascorbic acid decomposition in acerola puree depends on the operating conditions and is raised with the voltage input. At low voltage gradients, total ascorbic acid decomposition is comparable with the traditional heating method (Mercali, Jaeschke, Tessaro, & Marczak, 2012).

This voltage magnitude dependence is also found when studying anthocyanin reduction. In the case of grape juice, enzyme activity, i.e., phenol oxidase activity, increases with increase in temperature during resistance heating, until a critical temperature is attained, and thereafter it is dropped, this temperature depends on voltage gradients (Icier, Yildiz, & Raysal, 2008). On the basis of such results, the researchers suggested that the deactivation phenol oxidase occurs differently by ohmic as well as conventional heating. In orange juice, both ohmic and conventional heating leads to a comparable pectin esterase deactivation (by 90%–98%) (Leizerson & Shimoni, 2005b). In the case of orange juice, pectin esterase (pectin methylesterase) has been degraded significantly by ohmic heating as compared to conventional heat treatment (Demirdoven & Baysal, 2014). In the case of pea puree, during ohmic heat treatment at shorter processing times under certain conditions the peroxidase activity decreases as compared to conventional heat treatment (Icier, Yildiz, & Raysal, 2006). In the case of carrot pieces, heating methods does not affect peroxidases significantly (Lemmens et al., 2009). Electrical conductivity varies significantly between strawberry-based products. Electrical conductivity increases with field strength in the case of strawberry jelly and fresh strawberries but not in strawberry pulp, possibly due to the presence of texturizing agents. This parameter declines with the rise in sugar and solids concentration.

In addition to this, influence of various heating methods (conventional heating, microwave, and ohmic heating) on textural characteristics of red beet pieces, golden carrot, and carrot has been examined and compared with microwave and conventional heating. The samples exposed to various operating conditions have been observed for textural changes using texture profile analysis at varying operating times. It has been observed that resistance heating increases rate of softening but the final hardness of the treated samples processed with resistance heating decreases more than other samples treated with either microwave or conventional heating methods.

3.5 Applications of ohmic heating in fruits

The nutritional value of most vegetables and fruits is changed during conventional heat treatment, which dictates the study for other heating techniques resulting in higher quality products. Minimal ohmic heating results in significant retention of food quality. Further, there is considerable reduction in vacuum drying time when minimally treated with ohmic heating, which may have a vital role in product quality and economy.

3.5.1 Pineapple and pomegranate

Electrical conductivity of pineapple has been studied over temperature range of 25–140°C by Sarang et al. (2008). During ohmic heat treatment, electrical conductivity and viscosity of pineapple juice was calculated from 25°C to 70°C temperature range by Singh, Singh, and PS (2008). The influence of resistance heat treatment on rate of heating, electrical conductivity, pH of juice of pomegranate, and system performance had been reported by Darvishi et al. (2013).

3.5.2 Acerola

Mercali et al. (2012) reported the ascorbic acid reduction in acerola pulp occurs due to conventional and ohmic heating. Temperature dependence on electrical conductivity for acerola pulp has been calculated during operation and construction of resistance heating system by Sarkis et al. (2013). Mercali et al. (2014) reported that color changes and vitamin C reduction in acerola pulp occurs due change in electric field frequency during ohmic heating.

3.5.3 Orange

During ohmic heating, electrical conductivity and viscosity of orange juice has been calculated from 25°C to 70°C temperature by Singh et al. (2008). The importance of particle concentration in whole orange juices on the electrical conductivity of double phase system during resistance heating has been evaluated by Palaniappan and Sastry (1991a). Likewise, orange juice having spores of Bacillus subtilis has been studied with the help of alternating current electric field (Uemura & Isobe, 2003). Lima, Heskitt, Burianek, Nokes, and Sastry (1999) determined vitamin C reduction kinetics using ohmic heat treatment to heat orange juice with an electric field of 18.2V/cm for 30min at 90°C. Leizerson and Shimoni (2005) studied the influence of ultrahigh-temperature continuous resistance heating on orange juice with respect to traditional pasteurization method. Vikram, Ramesh, and Prapulla (2005) calculated the kinetics of vitamin C reduction during resistance heating of orange juice by introducing electric field strength with 42V/cm.

3.5.4 Apple

Lima, Heskitt, and Sastry (1999) determined the effect of hot-air drying rate and ohmic heating frequency on apple juice yield. Lima, Heskitt, and Sastry (1999) and Wang and Sastry (2000) evaluated the extraction rate of resistance heat-treated tissue of apple and untreated samples, and also evaluated the influence of frequency on the yield of extraction. The influence of resistance heat treatment on electrical conductivity of golden and red and apple over temperature range of 25–140°C has been studied by Sarang et al. (2008). It has been found that for red delicious apples, ohmic heating increases the rate of extraction of apple juice (Wang, 1995). Icier and Ilicali (2004) studied the impact of concentration on the rate of resistance heat treatment of apple juice. The influence of resistance heat treatment on extraction yield of juice from apple tissues has been studied by Praporscic, Lebovka, Ghnimi, and Vorobiev (2006). Singh et al. (2008) studied the influence of resistance heat treatment on electrical conductivity and viscosity of apple juices over temperature range of 25–70°C. Lima, Elizondo, and Bohuon (2010) studied the influence of temperature during resistance heat treatment on vitamin C reduction of ground cashew apples. Jakob et al. (2010) determined the deactivation kinetics of pectin methylesterase of fresh apple juice during ohmic heating.

3.5.5 Peach and pear

The rate of ohmic heating on peaches has been determined at an electric field strength of 60 V/cm, and frequencies ranging from 50 Hz to 1 MHz. Electrical conductivity of pear and peach has been calculated at temperatures ranging from 25°C to 140°C by Sarang et al. (2008). Icier and Ilicali (2005) studied the electrical conductivity of peach puree during ohmic heat treatment. Moreno et al. (2011) determined the influence of ohmic heat treatment on microstructure and osmotic dehydration kinetics of pears.

3.5.6 Apricot

The influence of resistance heat treatment on the shelf-life and quality of apricots in syrup using a continuous pilot scale ohmic unit has been evaluated by Pataro, Donsi, and Ferrari (2011), while Icier and Ilicali (2005) studied the impact of resistance heating on electrical conductivity of apricot puree. Both the studies reported the increased shelf stability and quality of apricots.

3.5.7 Blueberry

Temperature dependence on electrical conductivity for blueberry pulp has been calculated during operation and construction of resistance heating system by Sarkis et al. (2013). Marczak, Tessaro, Jaeschke, and Sarkis (2013) studied the reduction of anthocyanin content in blueberry pulp after heating using conventional and resistance heating. Brownmiller, Howard, and Prior (2008) evaluated anthocyanin reduction levels in blueberries during ohmic heating.

3.5.8 Quince and sour cherry

Bozkurt and Icier (2009) investigated the influence of resistance heating on the rheological parameters of quince nectar. Icier and Ilicali (2004) studied the influence of concentration on rate of resistance heating on cherry juice. Garzon and Wrolstad (2002) studied the influence of solid content on total anthocyanin reduction in sour cherries during ohmic heating.

3.5.9 Strawberry

Castro, Teixeira, Salengke, Sastry, and Vicente (2003) studied the influence of sugar content and temperature on electrical conductivity of strawberry samples. Moreno et al. (2012) showed the impact of resistance heat treatment on microstructure and osmotic dehydration kinetics of strawberries. The influence of ohmic heat treatment on electrical conductivity of strawberry at temperature range of 25–140°C has been reported by Sarang et al. (2008). Castro, Macedo, Teixeira, and Vicente (2004) showed the impact of various heating processes and field strength on electrical conductivity of strawberry samples and also investigated the vitamin C reduction kinetics. Cemeroglu, Velioglu, and Isik (1994) and Garzon and Wrolstad (2002) studied the influence of solid content on total anthocyanin depletion in strawberries during ohmic heat treatment. The application of ohmic heating has various advantages on the mass transference acceleration.

3.5.10 Grape and guava

During resistance heating of grape juice, the temperature, holding time, and voltage gradient have various effects on the polyphenoloxidase activity (Icier et al., 2008). Srikalong, Makrudin, Sampavamontri, and Kovitthaya (2011) studied the impact of resistance heating on sensory properties and mechanical extraction of guava juice.

3.6 Applications of ohmic heating in vegetables

3.6.1 Carrot

Palaniappan and Sastry (1991a, 1991b) showed the impact of voltage applied and insoluble solids on electrical conductivity of prepasteurized carrot juices during resistance heat treatment. Jakob et al. (2010) investigated the deactivation kinetics of peroxidase during resistance heat treatment of carrot. In addition, the impact of resistance and traditional heating on textural parameters of carrot pieces has also been examined (Farahnaky, Azizi, & Gavahian, 2012). Zareifard, Ramaswamy, Trigui, and Marcotte (2003) also examined the impact of resistance heat treatment and electrical conductivity on two-phase (solid phase containing carrot puree and a liquid phase) food systems.

3.6.2 Broccoli and beetroot

Using alternating electric field, the diffusion of beet dye from beetroot into the solution has been observed by Lima, Heskitt, and Sastry (2001) and Halden, De Alwis, and Fryer (1990). Halden et al. (1990) also showed the temperature impact on electrical conductivity of beetroot when treated with conventional and ohmic heat treatment. Mizrahi (1996) observed the influence of blanching on the removal of soluble solids of beetroot using ohmic heating. Similarly, Farahnaky et al. (2012) observed the influence of conventional and ohmic heating on textural properties of pieces of red beet. The influence of electrical processing on mass transfer in beetroot has been investigated by Fryer, Miri, and Parral (2012). Kulshrestha and Sastry (2003) also showed the influence of electric fields (as low as 10 Hz) on beet cell membrane permeability. During ohmic heating of broccoli, the inactivation kinetics of peroxidase has been carried out by Jakob et al. (2010).

3.6.3 Radish, turnip, and tomato

Imai et al. (1995) showed the influence of the electric field frequency on rate of heating of radish in the frequency range 50 Hz–10 kHz. Electrical conductivity and viscosity of tomato juice has been calculated over a temperature range (25–70°C) during resistance heat treatment by Singh et al. (2008). Palaniappan and Sastry (1991a, 1991b) reported the impact of applied voltage and insoluble solids on electrical conductivity of prepasteurized tomato concentrate during resistance heat treatment. Palaniappan and Sastry (1991a) examined the importance of concentration of particle in tomato concentrate on the electrical conductivity of a double phase system during resistance heat treatment. Lima, Heskitt, Burianek, et al. (1999) and Lima, Heskitt, and Sastry (1999) reported the impact of electrical conductivity and temperature of tissue of turnip at frequency (4, 10, 25, and 60 Hz) during resistance heat treatment.

3.6.4 *Cabbage and cauliflower*

Stabilization and processing of cauliflower by ohmic heat treatment has been examined by Goullieux, Zuber, and Godereaux (2001). Eliot, Goullieux, and Pain (1999) observed the effect of precooking by resistance heat treatment on the firmness of cauliflower. Volden et al. (2008) observed vitamin C reduction kinetics of cabbage at 95°C during resistance heat treatment. Cauliflower, being a brittle material, cannot withstand conventional heating. The processing feasibility of cauliflower by resistance heat treatment has been examined. Florets of cauliflower were sterilized in a 10 kW APV continuous resistance heating pilot plant with processing conditions and configurations of pretreatments. The stability of the end product has been investigated and textural parameters have been measured by mechanical measurements. Ohmic heating provides food material of appealing appearance, with remarkable firmness characteristics, and a higher amount of particles >1 cm. Stability at 25°C and 37°C has been observed, and in one case, the material remained stable at 55°C. Precooking of cauliflower at low temperature, high flow rate, and adequate electrical conductivity of florets appeared to be ideal conditions. The aim of applying resistance heating is to process brittle food products.

3.6.5 *Potato and pea*

Color changes and peroxidase deactivation by ohmic blanching of pea puree has been investigated by Icier et al. (2006). Zhong and Lima (2003) studied the impact of ohmic heat treatment on rate of vacuum drying on tissue of sweet potato. The resistance heating effect on cell membranes of potato has been investigated by measuring dielectric spectra from 100 Hz to 20 kHz. Sastry and Palaniappan (1992a, 1992b) reported the performance of resistance heating on particle-liquid mixtures using sodium phosphate solutions for potato cubes. Jakob et al. (2010) studied the deactivation kinetics of peroxidase by resistance heating of potato. Resistance heating is used to increase the rate of drying of potato cylinders as studied by Wang (1995). The impact of ohmic heating on yield of juice from tissue of potato has been reported by Praporscic et al. (2006).

4 Conclusion

In conclusion, ohmic heating and microwave heating is an alternative heating process where the food material is thermally processed because of intrinsic resistance. Compared to conventional heat treatment, microwave heating shows lower consumption of energy, with materials having improved sensory attributes. The recovery rate and final moisture content is significantly decreased due to hot air microwave heating with rise in microwave power, which in turn raises the rate of dehydration, shrinkage of the samples, and rate of rehydration. Microwave cooking is capable of preserving antioxidant activity, bioactive constituents in vegetables, and enhancing the in vitro protein digestibility of foods by considerably decreasing antinutritional factors, moreover, sufficient degradation in nutrients may occur if processing with ample amounts of water. Microwave sterilization is efficiently used to achieve microbial safety of foods. It also achieves minimum color changes, antioxidant activity, and bioactive constituents due to the deactivation of enzyme activity and less processing time. In addition, the resistance heating rate depends largely on the applied voltage gradient, temperature, size of particle, frequency, and electrolytes concentration. It rises linearly with increase in concentration of ionic constituents, temperature, and voltage gradient. The influence of ohmic heat treatment on nutritional value and quality of vegetables and fruits has been studied. During ohmic heating, applied electrical fields cause rapid deactivation of microorganisms and enzymes, resulting in enhanced shelf-life.

References

Ahmed, N., Singh, J., Chauhan, H., Gupat, P., Anjum, A., & Kaur, H. (2013). Different drying methods: Their application and recent advances. *International Journal of Food Nutrition and Safety, 4*(1), 34–42.

Akdaş, Z. Z., & Bakkalbaşı, E. (2016). Influence of different cooking methods on color, bioactive compounds, and antioxidant activity of kale. *International Journal of Food Properties, 20*(4), 877–887.

Antunes, M. D. C., Dandlen, S., Cavaco, A. M., & Miguel, G. (2010). Effects of postharvest application of 1-mcp and postcutting dip treatment on the quality and nutritional properties of fresh-cut kiwifruit. *Journal of Agricultural and Food Chemistry, 58*(10), 6173–6181.

Armesto, J., Gomez-Limia, L., Carballo, J., & Martínez, S. (2016). Effects of different cooking methods on some chemical and sensory properties of Galega kale. *International Journal of Food Science & Technology, 51*(9), 2071–2080.

Arroqui, C., Rumsey, T. R., Lopez, A., & Virseda, P. (2002). Losses by diffusion of ascorbic acid during recycled water blanching of potato tissue. *Journal of Food Engineering, 52*, 25–30.

Assiry, A. M., Gaily, M. H., Alsamee, M., & Sarifudin, A. (2010). Electrical conductivity of seawater during ohmic heating. *Desalination, 260*, 9–17.
Ayadi, M. A., Leuliet, J. C., Chopard, F., Berthou, M., & Lebouche, M. (2005). "Experimental study of hydrodynamics in a flat ohmic cell- impact on fouling by dairy products". *Journal of Food Engineering, 70*, 489–498.
Aydogdu, A., Sumnu, G., & Sahin, S. (2015). Effects of microwave-infrared combination drying on quality of eggplants. *Food and Bioprocess Technology, 8*(6), 1198–1210.
Basak, T., & Ayappa, K. G. (2002). Role of length scales on microwave thawing dynamics in 2D cylinders. *International Journal of Heat and Mass Transfer, 45*, 4543–4559.
Biss, C. H., Combes, S. A., & Skudder, P. J. (1989). The development and application of ohmic heating for the continuous processing of particulate foodstuffs. In R. W. Field & J. A. Howell (Eds.), *Processing engineering in the food industry* (pp. 17–27).
Bozkurt, H., & Icier, F. (2009). Rheological characteristics of quince nectar during Ohmic heating. *International Journal of Food Properties, 12*, 844–859.
Brownmiller, C., Howard, L. R., & Prior, R. L. (2008). Processing and storage effects on monomeric anthocyanins, percent polymeric color, and antioxidant capacity of processed blueberry products. *Journal of Food Science, 73*(5), H72–H79.
Buffler, C. R. (1993). *Microwave cooking and processing: Engineering fundamentals for the food scientist.* 1st ed. New York: Van Nostrand Reinhold.
Castro, I., Macedo, B., Teixeira, J. A., & Vicente, A. A. (2004). The effect of electric field on important food-processing enzymes: Comparison of inactivation kinetics under conventional and ohmic heating. *Journal of Food Science, 69*(9), C696–C701.
Castro, I., Teixeira, J. A., Salengke, S., Sastry, S. K., & Vicente, A. A. (2003). The influence of field strength, sugar and solid content on electrical conductivity of strawberry products. *Journal of Food Process Engineering, 26*, 17–29.
Cemeroglu, B., Velioglu, S., & Isik, S. (1994). Degradation kinetics of anthocyanins in sour cherry juice and concentrate. *Journal of Food Science, 59*(6), 1216–1218.
Chamchong, M., & Datta, A. K. (1999). Thawing of foods in a microwave oven: I. Effect of power levels and power cycling. *Journal of Microwave Power and Electromagnetic Energy, 34*, 9–21.
Chen, Z., Li, Y., Wang, L., Liu, S., Wang, K., Sun, J., et al. (2016). Evaluation of the possible non-thermal effect of microwave radiation on the inactivation of wheat germ lipase. *Journal of Food Process Engineering*, 1–11.
Cohen, J. S., & Yang, T. C. S. (1995). Progress in food dehydration. *Trends in Food Science & Technology, 6*, 20–25.
Contreras, C., Martín, M. E., Martínez-Navarrete, N., & Chiralt, A. (2005). Effect of vacuum impregnation and microwave application on structural changes which occurred during air-drying of apple. *Lebensmittel-Wissenschaft und-Technologie, 38*, 471–477.
Contreras, C., Martín-Esparza, E., & Martínez-Navarrete, N. (2012). Influence of drying method on the rehydration properties of apricot and apple. *Journal of Food Process Engineering, 35*, 178–190.
Darvishi, H., Khostaghaza, M. H., & Najafi, G. (2013). Ohmic heating of pomegranate juice: Electrical conductivity and pH change. *Journal of the Saudi Society of Agricultural Sciences, 12*, 101–108.
Datta, A. K., & Anantheswaran, R. C. (2000). *Handbook of microwave technology for food applications.* New York: Marcel Dekker Inc.
Datta, A. K., & Davidson, P. M. (2000). Microwave and radio frequency processing. *Journal of Food Science, 65*, 32–41.
Davies, L. J., Kemp, M. R., & Fryer, P. J. (1999). The geometry of shadows: Effects of inhomogeneities in electrical field processing. *Journal of Food Engineering, 40*, 245–258.

de Alwis, A. A. P., Halden, K., & Fryer, P. J. (1989). Shape and conductivity effects in the ohmic heating of foods. *Chemical Engineering Research and Design, 67*, 159–168.

De La Vega-Miranda, B., Santiesteban-Lopez, N. A., Lopez-Malo, A., & SosaMorales, M. E. (2012). Inactivation of Salmonella typhimurium in fresh vegetables using water-assisted microwave heating. *Food Control, 26*(1), 19–22.

de Lima, A. C. S., da Rocha Viana, J. D., de Sousa Sabino, L. B., da Silva, L. M. R., da Silva, N. K. V., & de Sousa, P. H. M. (2017). Processing of three different cooking methods of cassava: Effects on in vitro bioaccessibility of phenolic compounds and antioxidant activity. *LWT-Food Science and Technology, 76*, 253–258.

Demirdoven, A., & Baysal, T. (2014). Optimization of ohmic heating applications for pectin methylesterase inactivation in orange juice. *Journal of Food Science and Technology, 51*, 1817–1826.

Dolinsky, M., Agostinho, C., Ribeiro, D., Rocha, G. D. S., Barroso, S. G., Ferreira, D., et al. (2015). Effect of different cooking methods on the polyphenol concentration and antioxidant capacity of selected vegetables. *Journal of Culinary Science & Technology, 14*(1), 1–12.

dos Reis, L. C. R., de Oliveira, V. R., Hagen, M. E. K., Jablonski, A., Flôres, S. H., & de Oliveira Rios, A. (2015). Carotenoids, flavonoids, chlorophylls, phenolic compounds and antioxidant activity in fresh and cooked broccoli (Brassica oleracea var. Avenger) and cauliflower (Brassica oleracea var. Alphina F1). *LWT-Food Science and Technology, 63*(1), 177–183.

Drouzas, A. E., & Schubert, H. (1996). Microwave application in vacuum drying of fruits. *Journal of Food Engineering, 28*, 203–209.

Eliot, S. C., Goullieux, A., & Pain, J. P. (1999). Processing of cauliflower by ohmic heating: Influence of precooking on firmness. *Journal of the Science of Food and Agriculture, 79*, 1406–1412.

Farahnaky, A., Azizi, R., & Gavahian, M. (2012). Accelerated texture softening of some root vegetables by ohmic heating. *Journal of Food Engineering, 113*, 275–280.

Fryer, B. P. J., Miri, T., & Parral, G. P. (2012). The effect of electrical processing on mass transfer in beetroot and model gels. *Journal of Food Engineering, 112*, 208–217.

Gally, T., Rouaud, O., Jury, V., Havet, M., Oge, A., & Le-Bail, A. (2017). Proofing of bread dough assisted by ohmic heating. *Innovative Food Science and Emerging Technologies, 39*, 55–62.

Garzon, G. A., & Wrolstad, R. E. (2002). Comparison of the stability of pelargonidin based anthocyanins in strawberry juice and concentrate. *Journal of Food Science, 67*(4), 1288–1299.

Ghnimi, S., Malaspina, N. F., & Zaıd, I. (2009). *Proce ́de ́ de chauffage des fluides visqueux etencrassant a ' jet de fluide avec capteur radar.* [Patent EP09151169].

Goullieux, A., & Pain, J.-P. (2005). Ohmic heating. In *Emerging technologies for food processing* (pp. 469–505). London: Academic Press.

Goullieux, A., Pain, J. P., & Baudez, P. (1997). Anewpilot-scalebatchohmic heater. In *Proceedings of the 7th international congress on engineering and food, part I (pp. C29–C32)*. Sheffield: Sheffield Academic Press.

Goullieux, A., Zuber, F., & Godereaux, S. C. E. (2001). Processing and stabilisation of cauliflower by ohmic heating technology. *Innovative Food Science and Emerging Technologies, 2*, 279–287.

Gowen, A., Abu-Ghannam, N., Frias, J., & Oliveira, J. (2006). Optimisation of dehydration and rehydration properties of cooked chickpeas (Cicer arietinum L.) undergoing microwave—Hot air combination drying. *Trends in Food Science and Technology, 17*(4), 177–183.

Gupta, M., & Wong, W. L. E. (2007). *Microwaves and metals*. Singapore: John Wiley & Sons (Asia) Pte. Ltd.

Halden, K., De Alwis, A. A., & Fryer, P. J. (1990). Changes in the electrical conductivity of foods during ohmic heating. *International Journal of Food Science and Technology, 25*, 9–35.

Herting, G., Wallinder, I. O., & Leygraf, C. (2008). Corrosion-induced release of chromium and iron from ferritic stainless steel grade AISI 430 in simulated food contact. *Journal of Food Engineering, 87*, 291–300.

Horuz, E., & Maskan, M. (2015). Hot air and microwave drying of pomegranate (Punica granatum L.) arils. *Journal of Food Science and Technology, 52*(1), 285–293.

Icier, F, & Baysal, T. (2004a). Dielectrical properties of food materials—1: Factors affecting and industrial uses. *Critical Reviews in Food Science and Nutrition, 44*, 465–471.

Icier, F., & Baysal, T. (2004b). Dielectric properties of food materials-2: Measurement techniques. *Critical Reviews in Food Science and Nutrition, 44*, 473–478.

Icier, F., & Ilicali, C. (2004). Electrical conductivity of apple and sourcherry juice concentrates during ohmic heating. *Journal of Food Process Engineering, 27* (3), 159–180.

Icier, F., & Ilicali, C. (2005). Temperature dependent electrical conductivities of fruit purees during ohmic heating. *Food Research International, 38*, 1135–1142.

Icier, F., Yildiz, H., & Raysal, T. (2006). Peroxidase inactivation and color changes during ohmic heating of pea puree. *Journal of Food Engineering, 74*(3), 424–429.

Icier, F., Yildiz, H., & Raysal, T. (2008). Polyphenoloxidase deactivation kinetics during ohmic heating of grape fruit. *Journal of Food Engineering, 85*(3), 410–417.

Imai, T., Uemura, K., Ishida, N., Yoshizaki, S., & Noguchi, A. (1995). Ohmic heating of Japanese white radish Rhaphanus sativus L. *International Journal of Food Science and Technology, 30*, 461–472.

Jakob, A., Bryjak, J., Wojtowicz, H., Illeova, V., Annus, J., & Polakovic, M. (2010). Inactivation kinetics of food enzymes during ohmic heating. *Food Chemistry, 123*(2), 369–376.

Kaur, N., & Singh, A. (2016). Ohmic heating: Concept and applications—A review. *Critical Reviews in Food Science and Nutrition, 56*, 2338–2351.

Kim, H. J., Choi, Y. M., Yang, T. C. S., Taub, I. A., Tempest, P., Skudder, P., et al. (1996). Validation of ohmic heating for quality enhancement of food products. *Food Technology*, 253–261.

Kulshrestha, S., & Sastry, S. (2003). Frequency and voltage effects on enhanced diffusion during moderate electric field (MEF) treatment. *Innovative Food Science and Emerging Technologies, 4*(2), 189–194.

Latorre, M. E., Bonelli, P. R., Rojas, A. M., & Gerschenson, L. N. (2012). Microwave inactivation of red beet (Beta vulgaris L. var. conditiva) peroxidase and polyphenoloxidase and the effect of radiation on vegetable tissue quality. *Journal of Food Engineering, 109*, 676–684.

Leadley, C. (2008). Novel commercial preservation methods. In G. Tucker (Ed.), *Food biodeterioration and preservation*. Oxford: Blackwell Pub.

Lebovka, N. I., Praporscic, I., Ghnimi, S., & Vorobiev, E. (2005). Does electroporation occur during the ohmic heating of food? *Journal of Food Science, 70*, E308–E311.

Leizerson, S., & Shimoni, E. (2005). Effect of ultrahigh-temperature continuous ohmic heating treatment on fresh orange juice. *Journal of Agricultural and Food Chemistry, 53*(9), 3519–3524.

Leizerson, S., & Shimoni, E. (2005b). Stability and sensory shelf life of orange juice pasteurized by continuous ohmic heating. *Journal of Agricultural and Food Chemistry, 53*, 4012–4018.

Lemmens, L., Tiback, E., Svelander, C., Smout, C., Ahrne, L., Langton, M., et al. (2009). Thermal pretreatments of carrot pieces using different heating techniques: Effect on quality related aspects. *Innovative Food Science & Emerging Technologies, 10*, 522–529.

Lima, J. R., Elizondo, N. J., & Bohuon, P. (2010). Kinetics of ascorbic acid degradation and colour change in ground cashew apples treated at high temperatures (100-180°C). *International Journal of Food Science and Technology, 45*, 1724–1731.

Lima, M., Heskitt, B. F., Burianek, L. L., Nokes, S. E., & Sastry, S. K. (1999). Ascorbic acid degradation kinetics during conventional and ohmic heating. *Journal of Food Processing & Preservation, 23*(5), 421–443.

Lima, M., Heskitt, B., & Sastry, S. (1999). The effect of frequency and wave form on the electrical-conductivity temperature profiles of turnip tissue. *Journal of Food Process Engineering, 22*, 41–54.

Lima, M., Heskitt, B. F., & Sastry, S. K. (2001). Diffusion of beet dye during electrical and conventional heating at steady-state temperature. *Journal of Food Process Engineering, 24*, 331–340.

Ling, B., Tang, J., Kong, F., Mitcham, E. J., & Wang, S. (2014). Kinetics of food quality changes during thermal processing: A review. *Food and Bioprocess Technology, 8* (2), 343–358.

Lombrana, J. I., Rodriguez, R., & Ruiz, U. (2010). Microwave-drying of sliced mushroom. Analysis of temperature control and pressure. *Innovative Food Science & Emerging Technologies, 11*, 652–660.

Lu, Y., Turley, A., Dong, X., & Wu, C. (2011). Reduction of Salmonella enterica on grape tomatoes using microwave heating. *International Journal of Food Microbiology, 145*(1), 349–352.

Lv, W., Zhang, M., Bhandari, B., Yang, Z., & Wang, Y. (2017). Analysis about drying properties and vacuum impregnated qualities of edamame (glycine max (L.) Merrill). *Drying Technology,*. https://doi.org/10.1080/07373937.2016.1231201.

Marczak, L. D. F., Tessaro, I. C., Jaeschke, D. P., & Sarkis, J. R. (2013). Effects of ohmic and conventional heating on anthocyanin degradation during the processing of blueberry pulp. *LWT-Food Science and Technology, 51*, 79–85.

Marszałek, K., Mitek, M., & Skąpska, S. (2015). Effect of continuous flow microwave and conventional heating on the bioactive compounds, colour, enzymes activity, microbial and sensory quality of strawberry puree. *Food and Bioprocess Technology, 8*(9), 1864–1876.

Martín, M. E., Albors, A., Martínez-Navarrete, N., Chiralt, A., & Fito, P. (2000). Micro-structural changes in apple tissue subjected to combined air-microwave drying. In *Proceedings of the 12th international drying symposium IDS2000.* Elsevier Science, ISBN 0-444-50422-2. Paper no. 419 (CD-ROM).

Mercali, G. D., Jaeschke, D. P., Tessaro, I. C., & Marczak, L. D. F. (2012). Study of vitamin C degradation in acerola pulp during ohmic and conventional heat treatment. *LWT-Food Science and Technology, 47*, 91–95.

Mercali, G. D., Jaeschke, D. P., Tessaro, I. C., & Marczak, L. D. F. (2013). Degradation kinetics of anthocyanins in acerola pulp: Comparison between ohmic and conventional heat treatment. *Food Chemistry, 136*, 853–857.

Mercali, G. D., Schwartz, S., Marczak, L. D. F., Tessaro, I. C., & Sastry, S. (2014). Ascorbic acid degradation and color changes in acerola pulp during ohmic heating: Effect of electric field frequency. *Journal of Food Engineering, 123*, 1–7.

Metaxas, A. C., & Meredith, R. J. (1983). *Industrial microwave heating.* London: Peter Peregrinus Ltd.

Mizrahi, S. (1996). Leaching of soluble solids during blanching of vegetables by ohmic heating. *Journal of Food Engineering, 29*, 153–166.

Moreno, J., Simpson, R., Baeza, A., Morales, J., Munoz, C., Sastry, S., et al. (2012). Effect of ohmic heating and vacuum impregnation on the osmodehydration kinetics and microstructure of strawberries. *LWT-Food Science and Technology, 45*, 148–154.

Moreno, J., Simpson, R., Sayas, M., Segura, I., Aldana, O., & Almonacid, S. (2011). Influence of ohmic heating and vacuum impregnation on the osmotic dehydration kinetics and microstructure of pears. *Journal of Food Engineering, 104*, 621–627.

Mudgett, R. (1985). Dielectric properties of foods. In *Microwaves in the food processing industries* (pp. 15–37). New York: Academic Press.

Munoz, S., Achaerandio, I., Yang, Y., & Pujola, M. (2017). Sous vide processing as an alternative to common cooking treatments: Impact on the starch profile, color, and shear force of potato (Solanum tuberosum L.). *Food and Bioprocess Technology*, 1–11.

Nawirska-Olszanska, A., Stępień, B., & Biesiada, A. (2017). Effectiveness of the fountain microwave drying method in some selected pumpkin cultivars. *LWT-Food Science and Technology, 77*, 276–281.

Nayak, B., Liu, R. H., & Tang, J. (2015). Effect of processing on phenolic antioxidants of fruits, vegetables, and grains—A review. *Critical Reviews in Food Science and Nutrition, 55*(7), 887–919.

Nelson, S. O., Forbus, W. R., Jr., & Lawrence, K. C. (1994). Microwave permittivities of fresh fruits and vegetables from 0.2 to 20GHz. *Transactions of the ASAE, 37*(1), 183–189.

Noguchi, A. (2004). Potential of ohmic heating and high pressure cooking for practical use in soy protein processing. In *Proceedings VII world soybean research conference, IV international soybean processing and utilization conference. III Congresso Brasileiro de Soja Brazilian soybean congress, Foz do Iguassu, PR, Brazil*, (pp. 1094–1102).

Nott, K. P., & Hall, L. D. (1999). Advances in temperature validation of foods. *Trends in Food Science & Technology, 10*, 366–374.

Orsat, V., Changrue, V., & Raghavan, G. S. V. (2006). Microwave drying of fruits and vegetables. *Stewart Postharvest Review, 6*, 4–9.

Oszmianski, J., Wojdylo, A., & Kolniak, J. (2009). Effect of L-ascorbic acid, sugar, pectin and freeze-thaw treatment on polyphenol content of frozen strawberries. *LWT-Food Science and Technology, 42*, 581–586.

Ozkan, I. A., Akbudak, B., & Akbudak, N. (2007). Microwave drying characteristics of spinach. *Journal of Food Engineering, 78*, 577–583.

Palaniappan, S., & Sastry, S. K. (1991a). Electrical conductivities of selected solid foods during ohmic heating. *Journal of Food Process Engineering, 14*(3), 221–236.

Palaniappan, S., & Sastry, S. K. (1991b). Electrical conductivity of selected juices: Influences of temperature, solids content, applied voltage, and particle size. *Journal of Food Process Engineering, 14*, 247–260.

Parker, T. R. (2001). *Method of and line for breadcrumb production*. [Patent US6399130].

Pataro, G., Donsi, G., & Ferrari, G. (2011). Aseptic processing of apricots in syrup by means of a continuous pilot scale ohmic unit. *LWT-Food Science and Technology, 44*(6), 1546–1554.

Pellegrini, N., Chiavaro, E., Gardana, C., Mazzeo, T., Contino, D., Gallo, M., et al. (2010). Effect of different cooking methods on color, phytochemical concentration, and antioxidant capacity of raw and frozen brassica vegetables. *Journal of Agricultural and Food Chemistry, 58*(7), 4310–4321.

Pereira, R. N., & Vincente, A. A. (2010). Environmental impact of novel thermal and nonthermal technologies in food processing. *Food Research International, 43*, 1936–1943.

Piasek, A., Kusznierewicz, B., Grzybowska, I., Malinowska-Panczyk, E., Piekarska, A., Azqueta, A., et al. (2011). The influence of sterilization with EnbioJet® microwave flow pasteurizer on composition and bioactivity of aronia and blueberried honeysuckle juices. *Journal of Food Composition and Analysis, 24*(6), 880–888.

Podsędek, A. (2007). Natural antioxidants and antioxidant capacity of Brassica vegetables: A review. *LWT-Food Science and Technology, 40*(1), 1–11.

Polny, J. R., & Thaddeus, J. (1996). *Apparatus for electroheating food employing concentric electrodes.* [Patent US5562024].

Prakash, S., Jha, S. K., & Datta, N. (2003). Performance evaluation of blanched carrots dried by three different driers. *Journal of Food Engineering, 62,* 305–313.

Praporscic, I., Ghnimi, S., & Vorobiev, E. (2005). Enhancement of pressing of sugar beet cuts by combined ohmic heating and pulsed electric field treatment. *Journal of Food Processing and Preservation, 29,* 378–389.

Praporscic, I., Lebovka, N. I., Ghnimi, S., & Vorobiev, E. (2006). Ohmically heated, enhanced expression of juice from apple and potato tissues. *Biosystems Engineering, 93*(2), 199–204.

Ramaswamy, H. S., Marcotte, M., Sastry, S., & Abdelrahim, K. (2014). *Ohmic heating in food processing.* Boca Raton: CRC Press, Taylor & Francis Group.

Ranesh, M. N. (1999). Food preservation by heat treatment. In R. S. Rahman (Ed.), *Handbook of food preservation* (pp. 95–172). New York: Marcel Dekker.

Reitler, W., & Rudolph, M. (1986). Einsatz von elektrischem Strom fuer die konduktive Erwaermung von Nahrungsmitteln. *Elektrowärme International, 44*(B6), 275–279.

Reznik, D. (1997). *Electroheating methods.* [Patent US63580].

Reznik, D. (2000). *Conical shaped electrolyte electrode for electroheating.* [Patent US6088509].

Ruan, R., Chen, P., Chang, K., Kim, H. J., & Taub, I. A. (1999). Rapid food particle temperature mapping during ohmic heating using FLASH MRI. *Journal of Food Science, 64*(6), 1024–1026.

Ryynanen, S. (1995). The electromagnetic properties of food materials: A review of the basic principles. *Journal of Food Engineering, 26,* 409–429.

Sakr, M., & Liu, S. (2014). A comprehensive review on applications of ohmic heating (OH). *Renewable and Sustainable Energy Reviews, 39,* 262–269.

Salazar-Gonzalez, C., San Martin-Gonzalez, M. F., Lopez-Malo, A., & Sosa-Morales, M. E. (2012). Recent studies related to microwave processing of fluid foods. *Food Bioprocess and Technology, 5,* 31–46.

Sarang, S., Sastry, S. K., & Knipe, L. (2008). Electrical conductivity of fruits and meats during ohmic heating. *Journal of Food Engineering, 87*(3), 351–356.

Sarimeseli, A. (2011). Microwave drying characteristics of coriander (Coriandrum sativum L.) leaves. *Energy Conversion and Management, 52*(2), 1449–1453.

Sarkis, J. R., Mercali, G. D., Tessaro, I. C., & Marczak, L. D. F. (2013). Evaluation of key parameters during construction and operation of an ohmic heating apparatus. *Innovative Food Science and Emerging Technologies, 18,* 145–154.

Sastry, S. K., Heskitt, B. F., Jun, S., & Somavat, R. (2013). *Ohmic heating packet.* Patent US2013/0062332.

Sastry, S. K., & Palaniappan, S. (1992a). Influence of particle orientation on the effective electrical resistance and ohmic heating rate of a liquid-particle mixture 1. *Journal of Food Process Engineering, 15,* 213–227.

Sastry, S. K., & Palaniappan, S. (1992b). Mathematical modelling and experimental studies on ohmic heating of liquid–particle mixtures in a static heater. *Journal of Food Process Engineering, 15,* 241–261.

Sensoy, I., & Sastry, S. K. (2004). Ohmic blanching of mushrooms. *Journal of Food Process Engineering, 27,* 1–15.

Shaheen, M. S., El-Massry, K. F., El-Ghorab, A. H., & Anjum, F. M. (2012). Microwave applications in thermal food processing. In W. Cao (Ed.), *The development and application of microwave heating* (pp. 3–16). Croatia: InTech.

Shirsat, N., Lyng, J. G., Brunton, N. P., & Mckenna, B. M. (2004). Conductivities and ohmic heating of meat emulsion batters. *Journal of Muscle Foods, 15*, 121–137.

Shynkaryk, M. V., Ji, T., Alvarez, V. B., & Sastry, S. K. (2010). Ohmic heating of peaches in the wide range of frequencies (50 Hz to 1 MHz). *Journal of Food Science, 75*, E493–E500.

Simpson, D. P. (1994). *Internal resistance ohmic heating apparatus for fluids, UK.* Patent no. GB2268671A.

Singh, M., Raghavan, B., & Abraham, K. O. (1996). Processing of marjoram (Marjona hortensis Moench.) and rosemary (Rosmarinus officinalis L.). Effect of blanching methods on quality. *Nahrung, 40*, 264–266.

Singh, H., Singh, S. P., & PS, T. (2008). Study on viscosity and electrical conductivity of fruit juices. *Journal of Food Science and Technology, 45*(4), 371–372.

Skudder, P. J., & Biss, C. H. (1987). Aseptic processing of food products using ohmic heating. *The Chemical Engineer, 433*, 26–28.

Sosa-Morales, M. E., Valerio-Junco, L., Lopez-Malo, A., & Garcia, H. S. (2010). Dielectric properties of foods: Reported data in the 21st century and their potential applications. *LWT-Food Science and Technology, 43*, 1169–1179.

Soysal, Y. (2004). Microwave drying characteristics of parsley. *Biosystems Engineering, 89*, 167–173.

Srikalong, P., Makrudin, T., Sampavamontri, P., & Kovitthaya, E. (2011). Effect of ohmic heating on increasing guava juice yield. In Vol. 7. *2nd international conference on biotechnology and food science.*

Stancl, J., & Zitny, R. (2010). Milk fouling at direct ohmic heating. *Journal of Food Engineering, 99*, 437–444.

Stirling, R. (1987). Ohmic heating-A new process for the food industry. *Power Engineering Journal, 6*, 365–371.

Taher, B. J., & Farid, M. M. (2001). Cyclic microwave thawing of frozen meat: Experimental and theoretical investigation. *Chemical Engineering and Processing, 40*, 379–389.

Tian, J., Chen, J., Lv, F., Chen, S., Chen, J., Liu, D., et al. (2016). Domestic cooking methods affect the phytochemical composition and antioxidant activity of purple-fleshed potatoes. *Food Chemistry, 197*, 1264–1270.

Tran, V. N., Stuchly, S. S., & Kraszewski, A. W. (1984). Dielectric properties of selected vegetables and fruits 0.1–10GHz. *Journal of Microwave Power, 19*(4), 251–258.

Turner, I. W., & Jolly, P. (1991). Combined microwave and convective drying of a porous material. *Drying Technology An International Journal, 9*, 1209–1270.

Uemura, K., & Isobe, S. (2003). Developing a new apparatus for inactivating Bacillus subtilis spore in orange juice with a high electric field AC under pressurized conditions. *Journal of Food Engineering, 56*(4), 325–329.

Umudee, I., Chongcheawchamnan, M., Kiatweerasakul, M., & Tongurai, C. (2013). Sterilization of oil palm fresh fruit using microwave technique. *International Journal of Chemical Engineering and Applications, 4*(3), 111–113.

Vadivambal, R., & Jayas, D. S. (2010). Non-uniform temperature distribution during microwave heating of food materials—A review. *Food and Bioprocess Technology, 3*, 161–171.

Vikram, V. B., Ramesh, M. N., & Prapulla, S. G. (2005). Thermal degradation kinetics of nutrients in orange juice heated by electromagnetic and conventional methods. *Journal of Food Engineering, 69*(1), 31–40.

Volden, J., Borge, G. I. A., Bentsson, G. B., Hansen, M., Thygesen, I. E., & Wicklund, T. (2008). Effect of thermal treatment on glucosinolates and antioxidant-related parameters in red cabbage. *Food Chemistry, 109*(3), 595–605.

Wachtel-Galor, S., Wong, K. W., & Benzie, I. F. F. (2008). The effect of cooking on Brassica vegetables. *Food Chemistry, 110*(3), 706–710.

Wang, W. C. (1995). *Ohmic heating of foods: Physical properties and applications.* Columbus, OH: The Ohio State University.

Wang, W. C., & Sastry, S. K. (1997a). Changes in electrical conductivity of selected vegetables during multiple thermal treatments. *Journal of Food Process Engineering, 20,* 499–516.

Wang, W. C., & Sastry, S. K. (1997b). Starch gelatinization in ohmic heating. *Journal of Food Engineering, 34,* 225–242.

Wang, W. C., & Sastry, S. K. (2000). Effects of thermal and electrothermal pretreatments on hot air drying rate of vegetable tissue. *Journal of Food Process Engineering, 23*(4), 299–319.

Wang, J., & Xi, Y. S. (2005). Drying characteristics and drying quality of carrot using a twostage microwave process. *Journal of Food Engineering, 68,* 505–511.

Wojdyło, A., Figiel, A., Lech, K., Nowicka, P., & Oszmianski, J. (2013). Effect of convective and vacuumemicrowave drying on the bioactive compounds, color, and antioxidant capacity of sour cherries. *Food and Bioprocess Technology, 7*(3), 829–841.

Xu, F., Zheng, Y., Yang, Z., Cao, S., Shao, X., & Wang, H. (2014). Domestic cooking methods affect the nutritional quality of red cabbage. *Food Chemistry, 161,* 162–167.

Yaghmaee, P., & Durance, T. D. (2001). Predictive equation for dielectric properties of NaCl, D-sorbitol and sucrose solutions and surimi at 2450 MHz. *Journal of Food Science, 67*(6), 2207–2211.

Yildiz, H., Icier, F., & Baysal, T. (2010). Changes in beta-carotene, chlorophyll and color of spinach puree during Ohmic heating. *Journal of Food Process Engineering, 33,* 763–779.

Zareifard, M. R., Ramaswamy, H. S., Trigui, M., & Marcotte, M. (2003). Ohmic heating behaviour and electrical conductivity of two-phase food systems. *Innovative Food Science and Emerging Technologies, 4,* 45–55.

Zell, M., Lyng, J. G., Cronin, D. A., & Morgan, D. J. (2009). Ohmic cooking of whole beef muscle–optimization of meat preparation. *Meat Science, 81,* 693–698.

Zell, M., Lyng, J. G., Morgan, D. J., & Cronin, D. A. (2009). Development of rapid response thermocouple probes for use in a batch ohmic heating system. *Journal of Food Engineering, 93,* 344–347.

Zhang, M., Tang, J., Mujumdar, A. S., & Wang, S. (2006). Trends in microwave-related drying of fruits and vegetables. *Trends in Food Science & Technology, 17,* 524–534.

Zhong, X., Dolan, K. D., & Almenar, E. (2015). Effect of steamable bag microwaving versus traditional cooking methods on nutritional preservation and physical properties of frozen vegetables: A case study on broccoli (Brassica oleracea). *Innovative Food Science & Emerging Technologies, 31,* 116–122.

Zhong, T., & Lima, M. (2003). The effect of ohmic heating on vacuum drying rate of sweet potato tissue. *Bioresource Technology, 87*(3), 215–220.

Zielinska, M., Sadowski, P., & Błaszczak, W. (2015). Freezing/thawing and microwave-assisted drying of blueberries (Vaccinium corymbosum L.). *LWT-Food Science and Technology, 62*(1), 555–563.

Zitoun, K. B. (1997). *Continuous flow of solid-liquid food mixtures during ohmic heating: Fluid interstitial velocities, solid area fraction, orientation and rotation.* Dissertation Abstracts International, B, Vol. 57 (7).

Zuber, F. (1997). Ohmic heating: A new technology for stabilising ready-made dishes. *Viandes et Produits Carnes, 18*(2), 91–95.

Further reading

Benlloch-Tinoco, M., Igual, M., Rodrigo, D., & Martínez-Navarrete, N. (2013). Comparison of microwaves and conventional thermal treatment on enzymes activity and antioxidant capacity of kiwifruit puree. *Innovative Food Science and Emerging Technologies, 19*, 166–172.

Du, G., Li, M., Ma, F., & Liang, D. (2009). Antioxidant capacity and the relationship with polyphenol and vitamin C in Actinidia fruits. *Food Chemistry, 113*(2), 557–562.

Goldblith, S. A. (1966). Basic principles of microwaves and recent developments. *Advances in Food Research, 15*, 277.

Lin, T. M., Durance, D. T., & Scaman, C. H. (1998). Characterization of vacuum microwave, air and freeze dried carrot slices. *Food Research International, 31* (2), 111–117.

Rohini, K. P., & Ms. Chatali, S. P. (2017). Microwave fruit and vegetables drying. *International Advanced Research Journal in Science, Engineering and Technology, 4*, 82–84.

Rohini, K. P., & Uttam, L. B. (2018). Microwave system for fruit drying. *International Journal of Electrical and Electronics Research, 6*, 1–5.

Torringa, E. M., Van Dijk, E. J., & Bartels, P. V. (1996). Microwave puffing of vegetables: Modeling and measurements. In *Proceedings of the 31st microwave power symposium* (pp. 16–19). Manassas: International Microwave Power Institute.

Wang, W. C., Chen, J. I., & Hua, H. H. (2002). Study of liquid-contact by ohmic heating. In *2002 IFTAnnual meeting book of abstracts, paper 91F-4*. Chicago, IL: Institute of Food Technologists.

Wang, W. C., & Sastry, S. K. (1993). Salt diffusion into vegetable tissues as a pretreatment for ohmic heating: Electrical conductivity profiles and vacuum infusion studies. *Journal of Food Engineering, 20*, 299–309.

14

Cold plasma processing of fresh-cut fruits and vegetables

Cherakkathodi Sudheesh, Kappat Valiyapeediyekkal Sunooj
Department of Food Science and Technology, Pondicherry University, Puducherry, India

Fresh-cut produce (fruits and vegetables) can be minimally processed, lightly processed, partially processed, or freshly processed. Fresh-cut fruits and vegetables are one of the bestselling commodities in grocery stores over the last 10 years. As per International Fresh Cut Produce Association (IFPA), fresh-cut produce are “any fruit or vegetable or combination thereof that has been physically altered from its original form, but remains in a fresh state.” Fresh-cut produce is trimmed, peeled, and/or cut into wholly functional forms before packing without the loss of sensory attributes and higher nutritional values.

1 Processing of fresh-cut fruits and vegetables

The processing of fresh-cut fruits and vegetables includes sorting, grading, peeling, trimming, deseeding, and cutting in to specific shape. The flow chart for processing of fresh-cut fruits and vegetables is given in Fig. 1. The quality of fresh-cut fruits and vegetables are decided by consumer-based parameters such as freshness, color, texture, flavor, nutritional prominence, etc.

Fresh-cut produce undergoes more rapid deterioration than unprocessed raw materials. This results from the damage caused during the processing (cutting, peeling, slicing, deseeding, etc.) adopted for the fresh-cut produce. The minimal processing adopted for fresh-cut produce reduces its shelf-life and they exhibit some serious symptoms. The major storage symptoms shown by the fresh-cut produce after processing stages are tissue softening, nutritional loss, dehydration, higher respiration, higher

Fresh-cut Fruits and Vegetables. https://doi.org/10.1016/B978-0-12-816184-5.00014-8

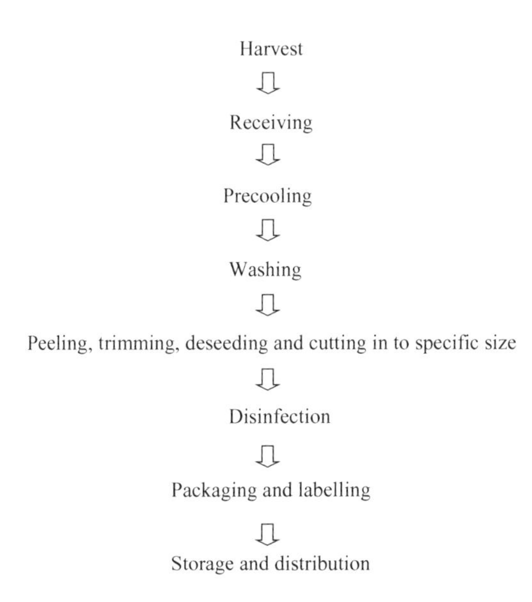

Fig. 1 Flow chart for fresh-cut process of fruits and vegetables. From FAO, (2010). Processing of fresh-cut tropical fruits and vegetables: A technical guide.

ethylene production, enzymatic browning of cut surfaces, and microbial spoilage. The pathogenic microflora grown in the cut surfaces of fruits and vegetables can also create severe health problems in consumers. The commonly grown microflora on the fresh-cut produces are *Salmonella* species, *Escherichia coli, Listeria monocytogenes*, and *Shigella* (Ma, Zhang, Bhandari, & Gao, 2017).

Traditional preservation methods for fresh-cut fruits and vegetables can be classified in to three categories: physical preservation, chemical preservation, and bio preservation. Physical preservation technique is related to controlling the parameters such as temperature, humidity, pressure, and gas composition, etc. (Krasaekoopt & Bhandari, 2010). Low temperature preservation is an efficient preservation technique for fresh-cut fruits and vegetables. It controls the growth of spoilage-causing microorganisms. Cold preservation technique consumes more energy. During storage, certain undesirable changes will occur in fresh-cut produce as the result of changes in the optimal storage temperature. Chemical based preservation technique is related to the use of additives like natural and synthetic ones in the fresh-cut fruits and vegetables (Meireles, Giaouris, & Simões, 2016), but consumers prefer the fresh produce be preserved by using natural additives. Bio preservation technique exploits the antimicrobial potential of natural microorganisms. Antimicrobial agents produced by the microorganisms also have a significant role in the preservation of fresh-cut produce. Bacteriophages, bacteriocins, and bio protective microorganisms perform effective roles in fresh-cut produce preservation (Ma et al., 2017).

2 Quality changes in fresh-cut fruits and vegetables

The quality of fresh-cut fruits and vegetables is decided based on sensorial, textural, and safety points of view. When we consider the fresh-cut produce on the retail market, consumers give more importance to the parameters such as color, flavor, texture, and appearance of the fresh-cut produce for its acceptance. Changes of these parameters will influence the consumer acceptance and shelf-life of the products. Changes occurring in the fresh-cut produce during storage period are given in Table 1.

2.1 Enzymatic browning

Enzymatic browning is the development of a brown color on the cut surfaces of fruits and vegetables. Browning is caused by the interaction of polyphenol oxidase (PPO) with the phenolic substrate. During the processing of fresh-cut produce, tissue damage occurs by cutting and it facilitates browning reaction

Table 1 Quality changes of fresh-cut produce during storage

Quality changes	Fresh-cut produce
Enzymatic browning	Apple
Enzymatic browning	Apricot
Enzymatic browning, decaying	Mango
Juice leakage, softening	Water melon
Enzymatic browning	Lettuce
Enzymatic browning	Potato
Enzymatic browning, loss of moisture, lignin formation	Carrot
Chlorophyll degradation	Broccoli
Enzymatic browning, tissue softening	Eggplant
Softening, fermentation	Cucumber
Softening, fermentation	Onion

From Ma, L., Zhang, M., Bhandari, B., Gao, Z. (2017). Recent developments in novel shelf life extension technologies of fresh-cut fruits and vegetables. Trends in Food Science & Technology, *64, 23–38. doi:10.1016/j.tifs.2017.03.005.*

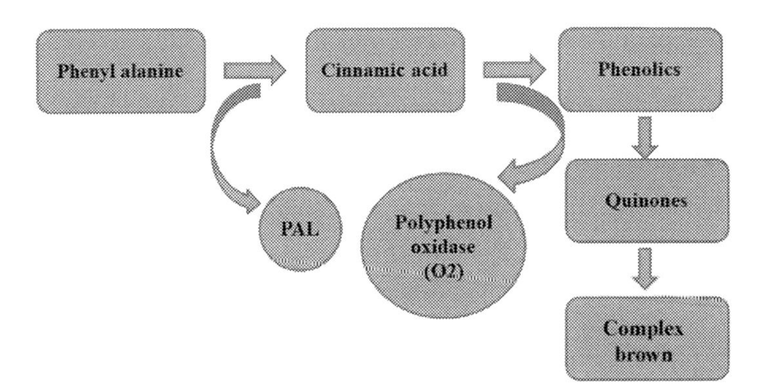

Fig. 2 Phenolic metabolism of browning reaction. From FAO, (2010). Processing of fresh-cut tropical fruits and vegetables: A technical guide.

by increasing the ethylene production and respiration. Phenyl alanine ammonia lyase (PAL) (stimulated by the ethylene production) is an enzyme that catalyzes the development of phenolic compounds. Polyphenolic compounds will convert in to the brown compounds in the presence of oxygen. The enzyme polyphenol oxidase acts as catalyst for this reaction (Fig. 2). Higher respiration rate of fresh-cut produce decreases the moisture, carbohydrates, vitamins, and organic acids. It will affect the flavor and aroma of the fresh-cut produce. Cell wall and membrane degradation also affects the loss of water content. The microbial growth on the cut surfaces will increase due to the higher availability of sugar content and it increases the spoilage level of fresh-cut produce. The factors that affect the rate of enzymatic browning are the amount of polyphenolic compounds, active polyphenol oxidase, temperature, pH, presence of oxygen and antioxidant components, etc. (James & Ngarmsak, 2010; Ma et al., 2017).

2.2 Other color changes

Carotenoids are the lipid soluble pigments present in fruits and vegetables. They provide yellow color to the fruits and vegetables. These pigments undergo oxidative degradation in the presence of an enzyme called lipoxygenase (e.g., broccoli and spinach). Development of yellow color on the green leafy vegetables reduces the quality and storage stability. Translucent appearance on the cut or peeled carrot as the result of dehydration is referred to as white blush and it reduces the consumer acceptability.

2.3 Flavor changes

The flavor part of fresh-cut fruits and vegetables includes sweetness, bitterness, acidity, and astringency. The specific flavor of fruits and vegetables depend on the presence of components like sugar, organic acids, and phenolic and volatile compounds. After cutting the fruits and vegetables, flavor components will be lost by the higher rate of enzymatic reaction. Genetic factors, maturity, and postharvest treatments are other factors that affect the flavor of fresh-cut fruits and vegetables. The aroma of fruits has higher threshold perception than vegetable aroma. It is caused by the presence of high volatile and aliphatic compounds (nitrogen and sulfur compounds) present in the fruits.

The development of off flavor in fresh-cut produce is also due to the presence of certain acids, alcohols, and carbon dioxide (CO_2) gases. Lactic acid bacteria and *Pseudomonas* are the major bacterial species responsible for the off flavor development. Lipase enzyme and break down of amino acids by bacteria will produces certain off flavor compounds in the fresh-cut produce.

2.4 Changes in nutritional quality

The storage conditions of fresh-cut produce affects its nutritional status. The nutritional values of fruits and vegetables depend upon the phytochemical compounds. Both primary and secondary metabolites have a significant role in the nutritional standard of fruits and vegetables. Primary metabolites include vitamins, dietary fibers, and minerals. Carotenoids, flavonoids, and anthocyanin are secondary metabolites. These plant metabolites have a significant role in human and animal diets due to their bioactivity. The biological activity includes antioxidant, antiviral, antibacterial, and antiinflammatory functions. The bioactive components also have numerous roles in the prevention of certain chronic disorders such as hypertension, coronary heart disorder, stroke, and osteoporosis.

2.5 Textural changes

Higher amounts of moisture will leach out at a rapid rate from the cut surfaces of fresh-cut produce during storage. Fresh-cut produce will be firm, crisp, and crunchy in texture. The loss of moisture content leads to the shriveling and wilting of fresh-cut produce. It reduces the crisp and firm texture of the product. Presence of certain enzymes (pectin methyl esterase (PME) and polygalacturonase (PG)) will lead to the dissolution of the peptic cell, reducing the resistance against pressure (Barbagallo, Chisari, & Caputa, 2012). Softening of tissues during the storage period is correlated with degradation of the peptic cell. The fertilizer used during the farming of fruits vegetables has a major role in the texture of products. Higher amounts of potassium and calcium in the fertilizer improve the firmness of fresh-cut produce. Higher amounts of nitrogen reduces the textural properties.

2.6 Quality changes by the microorganisms

Microorganisms associated with fresh-cut fruits and vegetables are less in amount as compared with unprocessed raw fruits and vegetables. Washing of produce by chlorinated water during the processing will reduce the microbial load. Cutting, shredding, and peeling procedures will increase the amount of mesophilic aerobic bacteria. The microbial growth in fresh-cut produce depends on the types of produce and storage conditions. The neutral pH of vegetables contributes to favorable conditions for the growth of bacteria; presence of black or blue color in the fresh-cut produce indicates its spoilage. Cut fruits are spoiled by the growth of yeast and lactic acid bacteria.

The quality of fresh-cut produce could be increased by adopting novel technologies. The novel technologies such as physical technologies (modified atmospheric package, pressurized inert gases, pulsed light, electron beam irradiation, UV light, and cold plasma treatment) and chemical technologies (ozone, acid hydrolyzed water, bacteriocins, bio protective microorganisms) have significant roles in increasing the shelf-life of fresh-cut produce (Ma et al., 2017). Among the physical technologies, cold plasma is a novel and green food processing technology.

3 Cold plasma

The term plasma was introduced in the 1920s by Irving Langmuir. It is a partially or fully ionized state of gas. The phase change from solid to liquid and thereafter to gas is based on the energy input. With increase in the energy given to gaseous state, it will get converted to the system containing ionized particles. This can be called the plasma phase of matter.

3.1 Generation of cold plasma

Plasma is the fourth state of matter and contains electrons, free radicals, neutral atoms, excited molecules, and positive and negative ions. Based on the temperature, plasma can be classified as high temperature plasma and low temperature plasma (cold plasma). High temperature plasma can be created by the use of higher temperature and all the particles are maintained in thermal equilibrium condition. It is not recommended for food processing. Cold plasma, further divided as low pressure plasma and atmospheric pressure plasma, uses atmospheric temperature for plasma generation (Thirumdas, Trimukhe, Deshmukh, & Annapure, 2017). Plasma can be produced by exposing different gases (O_2, N_2, H_2, He, Ne, Ar, NH_3, CH_4, CF_4) to electric fields (between two electrodes) as alternating current (higher frequency) or direct current, magnetic field, thermal, microwave, and radio frequency. It leads to the formation of plasma species like electrons, ions, and free radicals, which make collisions with higher kinetic energy (Thirumdas, Sarangapani, & Annapure, 2014). The net charge of plasma system is zero. The changes caused by the plasma depends on the factors like feed gas composition, humidity, surrounding phase, power, and voltages (Misra, Pankaj, Segat, & Ishikawa, 2016; Sudheesh et al., 2019).

3.2 Plasma categories

Based on the plasma generation, it can be classified as corona discharge plasma, microwave discharge plasma, dielectric barrier discharge plasma (DBD), radio frequency plasma, and gliding arc discharge plasma. DBD can be produced at low pressure and atmospheric pressure. DBD was historically known as silent discharge plasma. A dielectric layer (glass, quartz, ceramic, etc.) is kept between the anode and cathode, an AC voltage (amplitude 1–100 kV) with frequency of few Hz to MHz is applied. When potential difference is created between the electrodes, current will continue to flow through the discharge resulting in direct current discharge. Radiofrequency discharge plasma can be produced by applying AC voltage between the electrodes (anode and cathode). The frequency used for the generation of radiofrequency plasma is in the range of 1 kHz to 10^3 MHz (13.56 MHz is the most commonly used frequency) (Thirumdas et al., 2014).

3.3 Functional properties of cold plasma

Cold plasma has a wide range of applications in the food industry, like increasing shelf-life, enhancing the rate of seed germination, microbial inactivation, alteration of enzyme activity (functionalization or inactivation), adjusting the hydrophilic or hydrophobic properties, etching or deposition of thin films and starch modifications, etc. (Thirumdas et al., 2017).

4 Effect of plasma on fresh-cut produce

Nowadays, cold plasma has a significant role in the processing of fresh-cut fruits and vegetables. The spoilage of fresh-cut produce by microbes, enzymes, and storage conditions can be reduced by the effect of cold plasma processing.

4.1 Effect of plasma on the microbial cell

The plasma species will interact with food material and inactivate or destruct the microbial cells. The interaction between microbial cells and plasma species is the key factor for the effect of cold plasma. The oxidative effect of plasma species lead to the destruction of microbial cell walls. The moisture content has an important role in the microbial destruction. Increase in moisture content will increase the activity of plasma species on the microbial cells. Hence, moist microorganisms get destroyed more easily than drier organisms (Dobrynin, Fridman, Friedman, & Fridman, 2009). Plasma species will damage the DNA molecules present in the chromosome of the microbial cells. Hydrogen peroxide, hydroxyl ion, and superoxide anion are the major reactive oxygen species formed during plasma processing. The water molecules present in the microbial cells undergo ionization and are converted to highly reactive free radicals (hydroxyl free radical) (Fig. 3) and cause damage in the DNA molecules in the nucleus of microbial cells. The formation of malondialdehyde (MDA) by plasma treatment have a significant role in the destruction of microbial cells. Hydroxyl free radicals (OH^-) formed in the hydration layer of the DNA molecules make more damages in the microbial cell as compared to other free radicals. About 90% of DNA damages are mainly due to the hydroxyl free radicals. The oxidation of cellular components such as DNA, chromosomes, nucleus, and cell membrane by the hydroxyl free radicals leads to its destruction (Thirumdas et al., 2014).

Various kinds of reactive oxygen species will affect the microbial cells. The singlet stage of oxygen creates more damage to the

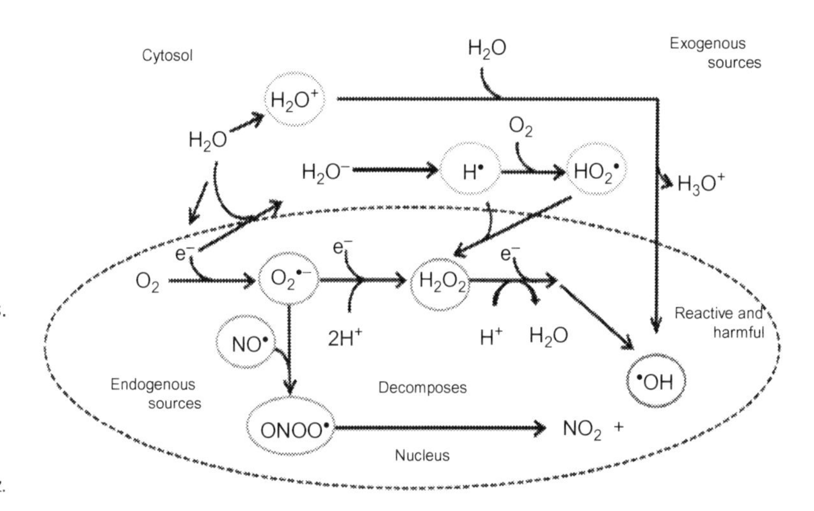

Fig. 3 Dissociation of water molecules into reactive species. From Thirumdas, R., Sarangapani, C., Annapure, U. S. (2014). Cold plasma: A novel non-thermal technology for food processing. *Food Biophysics*, 10, 1–11. doi:10.1007/s11483-014-9382-z.

microbial cells. The atomic oxygen is more reactive than molecular oxygen. The singlet oxygen stage is more susceptible to lipid bi layer. Hence, it makes more damages on the lipid, protein, etc. in the microbial cell. The degradation of lipid bi layer leads to the improper transportation of molecules across the cell membrane. During the plasma processing, plasma species will be bombarded on the microbial cells and make damages on the cellular components. These damages will depend on the interaction between the plasma species like free radicals, positive and negative ions, etc. with cell components. The charges accumulated on the surface of microbial cells create electrostatic force on the cell membrane, causing it to burst. This can be called electropermeabilization. Pulsed electric field treatment (PEF) also uses the same principle for cell degradation.

Microbial reduction of fresh-cut produce such as strawberry, lettuce, cherry, potato, tomato, and cabbage are successfully attained by cold plasma treatment. Microwave powered cold plasma treatment (Fig. 4A) showed 0.3–2.1 log CFU/g reduction of *L. monocytogenes* on cabbage and 1.5 log CFU/g reduction of *Salmonella typhimurium* on lettuce and cabbage (Lee, Kim, Chung, & Min, 2015). The effect of treatment time on the inactivation of *L. monocytogenes* on cabbage is revealed in Fig. 5. While increasing the treatment time, the number of *L. monocytogenes* decreased and a leaner relationship was observed between the level of microbial reduction (log reduction) and treatment time. Jet plasma treatment also showed 2.72, 1.76, and 0.94 log CFU/g reductions of *Salmonella typhimurium* on strawberry, lettuce, and potato, respectively (Fernández, Noriega, & Thompson, 2013). DBD treatment (Fig. 4B) on strawberry reduced 12.85% of mesophilic count and 44%–95% of yeast and mold count. It does not create any changes in the respiration rate and firmness (Fig. 6A and B) (Misra et al., 2014). Some other effects of cold plasma on fresh-cut fruits and vegetables are given in Table 2. The effect of atmospheric pressure cold plasma (APCP) on *E. coli* is shown in Fig. 7. The first images (Fig. 7A and B) show native forms of *E. coli* with smooth surface and rod shape. The size of the bacteria is 2 μm. APCP treated cells showed significant damage on the bacterial cell surface. Fig. 7C showed complete destruction of cells. Plasma treatment leads to the complete loss of cell membrane surrounding the cytoplasmic components. Cell membrane and inner and outer surface of cells showed higher perforations after plasma treatment. The presences of dots surrounding the cells represents the cell debris released to the outside by the effect of plasma.

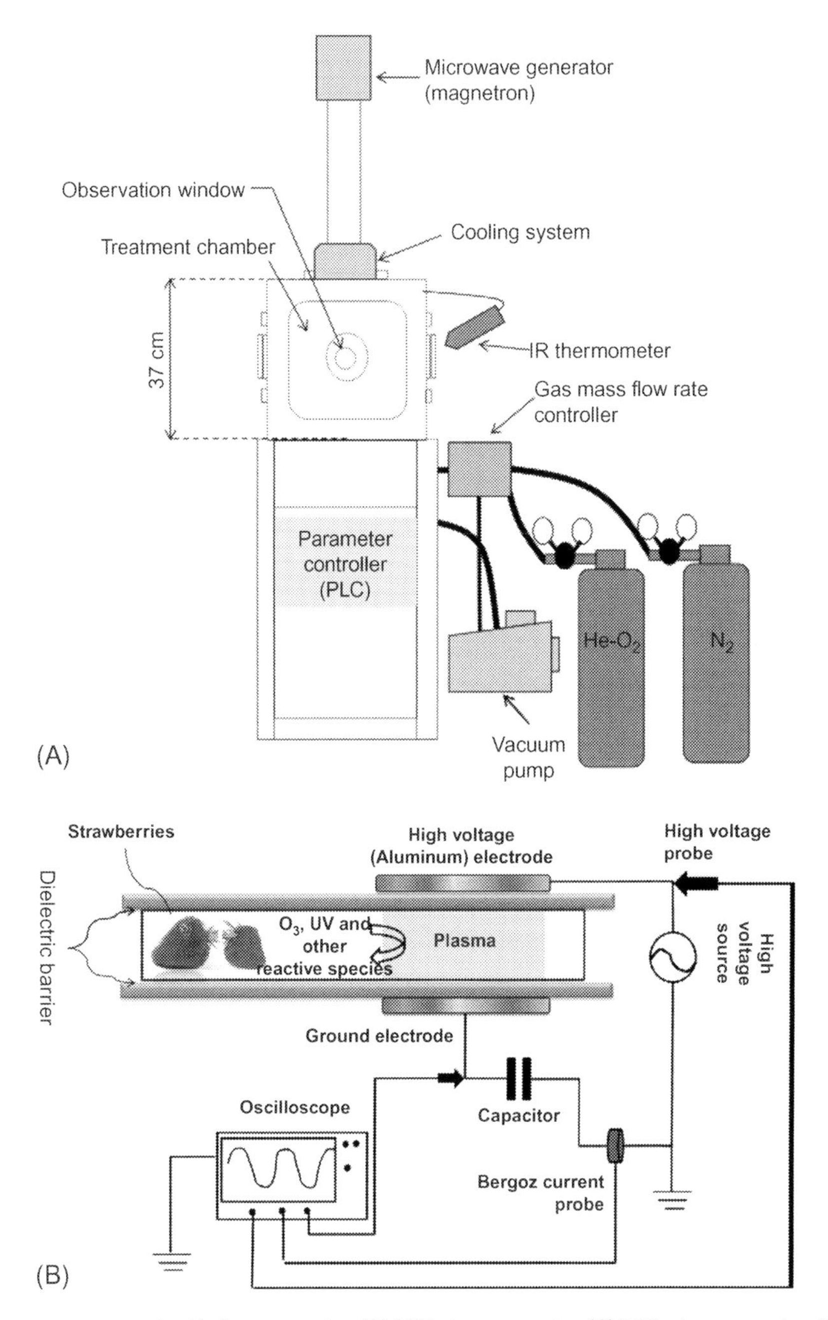

Fig. 4 (A) Microwave powered cold plasma reactor. (B) DBD plasma reactor. (C) DBD plasma reactor. (A) From Lee, H., Kim, J. E., Chung, M. S., Min, S.C. (2015). Cold plasma treatment for the microbiological safety of cabbage, lettuce, and dried figs. *Food Microbiology*, 51, 74–80. doi:10.1016/j.fm.2015.05.004; (B) From Misra, N. N., Patil, S., Moiseev, T., Bourke, P., Mosnier, J. P., Keener, K. M., Cullen, P. J. (2014). In-package atmospheric pressure cold plasma treatment of strawberries. *Journal of Food Engineering*, 125, 131–138. doi:10.1016/j.jfoodeng.2013.10.023; (C) From Tappi, S., Berardinelli, A., Ragni, L., Dalla Rosa, M., Guarnieri, A., Rocculi, P. (2014). Atmospheric gas plasma treatment of fresh-cut apples. *Innovative Food Science and Emerging Technologies*, 21, 114–122. doi:10.1016/j.ifset.2013.09.012.

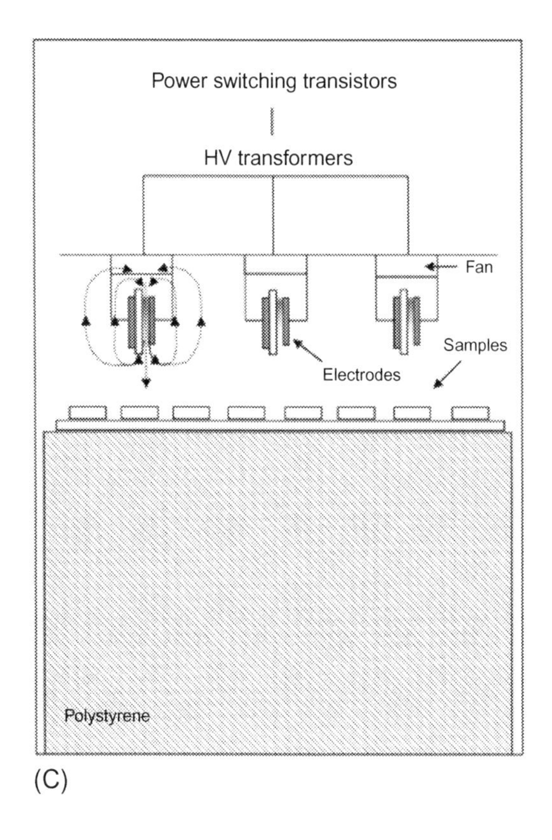

Fig. 4, Cont'd

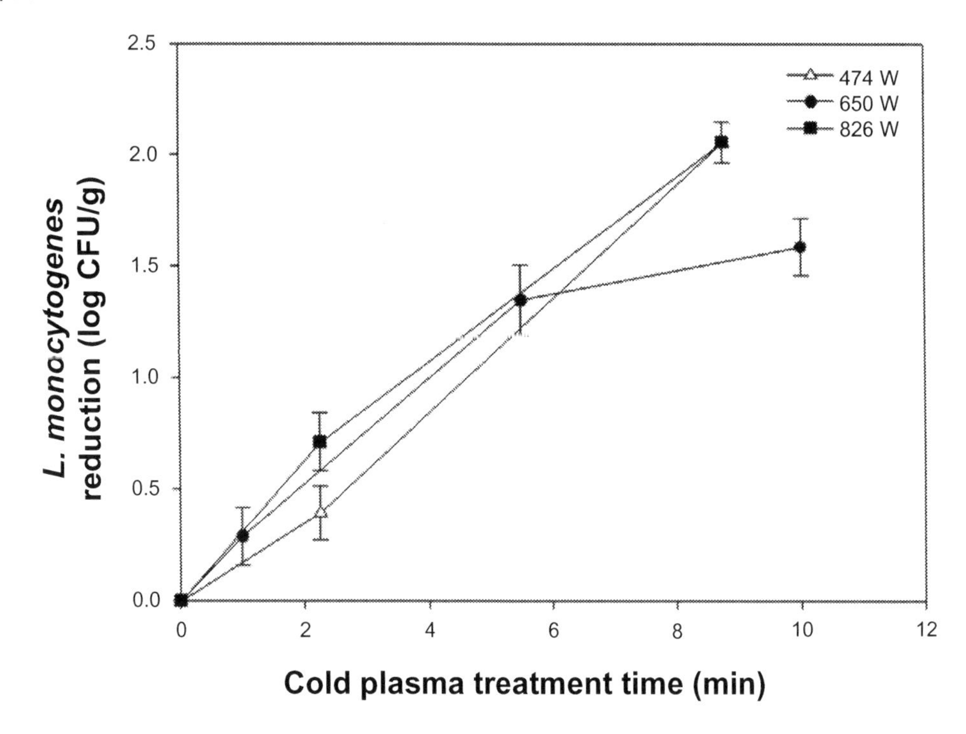

Fig. 5 Cold plasma treatment time (at different power levels) weds *L. monocytogenes* reduction. Helium oxygen mixture used for the development of plasma system at 667 Pa. From Lee, H., Kim, J. E., Chung, M. S., Min, S.C. (2015). Cold plasma treatment for the microbiological safety of cabbage, lettuce, and dried figs. *Food Microbiology*, 51, 74–80. doi:10.1016/j.fm.2015.05.004.

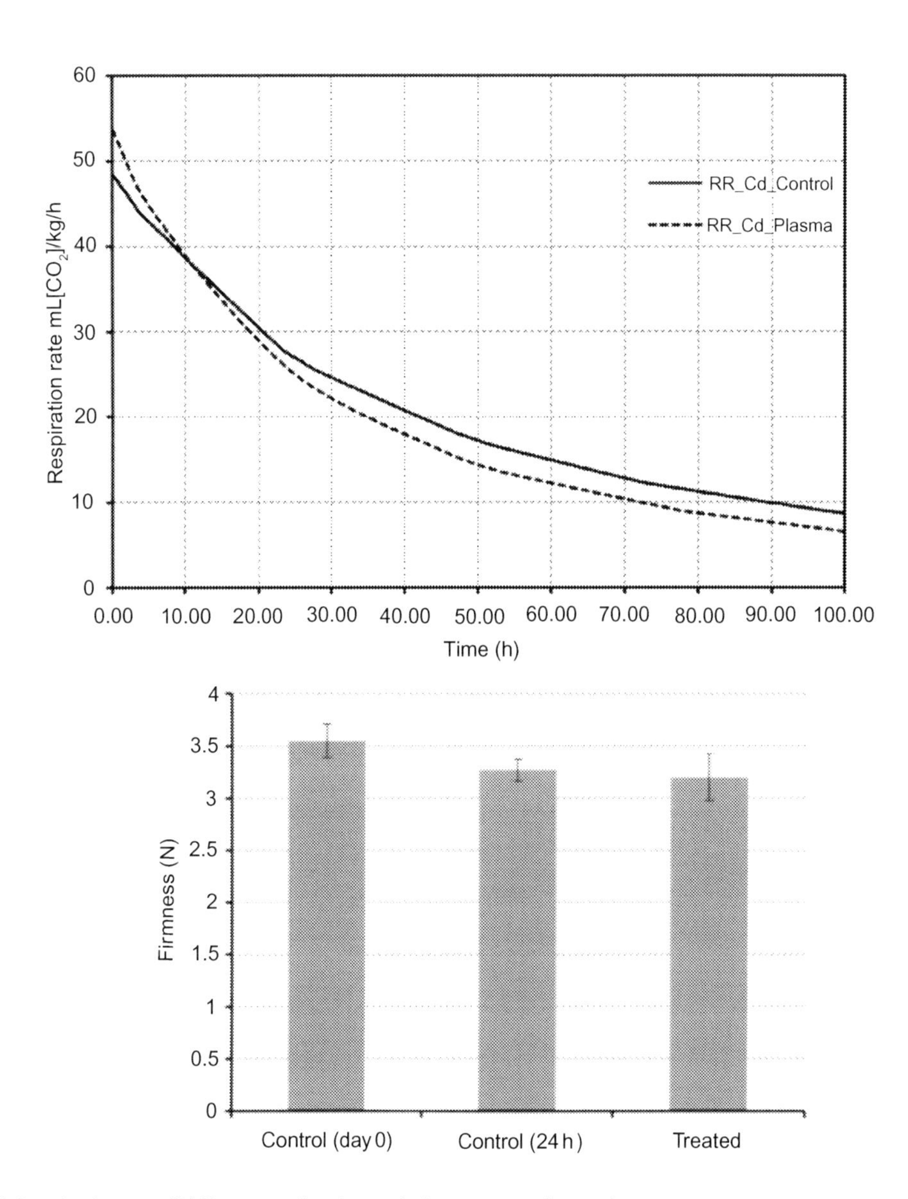

Fig. 6 (A) Respiration rate (B) firmness of native and plasma treated strawberry. From Misra, N. N., Patil, S., Moiseev, T., Bourke, P., Mosnier, J. P., Keener, K. M., Cullen, P. J. (2014). In-package atmospheric pressure cold plasma treatment of strawberries. *Journal of Food Engineering*, 125, 131–138. doi:10.1016/j.jfoodeng.2013.10.023.

Table 2 Application of cold plasma on fresh-cut produce

Products	Treatment	Effects	Reference
Cherry tomato	DBD plasma. 30 kV for 30, 60, 180s, and 300 s	Insignificant changes in the weight loss, pH, and firmness. Respiration rate and color does not change after DBD treatment	Misra, Keener, Bourke, Mosnier, and Cullen (2014)
Cherry tomato, strawberry	DBD plasma.120 kV for 10, 60, 120, and 300 s.	Treatment for 10, 60, and 120 s reduced *Salmonella*, *E. coli*, and *L. monocytogenes* to undetectable levels from 3.1, 6.3, and 6.7 $\log_{10}$ CFU/sample. Treatment for 300 s makes reduction by 3.5, 3.8, and 4.2 $\log_{10}$ CFU/sample of *Salmonella*, *E. coli*, and *L. monocytogenes*	Ziuzina, Patil, Cullen, Keener, and Bourke (2014)
Blueberry	Jet plasma. 549 W for 15, 30, 45, 60, 90, and 120 s	All treatments significantly reduced aerobic plate count (APC) from 0.8 to 1.6 log CFU/g and 1.5 to 2.0 log CFU/g as compared to control for 1–7 days (Fig. 10). The firmness was reduced significantly after 90 and 120 s treatments, the softening of berries after treatments attributed the collisions between berries and container (Fig. 9). A significant reduction in anthocyanins was observed after 90 s	Ramazzina et al. (2015)
Fresh-cut apple	DBD plasma.150 W for 30 and 120 min	Treatment slightly reduced (up to 10%) antioxidant activity and antioxidant content	Ramazzina et al. (2016)

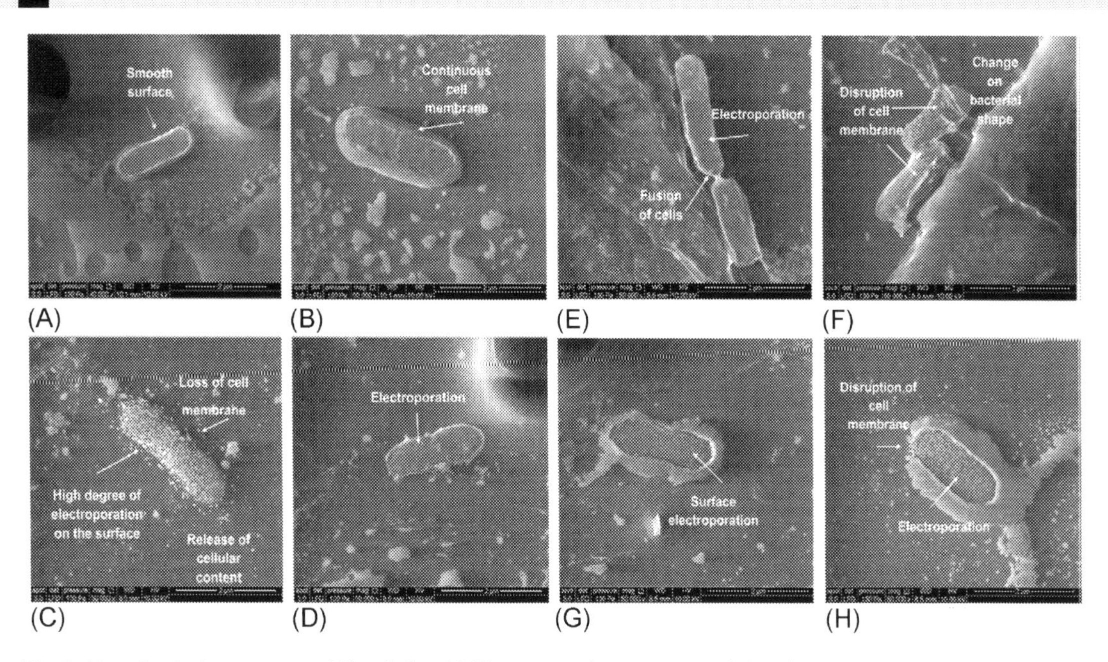

Fig. 7 Alteration in the structure of *E. coli* after APCP treatment (10 min, 12.83 kV); (A, B) control cells at 40 and 50 K times; (C) cytoplasmic membrane losses and release of cellular components after treatment, a very high degree of electroporation; (D) electroporation observed on treated cell surface; (E) minor degree of electroporation observed on treated cells, but fusion was observed between them; (F) the shape of cell was changed, cytoplasmic membrane interruption and release of cell contents, electroporation on the treated cell surface; (G) electroporation on the treated cell surface; (H) cytoplasmic membrane interruption and release of cell contents, electroporation on the treated cell surface. From Effect of atmospheric pressure cold plasma (APCP) on the inactivation of *Escherichia coli* in fresh produce. *Food Control 34* (2013) 149–157.

4.2 Effect of plasma on the enzymes of fresh-cut produce

The endogenous enzymes present in fruits and vegetables have a significant role in the browning reaction. Enzymatic browning reaction spoils the sensory qualities of fresh-cut produce. The endogenous enzymes such as polyphenol oxidase and peroxidize are the major enzymes that lead to the enzymatic browning of fresh-cut produce. The general methods used for the inactivation of enzymes are blanching, heating, and commercial sterilization. The nonconventional techniques used for the inactivation of enzymes are irradiation, pulsed electric field treatment and high-pressure processing, and cold plasma.

Among these techniques, cold plasma has a considerable role in enzyme inactivation. The enzyme trypsin undergoes inactivation (zero at 4 J/cm^2) by the plasma treatment (Dobrynin et al., 2009). Plasma species make breakage in peptide bonds of the trypsin protein and alters 3D structure of protein; it leads to its inactivation. The polyphenol oxidase activity of guava pulp and fruit reduced to 70% and 10% after treatment with cold plasma with 2 kV for 300 s (Thirumdas et al., 2014). Treatment of DBD plasma (Fig. 4C) on fresh-cut apples for 30 min decreases 65% of browning area, and also reduced the polyphenol oxidase activity from 12% to 58%. (Tappi et al., 2014). In the case of fresh-cut melons, 17% of POD and 7% PME activities were reduced after cold plasma treatments (Tappi et al., 2016). DBD treatment on fresh-cut kiwi fruits reduced darkened area formation without any changes occurring in the texture, antioxidant content, and antioxidant activity (Fig. 8). It improved color retention of kiwi fruit (Table 3) (Ramazzina et al., 2015).

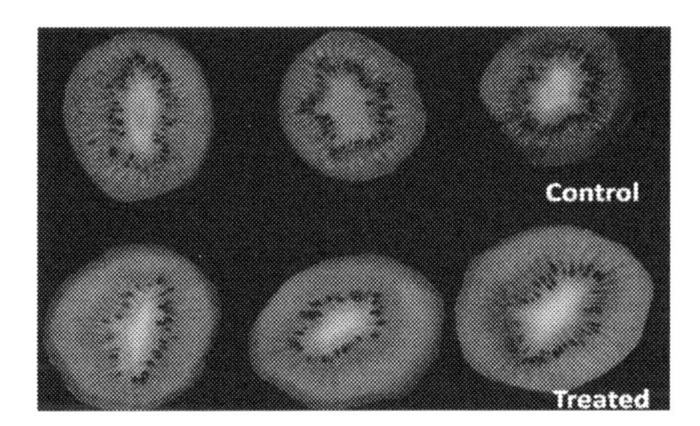

Fig. 8 Kiwi fruit slices subjected to 20 + 20 min DBD gas plasma treatment. Control sample stored at controlled conditions for 4 days. From Ramazzina, I., Berardinelli, A., Rizzi, F., Tappi, S., Ragni, L., Sacchetti, G., Rocculi, P. (2015). Effect of cold plasma treatment on physico-chemical parameters and antioxidant activity of minimally processed kiwifruit. *Postharvest Biology and Technology*, 107, 55–65. doi:10.1016/j.postharvbio.2015.04.008.

Table 3 Pericarp color and factorial ANOVA results of control (C) and treated (T) for 10+10 and 20+20 min fresh-cut kiwifruit immediately after the treatment and after 1 and 4 days of storage

Sample	Treatment time (min)	Storage time (day)	Lightness		Chroma		Hue angle (degrees)	
			Mean	*s.d.*	*Mean*	*s.d.*	*Mean*	*s.d.*
C	10+10	0	49.3^{a}	5.5	30.1^{a}	6.8	120.4^{a}	2.3
T	10+10	0	47.6^{ab}	6.2	27.1^{ab}	8.1	120.2^{a}	2.4
C	20+20	0	47.5^{ab}	5.7	29.0^{ab}	7.0	119.9^{a}	2.6
T	20+20	0	46.4^{abc}	4.8	26.2^{b}	6.3	120.7^{a}	3.5
C	10+10	1	44.7^{abce}	8.4	20.1^{cd}	4.5	111.5^{b}	7.1
T	10+10	1	46.2^{abc}	7.4	18.6^{cd}	3.8	111.6^{b}	3.7
C	20+20	1	43.0^{cde}	6.9	19.0^{cd}	4.6	112.5^{b}	3.3
T	20+20	1	48.1^{a}	9.3	19.7^{c}	5.1	111.4^{b}	2.9
C	10+10	4	39.9^{df}	6.4	18.4^{ab}	4.4	112.5^{b}	5.5
T	10+10	4	43.2^{bcde}	6.4	21.1^{c}	4.6	112.1^{b}	6.3
C	20+20	4	37.3^{f}	5.4	16.7^{d}	3.5	112.4^{b}	4.3
T	20+20	4	41.1^{def}	6.5	17.7^{ab}	4.7	111.1^{b}	3.3
F	Sample		10.1	...	1.11	n.s.	0.01	n.s.
F	S.t.		57.1	...	171	...	237	...
F	Sample × S.t.		8.00	...	8.32	...	0.75	n.s.
F	T.t		4571.48	...	6.83	...	0.99	n.s.
F	T.t. × S.t.			n.s.	2.44	n.s.	0.55	n.s.

C, control; *F*, F value; *n.s.*, not significant; *s.d.*, standard deviation; *S.t.*, storage time; *T.t.*, treatment time; *T*, treated. Data marked with the same letters (a–f) within each column are not significantly different at a $P < 0.05$ level.

From Ramazzina, I., Berardinelli, A., Rizzi, F., Tappi, S., Ragni, L., Sacchetti, G., Rocculi, P. (2015). Effect of cold plasma treatment on physico-chemical parameters and antioxidant activity of minimally processed kiwifruit. Postharvest Biology and Technology, *107, 55–65. doi:10.1016/j.postharvbio.2015.04.008.*

Fig. 9 Cold plasma treatment of blueberries in glass jars. From Atmospheric cold plasma inactivation of aerobic microorganisms on blueberries and effects on quality attributes. *Food Microbiology*, 46 (2015), 479–484.

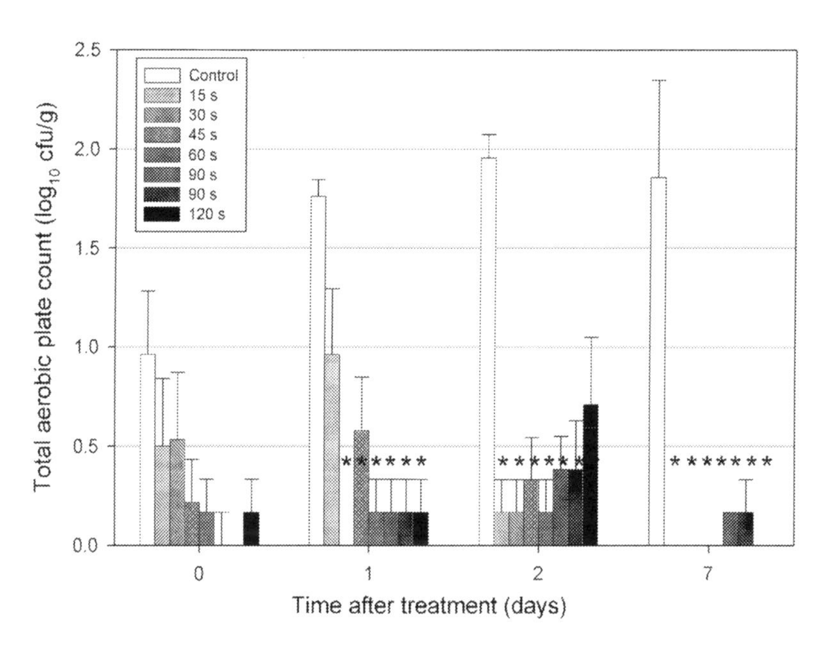

Fig. 10 Aerobic plate count of cold plasma treated blueberry at different storage days (1–7 days). From Atmospheric cold plasma inactivation of aerobic microorganisms on blueberries and effects on quality attributes. *Food Microbiology*, 46 (2015), 479–484.

References

Barbagallo, R. N., Chisari, M., & Caputa, G. (2012). Effects of calcium citrate and ascorbate as inhibitors of browning and softening in minimally processed 'Birgah' eggplants. *Postharvest Biology and Technology, 73*, 107–114. https://doi.org/10.1016/j.postharvbio.2012.06.006.

Dobrynin, D., Fridman, G., Friedman, G., & Fridman, A. (2009). Physical and biological mechanisms of direct plasma interaction with living tissue. *New Journal of Physics, 11*. https://doi.org/10.1088/1367-2630/11/11/115020.

Fernández, A., Noriega, E., & Thompson, A. (2013). Inactivation of *Salmonella enterica* serovar typhimurium on fresh produce by cold atmospheric gas plasma technology. *Food Microbiology, 33*, 24–29. https://doi.org/10.1016/j.fm.2012.08.007.

James, J. B., & Ngarmsak, T. (2010). Processing of fresh-cut tropical fruits and vegetables: A technical guide food. Rap Publication 2010/16. Food and Agriculture Organization of the United Nations, Regional Office for Asia and the Pacific, Bangkok.

Krasaekoopt, W., & Bhandari, B. (2010). Fresh-cut vegetables. In *Handbook of vegetables and vegetable process* (pp. 219–242). Wiley-Blackwell.

Lee, H., Kim, J. E., Chung, M. S., & Min, S. C. (2015). Cold plasma treatment for the microbiological safety of cabbage, lettuce, and dried figs. *Food Microbiology, 51*, 74–80. https://doi.org/10.1016/j.fm.2015.05.004.

Ma, L., Zhang, M., Bhandari, B., & Gao, Z. (2017). Recent developments in novel shelf life extension technologies of fresh-cut fruits and vegetables. *Trends in Food Science & Technology, 64*, 23–38. https://doi.org/10.1016/j.tifs.2017.03.005.

Meireles, A., Giaouris, E., & Simões, M. (2016). Alternative disinfection methods to chlorine for use in the fresh-cut industry. *Food Research International, 82*, 71–85. https://doi.org/10.1016/j.foodres.2016.01.021.

Misra, N. N., Keener, K. M., Bourke, P., Mosnier, J., & Cullen, P. J. (2014). In-package atmospheric pressure cold plasma treatment of cherry tomatoes. *Journal of Bioscience and Bioengineering, 118*, 177–182. https://doi.org/10.1016/j.jbiosc.2014.02.005.

Misra, N. N., Pankaj, S. K., Segat, A., & Ishikawa, K. (2016). Trends in food science & technology cold plasma interactions with enzymes in foods and model systems. *Trends in Food Science and Technology, 55*, 39–47. https://doi.org/10.1016/j.tifs.2016.07.001.

Misra, N. N., Patil, S., Moiseev, T., Bourke, P., Mosnier, J. P., Keener, K. M., et al. (2014). In-package atmospheric pressure cold plasma treatment of strawberries. *Journal of Food Engineering, 125*, 131–138. https://doi.org/10.1016/j.jfoodeng.2013.10.023.

Ramazzina, I., Berardinelli, A., Rizzi, F., Tappi, S., Ragni, L., Sacchetti, G., et al. (2015). Effect of cold plasma treatment on physico-chemical parameters and antioxidant activity of minimally processed kiwifruit. *Postharvest Biology and Technology, 107*, 55–65. https://doi.org/10.1016/j.postharvbio.2015.04.008.

Ramazzina, I., Tappi, S., Rocculi, P., Sacchetti, G., Berardinelli, A., Marseglia, A., et al. (2016). Effect of cold plasma treatment on the functional properties of fresh-cut apples. *Journal of Agricultural and Food Chemistry, 64*(42), 8010–8018. https://doi.org/10.1021/acs.jafc.6b02730.

Sudheesh, C., Sunooj, K. V., Sinha, S. K., George, J., Kumar, S., Murugesan, P., et al. (2019). Impact of energetic neutral nitrogen atoms created by glow discharge air plasma on the physico-chemical and rheological properties of kithul starch. *Food Chemistry, 294*, 194–202. https://doi.org/10.1016/j.foodchem.2019.05.067.

Tappi, S., Berardinelli, A., Ragni, L., Dalla Rosa, M., Guarnieri, A., & Rocculi, P. (2014). Atmospheric gas plasma treatment of fresh-cut apples. *Innovative Food*

Science and Emerging Technologies, 21, 114–122. https://doi.org/10.1016/j.ifset.2013.09.012.

Tappi, S., Gozzi, G., Vannini, L., Berardinelli, A., Romani, S., Ragni, L., et al. (2016). Cold plasma treatment for fresh-cut melon stabilization. *Innovative Food Science and Emerging Technologies, 33,* 225–233. https://doi.org/10.1016/j.ifset.2015.12.022.

Thirumdas, R., Sarangapani, C., & Annapure, U. S. (2014). Cold plasma: A novel non-thermal technology for food processing. *Food Biophysics, 10,* 1–11. https://doi.org/10.1007/s11483-014-9382-z.

Thirumdas, R., Trimukhe, A., Deshmukh, R. R., & Annapure, U. S. (2017). Functional and rheological properties of cold plasma treated rice starch. *Carbohydrate Polymers, 157,* 1723–1731. https://doi.org/10.1016/j.carbpol.2016.11.050.

Ziuzina, D., Patil, S., Cullen, P. J., Keener, K. M., & Bourke, P. (2014). Atmospheric cold plasma inactivation of Escherichia coli, Salmonella enterica serovar Typhimurium and Listeria monocytogenes inoculated on fresh produce. *Food Microbiology, 42,* 109–116. https://doi.org/10.1016/j.fm.2014.02.007.

Author Index

Note: Page numbers followed by *t* indicate tables.

C

D

E

H

I

J

O

P

S

T

U

V

W

Subject Index

Note: Page numbers followed by *f* indicate figures and *t* indicate tables.

Q

Printed in the United States
By Bookmasters